P9-EDR-121

ABOUT ISLAND PRESS

Island Press is the only nonprofit organization in the United States whose principal purpose is the publication of books on environmental issues and natural resource management. We provide solutions-oriented information to professionals, public officials, business and community leaders, and concerned citizens who are shaping responses to environmental problems.

In 1994, Island Press celebrated its tenth anniversary as the leading provider of timely and practical books that take a multidisciplinary approach to critical environmental concerns. Our growing list of titles reflects our commitment to bringing the best of an expanding body of literature to the environmental community throughout North America and the world.

Support for Island Press is provided by Apple Computer, Inc., The Bullitt Foundation, The Geraldine R. Dodge Foundation, The Energy Foundation, The Ford Foundation, The W. Alton Jones Foundation, The Lyndhurst Foundation, The John D. and Catherine T. MacArthur Foundation, The Andrew W. Mellon Foundation, The Joyce Mertz-Gilmore Foundation, The National Fish and Wildlife Foundation, The Pew Charitable Trusts, The Pew Global Stewardship Initiative, The Rockefeller Philanthropic Collaborative, Inc., and individual donors.

ABOUT THE SCHOOL OF NATURAL RESOURCES & ENVIRONMENT AT THE UNIVERSITY OF MICHIGAN

The School of Natural Resources and Environment at The University of Michigan (SNRE) is a research-oriented professional school focusing on the development of policies, management programs, and landscape designs that promote the conservation, protection, and sustained use of natural resources. The first school of its kind in the world, SNRE trains practitioners and researchers to solve complex natural resource and environmental problems ranging from the local to the international levels.

One of the core areas of research, study, and action at SNRE, "Ecosystems: Conservation, Management, and Restoration," includes work under way in topics ranging from landscape ecosystem classification and prairie restoration to endangered species policy and dispute resolution. Academic programs include studies in resource policy and behavior, resource ecology and management, and landscape architecture. Interdisciplinary perspectives are important and are fostered through several joint or dual degrees with other University programs, including the Michigan Business School, the Law School, the Center for Russian and Far Eastern Studies, the School of Public Policy, and the Department of Urban Planning.

ABOUT THE WILDERNESS SOCIETY

In 1935, a group of foresters, biologists, economists, and other professionals founded The Wilderness Society, declaring "all we desire to save from invasion is that extremely minor fraction of outdoor America which yet remains free from mechanical sights and sounds and smells." Nearly 30 years later, Wilderness Society staff and members succeeded in drafting and convincing Congress to pass the Wilderness Preservation Act of 1964, which today protects some 100 million acres of America's most precious wildlands.

The mission of The Wilderness Society today is to preserve wilderness and wildlife, protect America's prime forests, parks, rivers, deserts, and shorelands, and foster an American land ethic. Fulfilling that mission involves three main strategies: ecological and economic research and analysis; advocacy and litigation to influence public land policy; and public education including publishing monographs, newsletters, and a semi-annual magazine.

ECOSYSTEM MANAGEMENT IN THE UNITED STATES

ECOSYSTEM MANAGEMENT
IN THE UNITED STATES
AN ASSESSMENT OF CURRENT EXPERIENCE

Steven L. Yaffee
Ali F. Phillips
Irene C. Frentz
Paul W. Hardy
Sussanne M. Maleki
Barbara E. Thorpe

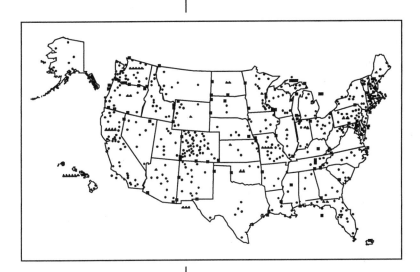

A collaborative effort of

THE UNIVERSITY OF
MICHIGAN

and

THE WILDERNESS SOCIETY

ISLAND PRESS
Washington, D.C. □ Covelo, California

ISLAND PRESS is a trademark of The Center for Resource Economics.

Library of Congress Cataloging-in-Publication Data

Ecosystem management in the United States: an assessment of current
    experience/Steven L. Yaffee . . . [et al.].
        p.   cm.
    "A collaborative effort of the University of Michigan and the
Wilderness Society."
    Includes bibliographical references and index.
    ISBN 1-55963-502-9 (pbk.)
    1. Ecosystem management—United States.   I. Yaffee, Steven Lewis.
II. University of Michigan.   III. Wilderness Society
QH76.E336   1996
333.7'0973—dc20   96-21825
                                                                CIP

Printed on recycled, acid-free paper

Manufactured in the United States of America

10 9   8   7   6   5   4   3   2   1

CONTENTS

FIGURES

FOREWORD

Two years ago, cooperative ecosystem management (CEM) was, to The Wilderness Society and almost everyone else we asked, an unknown quantity. Basic principles of ecosystem management—everything is connected to everything else, save all the parts, think like a mountain—had been with the modern conservation movement from its beginning, but the specifics of ecosystem management were elusive. Spurred by a growing recognition that many difficult environmental problems, such as endangered species recovery and water quality protection, are really ecosystem problems, scientists, resource management professionals, and others were beginning (again) to work ecological principles into public land management policy and practice.

Parallel to this scientific renaissance in land management was a great deal of administrative attention to finding new ways of resolving natural resource conflicts before they turn into "trainwrecks." All parties sought the answer to one fundamental question: Is there a way to sustain the complex natural systems on which we all ultimately depend without abdicating collective and individual responsibilities to the land and to each other, and without denying differences in values, traditions, and expectations? Cooperative ecosystem management seemed to hold the answer.

While scientists and managers attempted to translate ecological principles into practice, and economists and administrators searched for "win-win" strategies (the Holy Grail of natural resource management) to avoid "trainwrecks," little was really known about what ecosystem management looked like on the ground. Nor did we know how well cooperative, as opposed to competitive (or combative), approaches to ecosystem management were faring. There were some anecdotal reports—word of a river corridor project here or a habitat improvement project there—but by and large, discussion of cooperative ecosystem management was occurring in a virtual information vacuum.

To lessen that void, The Wilderness Society asked researchers at The University of Michigan's School of Natural Resources and Environment to conduct an objective survey of cooperative ecosystem management in America. Which agencies and organizations are attempting to implement CEM? Where? What do CEM projects look like? Why did they start? How are they doing? And what are they achieving?

The Wilderness Society initially sought to provide information that would better equip the U.S.D.A. Forest Service, National Park Service, Bureau of Land Management, and other public agencies to practice ecosystem management—that is, to consider appropriate spatial and temporal scales when taking management actions on lands under their jurisdiction. We recognized that this information would be crucial to sound land management decision making where agency decisions affect, and are affected by, actions taken by other agencies or entities affecting land outside of the direct federal control.

We also sought to find out for ourselves the kind of models CEM provides for our own bioregional work. We are interested in fostering truly *cooperative*, as opposed to *co-opted* or *compromised*, ecosystem management. Our goal—to establish and maintain a nationwide network of wildlands—won't be reached by "big-hug exercises," where dialogue among representatives of diverse interests leads only to fewer hard feelings or an endorsement of a failed *status quo*. It is essential, instead, to encourage informed interaction among various stakeholders in a context where achieving real improvements in American land management is possible. We sought an initial filter to help us sort out the "big hugs" from the real, hard work of ecosystem management and a rough guide to our own involvement in CEM projects in the future.

Finally, and at least as important, we intended to create a resource by which individuals, organizations, agencies, or corporations facing local environmental challenges could learn from the collective and individual experience of CEM projects nationwide. This book, together with the base of information that underlies it, is a substantial review of ecosystem management in the United States. It lists more than 600 CEM projects, with great detail provided on 105 selected projects. Peruse the "Assessment" to learn what CEM looks like on the ground, what makes CEM work, what gets in the way, what kinds of resources CEM requires, and where CEM projects can get outside help. Scan the "Catalog" following the "Assessment" to learn about particular projects and, possibly, find a model for your CEM project. The authors have listed contact information for all of the projects, so getting additional and up-to-the-minute guidance from people "doing" CEM will be that much easier.

What Professor Yaffee and his team at Michigan found is cause for hope. Across the country, cooperative ecosystem management projects encompassing diverse ecosystem types, with a mix of landownership patterns and managing entities, are addressing environmental stresses ranging from "simple" point source pollution to problems as complex as invasive exotic species and recreational overuse. The vast majority of the projects include ecosystem preservation or ecosystem restoration among their goals. Additionally, most of the projects employed 1) research and sound scientific information, and 2) stakeholder involvement as key strategies for meeting those goals. Conflict and trainwrecks down the track should be lessened by broader stakeholder participation from the beginning of CEM projects. Research, leading to a scientific basis for management actions, will help ensure that what the stakeholders agree to will result in real improvements in land and resource stewardship.

While CEM appears quite promising, it is in many respects a fledgling endeavor that will need plenty of care and feeding to really take flight. CEM is hard work. From gathering basic scientific, social, and economic data, to maintaining constructive working relationships among diverse stakeholders, doing CEM and doing it well will demand time, patience, and resources that many may find difficult to spare. Most of the CEM projects surveyed received at least some public investment in the form of staff time, technical assistance, or direct funding. Unfortunately, since the initial survey was completed (summer of 1995), some follow-up surveying reveals that federal support is in grave danger. Despite its purported preference for collaborative decision making at the local level, the current Congress has had a demoralizing effect on those who have struggled to make CEM bear fruit. As one federal CEM team leader expressed to the research team:

> We feel we are beginning to lose the impetus for ecosystem management [because of] budget problems and the [recent] furloughs. ... The pace is unrelenting. ... It just seems that [the work] is getting piled on more and more and more, and we have budget cuts, and you have to do more with less. We've been hearing "do more with less" for three to four years now. ... It's taking its toll, and I think it might end up affecting the ecosystem process in a negative way.

This concern reflects the research team's assessment that the human component—the folks doing the work, directing, participating in it—is the most important aspect of this promising new paradigm. Natural resource managers, project coordinators, and decision makers on the ground are struggling against imposing odds to make land management work. They have little direction and often face incredible political or administrative hurdles. Nevertheless, they are forging ahead. And those who are participating but not necessarily leading the effort—the landowners, other agency personnel, the diverse stakeholders who are coming to the table to talk—are taking risks as well, often going against the tide of what extremists and politicians are telling them. The success of CEM depends on our ability to create an environment in which they can do their important work.

These observations lead to some obvious policy recommendations. First, and perhaps most important, we need a sound national population policy that addresses the burdens that an exploding human population places on a shrinking resource base. All of these projects are doomed unless we slow the accelerating demand for raw materials and wild places. Second, we need to remove barriers to cooperation, such as those created by current interpretations of the Federal Advisory Committee Act, to put federal/non-federal collaboration on a firm legal footing. Likewise, we need to revisit anti-trust law to facilitate collaboration among private landowners toward sustainable land management. Third, we need to expand technical assistance to CEM cooperators, as sound scientific information and technical models are key to the success of CEM projects. And, finally, we need to commit ourselves as a society, through taxpayer support where needed, to help get projects off the ground and thoroughly implemented. Throughout this research, one message came through loud and clear: While CEM can be accomplished only on the ground, its success depends on the support of society at all levels, from individuals to the federal government.

So there is reason for hope—CEM represents a process by which more effective ecosystem protection can occur. But there is also reason for increased diligence and action. We must ensure that the progress CEM has already made and the lessons CEM holds are not lost, and the chance for a cleaner, wilder, healthier, and more sustainable future with them.

Gaylord Nelson
Counselor
The Wilderness Society

Washington, D.C.
April 1996

PREFACE

Ecosystem management has received a great deal of attention over the past five years. Scientific articles and conferences have highlighted the complexity of ecosystems, discussed measures of ecological integrity, and debated whether "pre-settlement conditions" should be the baseline for assessing management direction. Managers have tried to understand how human needs and demands can be integrated with ecological objectives. The Clinton Administration has viewed ecosystem management approaches as the light at the end of the spotted owl tunnel, and created an Interagency Task Force on Ecosystem Management, the National Biological Service, and the Council on Sustainable Development. Even the 104th Congress, while damning the Administration's ecosystem management initiatives, has embraced the concept of bottom-up, public-private collaboration that is featured in many ecosystem management efforts.

Yet the dialogue often has bogged down in arguments over "just what is ecosystem management" and "what are its goals." Some argue that an approach that is supported rhetorically by both environmental groups and commodity interests should be appraised with skepticism. Others suggest that when environmentalists hear the term "ecosystem management," they hear "ecosystem," while commodity groups hear "management." And like all things that promise to be a cure-all—Dr. Youngblood's Magical Ecosystem Management Elixir—we should be wary and ask hard questions.

Sometimes, however, things that sound good can actually be good. The intensity of the debate over ecosystem management at times has hidden a simple reality: People on the ground in places ranging from Soldotna, Alaska, to the Florida Everglades are already actively engaged in ecosystem management experiments. At times called "sustainable development," "sustainable community efforts," "place-based management," "bioregionalism," and "collaborative problem solving," these are management approaches that focus on long-term ecological and economic sustainability, involve stakeholders to a greater extent, and challenge long-standing boundaries established by geography, politics, and traditions. Few are managing at the ecosystem scale, but most are taking an ecosystem approach to resource management. While it is important to continue to sharpen our conceptualizations of ecosystem-based approaches, it is not necessary to reach consensus on a definition of ecosystem management before progress can be made.

Indeed, we should be careful not to let the theoretical and political debate keep us from addressing another underlying truth: The management approaches of the last fifty years are not sufficient to guide our society toward a sustainable future; we must find better ways. Conservation policy has produced short-term prosperity while eroding the resource base on which prosperity depends. The top-down management style of government agencies has disempowered local citizens, while changing fiscal realities made it less likely that a dominating government role could be maintained. Conflict, not consensus, and confusion, not clarity, have been the state of resource management in recent years—a state of affairs that has been costly for individuals, groups, and agencies of all political inclinations. Is there a better way?

When we began the research reported in this book, we did not know whether ecosystem-based approaches to resource management were the answer. At Michigan, where master's-degree students in natural resources and environment carry out interdisciplinary group projects as their thesis like capstone experience, a group of students was interested in ecosystem approaches. An earlier student group had synthesized what was known about the theory underlying ecosystem management and had drafted an ecosystem management framework. A 1994 conversation with Mark Shaffer, then at The Wilderness Society, pointed us to the task of collecting and analyzing the early experiences with on-the-ground experiments in ecosystem management. He argued that no one could resolve the theoretical and conceptual issues in debate without learning from experience. Only through a good dose of reality could the theory advance.

The more than one hundred cases summarized in this book provide that dose of reality, at least as it appears from an admittedly short implementation history. And somewhat unexpectedly, the story we tell is an optimistic one. People are succeeding in small and large ways, and doing things that they were unable to do just a few years ago. It is important to learn from their experience: their strategies and sources of progress; the obstacles they faced and how they were overcome. What surprised us was the exhilaration evident in the voices of people from all points on the resource management spectrum, and the level of interest they expressed in learning more from others' experiences. Not all of these experiences are successes, and most represent a lot of hard work. All have faced challenges, and some will not succeed over the long term. Nevertheless, it is clear that something different is going on here, and we should listen to the testimony provided by the pioneers in this important process of change.

HOW TO USE THIS BOOK

This book seeks to be the first practical and comprehensive guide to ecosystem management efforts for use by practitioners and decision makers alike. Just as ecosystems can be understood by looking at the landscape through a hierarchy of spatial scales, the book is designed to provide several levels of information. The "Assessment" offers our conclusions about the aggregate experience at 105 representative ecosystem management sites nationwide. It is relatively short, contains many graphs summarizing the site-based experience, and ends with a set of summary observations.

The "Catalog" provides a finer level of focus. It contains two-page descriptions on each of the 105 sites, describing the projects and project areas, the stresses that are evident on-site and the strategies employed to deal with them. Each write-up provides an assessment of the status of the effort, including factors that are facilitating and constraining progress. Contact information is provided for follow-up. To help readers access the vast set of experiences in the 105 case studies, we have provided several tools collected in a section entitled "Summary Information" at the beginning of the catalog. A map, lists of projects by state and region, date of origin, and other factors, can help you locate sites of particular interest. Matrices arraying projects by features such as outcomes, stresses, and involved organizations, give a visual overview of the data base and provide an additional way to zero in on interesting sites.

The "Appendixes" provide one more level of information. A list of 619 projects identified in the course of this study highlights the full range of nationwide efforts from which this sample was drawn. We have provided contact information for the sites where available. This list is an unparalleled source of baseline information for researchers and practitioners. Also included in the "Appendixes" is a list of documentary and electronic information for those interested in additional reading or research on the ecosystem management topic. A glossary and list of acronyms should help readers navigate through the wide array of experiences documented in the book.

Though additional events have no doubt occurred at the sites in the catalog since the research was completed (the site descriptions include activities through the end of 1995), we believe the aggregate picture they portray will be accurate for some time to come. They provide a rich set of lessons for those participating in ecosystem management projects as well as insights for those structuring resource management policies. We can all learn from these experiences and move forward.

ACKNOWLEDGMENTS

The research that led to this assessment was supported by funds provided by The Wilderness Society, The University of Michigan, and the U.S.D.A. Cooperative State Research Service. The authors gratefully acknowledge this support and note that the findings and conclusions in the book represent those of the authors and not necessarily those of the sponsors.

The research team wants to thank a variety of contributors to this effort. We greatly appreciate the advice, assistance, and support of: Greg Aplet, Julie Fox Gorte, Spencer Phillips, and Gaylord Nelson from TWS; Mark Shaffer of The Nature Conservancy; Arvid Thomsen and David Sumpter of HDR Engineering, Inc.; John Shepard of the Sonoran Institute; David Allan, Burton Barnes, Bob Grese, Kim Hall, Rachel Kaplan, Teri Malies, Jennifer Parody, Wayne Say, John Witter, and Julia Wondolleck from SNRE; Maia Enzer of American Forests; and Barbara Dean, Bill LaDue, and Barbara Youngblood of Island Press. In addition, a prior master's project report written by Sara Barth, Lynn Gooch, Jim Havard, David Mindell, Rachel Stevens, and Mark Zankel established a theoretical framework for ecosystem management that facilitated this work. Finally, we want to acknowledge the valuable assistance of the site contacts, whose experiences and observations form the heart of this analysis. Their enthusiasm for the on-the-ground efforts was unmistakable and contributed greatly to our sense of optimism about the future of ecosystem-based approaches to resource management.

Steven L. Yaffee
Ali F. Phillips
Irene C. Frentz
Paul W. Hardy
Sussanne M. Maleki
Barbara E. Thorpe

Ann Arbor, Michigan
April 1996

ECOSYSTEM MANAGEMENT
IN THE UNITED STATES

PART ONE

ASSESSMENT

INTRODUCTION

In the past two decades, increasing dissatisfaction with traditional resource management approaches, coupled with the development of new scientific information about ecosystems, has led to a search for new management strategies. Many people have argued that traditional approaches ignore important interconnections between geographic areas, such as when sedimentation or the introduction of invasive species on one land unit results in problems on an adjacent area. Land management approaches have often been overly focused on short-term goals such as commodity production at the expense of long-term ecological health, ultimately resulting in landscapes that are neither economically productive nor ecologically sound.

Management activities have often been inefficient as agencies and other groups operated in ignorance of each other's efforts. Sometimes efforts have been duplicative or conflicting; at other times, potential benefits of management partnerships have been lost. Traditional approaches have often neglected the diversity of human interests in management of natural resources and have resulted in conflict-laden impasses, as when interest groups and agencies battled to a standstill in the spotted owl dispute in the Pacific Northwest. Issues such as management of salmon stocks in the Northwest, of the Everglades in the Southeast, and of the Yellowstone region in the Rocky Mountains seem to cry out for a different approach.

Ecosystem management has been proposed as a new approach to resource management, and a body of literature has developed describing various goals and methods of such an approach (see Appendix, Resource Guide). Most authors emphasize a land management approach that incorporates an understanding of ecological systems, considers extended time and spatial scales, and highlights interconnections between landscapes, ecological processes, and humans and other organisms. Summarizing much of this literature, Edward Grumbine set forth a definition of ecosystem management:

> Ecosystem management integrates scientific knowledge of ecological relationships within a complex sociopolitical and value framework toward the general goal of protecting native ecosystem integrity over the long term.[1]

Many government agencies have picked up on the evolving concept of ecosystem management and have developed definitions to guide their land management activities. For example, U.S.D.A. Forest Service Chief Jack Ward Thomas describes ecosystem management as:

> a holistic approach to natural resource management, moving beyond a compartmentalized approach focusing on the individual parts of the forest. It is an approach that steps back from the forest stand and focuses on the forest landscape and its position in the larger environment in order to integrate the human, biological and physical dimensions of natural resource management. Its purpose is to achieve sustainability of all resources.[2]

While there is an ongoing debate about the goals and practices of ecosystem management, practitioners are actively testing new land management approaches. By building on this experience, policy makers and practitioners can improve the

---

1. R. Edward Grumbine, "What Is Ecosystem Management?," *Conservation Biology*, 8, 1 (1994):27–38.
2. "New Directions for the Forest Service," statement of Jack Ward Thomas, Chief, Forest Service, U.S. Department of Agriculture, before the Subcommittee on National Parks, Forests, and Public Lands and the Subcommittee on Oversight and Investigations, Committee on Natural Resources, U.S. House of Representatives, February 3, 1994.

practice of resource management on both public and private lands and, in the process, help to create landscapes and patterns of development that are sustainable both economically and ecologically.

Recognizing the need to accumulate ideas from ground-level experience, a research team at The University of Michigan's School of Natural Resources and Environment, acting with the financial support of The Wilderness Society, set out to complete an assessment of current experience with ecosystem management. The mission of the project was to catalog activities under way across the United States and analyze the experiences of the people involved. The following is a brief assessment of what the research team learned.

In conducting the research, the team first identified a set of 619 sites where ecosystem management was being carried out (Figure 1). To make the inventory as inclusive as possible, a broad set of selection criteria was applied. Sites were included if project managers sought to extend management across property or political boundaries to incorporate ecological boundaries or if they sought to shift management priorities away from emphasis on a single resource or species to consider ecosystem processes and the landscape as a whole. Hence, many of these projects are not textbook cases of ecosystem management (assuming there is one textbook definition). Rather, they incorporate elements of ecosystem-based approaches to land management in a significant way. The team "beat the bushes" broadly to identify candidate sites, through letters and telephone calls to a host of public and private agencies and institutions as well as review of a variety of documents.

The 619 candidate sites identified through this process are remarkably diverse and are located throughout the United States. The sites include projects in every state, with the number of projects per state ranging from 1 to 53. Some states are ecosystem management hot spots because of either an intensive agency focus on ecosystem management or the variety and uniqueness of ecosystems found in the state. Examples of such states are California and Colorado, with more than 50 projects reported in each. Considerable activity was also evident in Washington (37 projects)

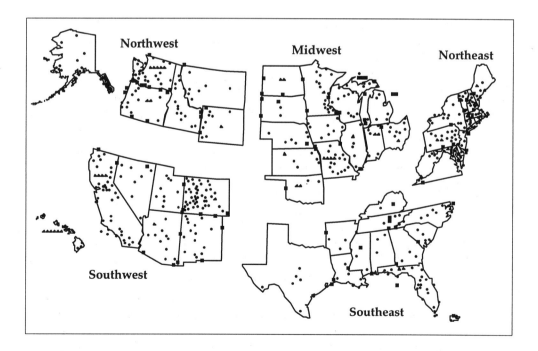

FIGURE 1. Location of the 619 "candidate" ecosystem management projects

and Minnesota (34 projects). Twenty-one to 29 projects were identified as under way in an additional eight states (Florida, Michigan, Missouri, New Mexico, New York, Oregon, Pennsylvania, and Wisconsin).

From these 619 candidate sites, 105 were chosen for analysis, taking into consideration their regional distribution and the types of agencies and groups involved. (Although the database contains 105 sites, respondents occasionally did not respond to all questions. Hence, *n*, the number of observations included in a particular element of the analysis, is often fewer than 105, as indicated in the figures.) For each of these sites, telephone interviews and mail surveys were conducted to obtain descriptive and evaluative information that was used in writing this assessment and the site descriptions that follow. The information about the sites was coded so that conclusions could be drawn about the number of sites evidencing different variables, such as project strategies or obstacles to progress. In the figures that are presented in the assessment, the data represents numbers of sites at which a factor was evident, as reported by site contacts and interpreted by the researchers. With the exception of the information about on-site stresses, most of the data represents answers to open-ended questions about projects and site characteristics. Hence, the figures probably underrepresent the actual percentage of projects evidencing a certain factor.

To the maximum extent possible, assessments by the individuals involved in these projects were used as the basis for evaluation; this assessment draws heavily on their words and insights. The participants' words are compelling, for they provide glimpses of the successes—both small and large—that lead many of those involved in these projects to be optimistic about the promise of ecosystem-based approaches to resource management.

Questions Addressed in the Assessment

The balance of this assessment summarizes the participants' experiences to date with ecosystem management. It is organized according to the following seven questions:

1. What characterizes the ecosystem management project areas?
2. What are the characteristics of the projects?
3. Why were the projects started?
4. What have the projects produced?
5. What has helped the projects to move forward?
6. What obstacles have the projects faced?
7. What do these experiences suggest for future ecosystem management efforts?

WHAT CHARACTERIZES THE
ECOSYSTEM MANAGEMENT PROJECT AREAS?

The 105 sites examined in the inventory exhibited a remarkable amount of diversity and mirrored the variety of ecosystems, landownership patterns, and land uses in the United States. Given the amount of interest expressed in ecosystem management by federal and state officials, one might expect that projects primarily involved public lands. In fact, projects in areas with predominantly private lands were more prevalent that those limited to activities on public lands. Human-induced stresses evident at these sites ranged from stream channelization to overgrazing by livestock and generally reflected prevailing land uses in different regions of the country. Some, such as urban land development and disruption of fire regime, were reported as a consistent stress throughout. These stresses have had a negative effect on ecological processes and composition, with many sites reporting the presence of rare or sensitive species along with other indicators of threats to ecological integrity.

Diverse Natural Features Reflecting the Variety of Ecosystems in the U.S.

Considerable variation in landforms and vegetation was evident in the project areas, reflecting the variety of ecosystems in the United States (Figures 2 and 3). Even projects located in the same state often displayed substantially different ecosystem characteristics. In addition to physiographic and vegetational differences, project areas also varied greatly in other ways, ranging from large-scale, rural, sparsely populated sites with few owners (as evident at the Mesa Creek Coordinated Resource Management Plan area in Colorado) to smaller-scale, urbanized, densely populated, fragmented sites with multiple owners (such as the Albany Pine Bush site in New York).

Presence of Federally Listed Threatened or Endangered Species
Respondents reported the presence of federally listed threatened or endangered species in 81 percent of the project areas. Commonly mentioned examples include

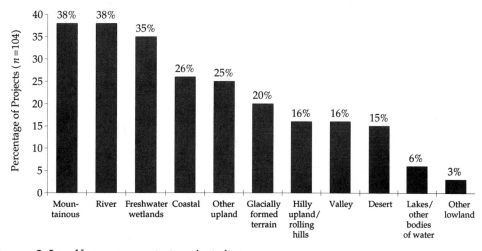

FIGURE 2. Landforms present at project sites

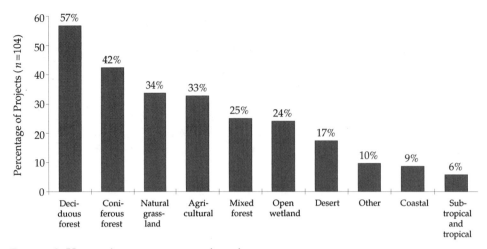

FIGURE 3. Vegetation present at project sites

the bald eagle (*Haliaeetus leucocephalus*), American peregrine falcon (*Falco peregrinus anatum*), gray wolf (*Canis lupus*), interior least tern (*Sterna antillarum*), and piping plover (*Charadrius melodus*). Some of the threatened or endangered species are endemic to particular areas and therefore were reported by only one or a few projects. For example, threatened or endangered species present in project areas in Hawaii, including the pendant Kihi fern (*Adenophorus periens*) and the Molokai creeper (*Parareomyza flammea*), do not occur outside the state.

Involvement of Both Public and Private Lands

In the Northeast, Southeast, and Midwest, more projects were located on private lands, whereas in the Northwest and Southwest, more projects were on public lands (Figure 4). Combining the data from these regions, 59 percent of the projects in the Northeast, Southeast, and Midwest involved predominantly privately owned lands;

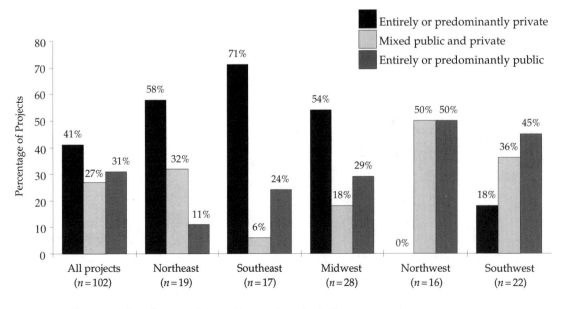

FIGURE 4. Landownership patterns at project sites

22 percent involved predominantly publicly owned lands (Figure 1 depicts how states are allocated into regions). This pattern stands in sharp contrast with the Northwest and Southwest, where only 11 percent of the project areas were privately owned and 47 percent are publicly owned. However, this difference is understandable when viewed in light of overall landownership patterns in both areas. Federal landholdings, especially those managed by the U.S.D.A. Forest Service and the Bureau of Land Management (BLM), are much more extensive in the Northwest and Southwest than in other parts of the country.

Existence of Numerous Anthropogenic Ecosystem Stresses

Nearly all project areas were experiencing one or more human-caused ecosystem stresses. While the importance of the stresses varied by region, differences generally can be attributed to regional differences in land use patterns. Respondents were asked to rank a list of fourteen stresses to the ecosystem on a scale of 1 to 5, where 1 indicated "not a stress" and 5 a "severe stress." A stress was considered significant to the ecosystem and project area if it rated 4 or 5. Figure 5 shows percentages of projects that ranked particular stresses as significant.

Although some stresses were clearly reported as significant more often than others (e.g., hydrologic alteration was rated as a 4 or 5 by 42 percent of the projects, whereas overfishing, overhunting, or overcollecting was rated as a 4 or 5 by only 13 percent), there was no clear winner of the title "worst stress." However, some interesting trends may be observed by considering regional patterns. A few stresses, including disruption of fire regime, seemed equally important throughout the country. On the other hand, some stresses were clearly more significant in certain areas. In most cases, these patterns probably reflect regional differences in land use, since most of the stresses are induced by human activities. Hence, grazing and range management were most often mentioned as significant stresses in the Southwest, the region where range management is most prevalent. Similarly, forest management was most often mentioned as a significant stress in the Northwest, and agricultural practices and land conversion for agricultural purposes were more frequently rated as significant stresses in the Southeast and Midwest.

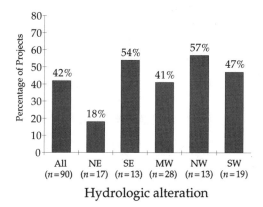
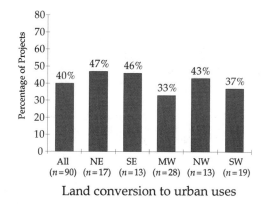

FIGURE 5. Anthropogenic ecosystem stresses at project sites: percentages of projects rating particular stresses as significant (presented in descending order of overall percentage)

Key: All = All sites; NE = Northeast; SE = Southeast; MW = Midwest; NW = Northwest; SW = Southwest

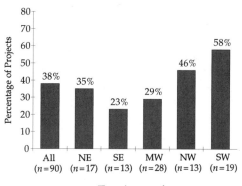

Exotic species

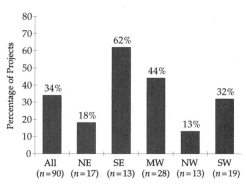

Agricultural Practices

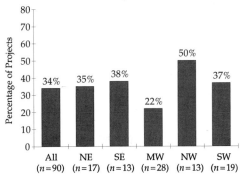

Roads or other infrastructure

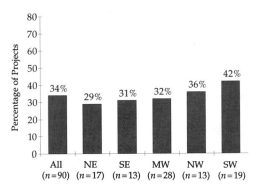

Disruption of fire regime

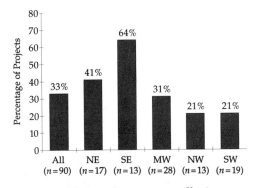

Non–point source pollution

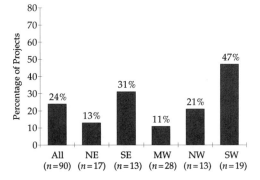

Grazing and range management

FIGURE 5. Anthropogenic ecosystem stresses at project sites *(continued)*

Key: All = All sites; NE = northeast; SE = southeast; MW = midwest; NW = Northwest; SW = Southwest

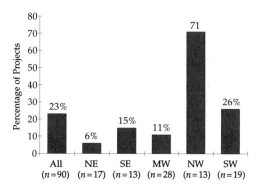

Timber/forest management

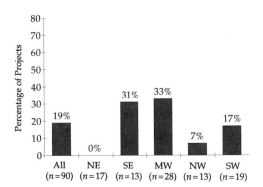

Land conversion to agricultural uses

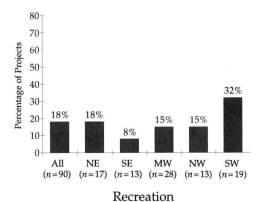

Recreation

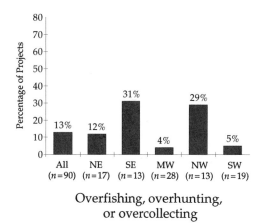

Point source pollution

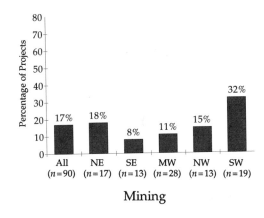

Mining

Overfishing, overhunting,
or overcollecting

FIGURE 5. Anthropogenic ecosystem stresses at project sites (*continued*)

Key: All = All sites; NE = northeast; SE = southeast; MW = midwest; NW = Northwest; SW = Southwest

WHAT ARE THE CHARACTERISTICS
OF THE PROJECTS?

Just as the project areas are diverse, so too are the approaches taken in the projects themselves. For example, statewide ecosystem management initiatives such as the Missouri Coordinated Resource Management project differed substantially from highly focused small-scale restoration projects such as the Camp Johnson Sandplain Restoration effort. Some activities, including the Applegate Partnership, were started by private citizens, while others, such as the Interior Columbia Basin Ecosystem Management Project, reflected the interests of high-level officials, including the President and the Congress. Nevertheless, some common patterns emerge: Most of the projects were fairly young and were initiated by a range of federal and state agencies and The Nature Conservancy. Once underway, the projects involved a full range of public and private organizations, with various groups playing different roles in site efforts. Project goals and strategies emphasized outcomes, such as ecological restoration, and the processes needed to eventually reach them, such as involving stakeholders in collaborative efforts.

Most in Planning or Early Implementation Stages

Although principles of ecosystem management have been discussed for several decades, most projects in the inventory were started only recently (Figure 6). As of 1995, more than one-third of the projects were still in the planning stage. Nearly two-thirds of the projects had initiated at least some implementation activities.

Initiated Primarily by Federal and State Agencies and The Nature Conservancy

The way a project is started often significantly affects how it unfolds. The differences in start-up among the projects often reflected the varying missions and styles of the organizations that initiated the projects. In the sample of sites included in this

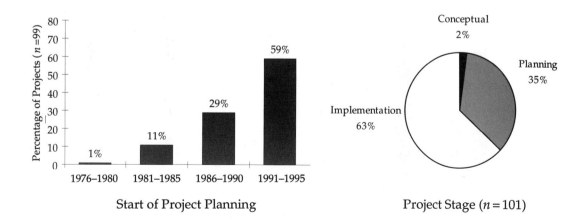

FIGURE 6. Dates of start-up and stages of projects

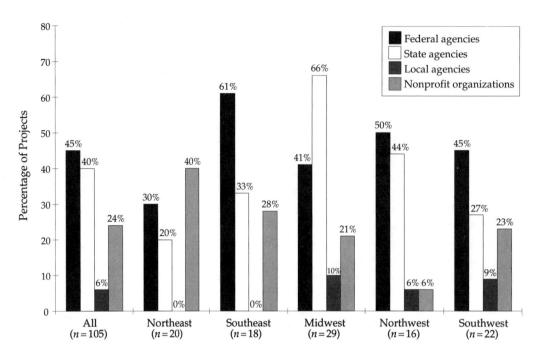

FIGURE 7. Initiating institutions by region (*Note*: Many projects were initiated by more than one agency or organization. Therefore, some percentages in Figures 6 and 7 add up to more than 100 percent.)

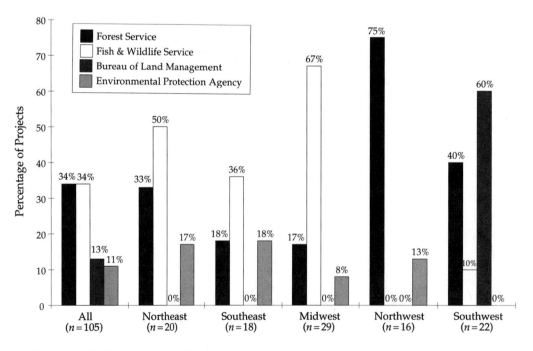

FIGURE 8. Federal agencies initiating projects by region

analysis, federal agencies initiated the largest portion of projects, followed closely by state agencies and then nonprofit organizations (Figure 7). This general pattern was found throughout the country. Of federal agencies that initiated projects, the Forest Service and the U.S. Fish and Wildlife Service (FWS) initiated the largest number, followed by the BLM and the Environmental Protection Agency (EPA). Not surprisingly, the Forest Service and the BLM tended to initiate more projects in the West, whereas the FWS started more in the East and Midwest (Figure 8). Other federal agencies initiating projects included the National Park Service, the Natural Resources Conservation Service (formerly the Soil Conservation Service), the Army Corps of Engineers, the Geological Survey, the National Biological Service, and the Tennessee Valley Authority. State natural resources agencies initiated most of the state-led programs; other state offices initiating projects included transportation and planning departments and governors' offices. The Nature Conservancy (TNC) initiated the vast majority of projects started by nonprofit organizations, reflecting the land stewardship emphasis of this organization.

A Full Range of Agencies and Organizations Involved After Project Initiation

Additional agencies and organizations became involved in the projects later in the planning process or during implementation. As a result, the number and diversity of involved groups go far beyond those of the initiating organizations. Public agencies from all levels of government (federal, state, regional, local, tribal) were involved, as well as private landowners, nonprofit organizations, industry, universities, and citizens (Figure 9). Levels of participation varied from organization to organization and from project to project. The median number of groups participating in a project was 11, with 82 percent of the projects involving more than 5 groups. Eighteen percent of the projects reported more than 20 participating groups, with four projects reporting more than 100.

The range of agencies and interests involved in the projects was wider than that of groups that initiated projects. Nevertheless, the patterns of involvement—which agencies and organizations were most commonly involved, and where—were roughly the same. The FWS and the Forest Service were the most commonly involved federal agencies, followed by the BLM, the Natural Resources Conservation

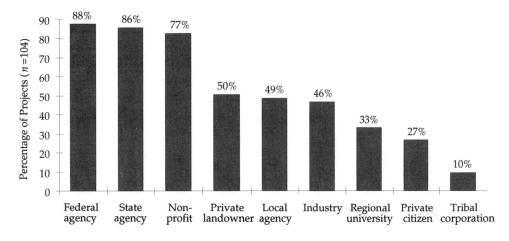

FIGURE 9. Agencies and organizations involved in the projects

Service, the National Park Service, the Environmental Protection Agency, the Army Corps of Engineers, and, to a lesser extent, other federal agencies. Of the national and local nonprofit groups that became involved, TNC participated most frequently. While few industrial or user groups initiated efforts, representatives from a variety of industrial sectors took part in projects once they were under way. Forest companies were involved in twenty-one projects, followed by mining interests participating at twelve sites. Other industries represented were agriculture and grazing, recreation, commercial fishing, real estate, and utilities.

Emphasis on Ecosystem Protection, Preservation, or Restoration and the Human Relationships Needed to Achieve Them

The five most commonly reported project goals were protecting or preserving the ecosystem, restoring the ecosystem, obtaining stakeholder support, maintaining or improving the local or regional economy, and providing guidelines for ecosystem management (Figure 10). Protecting or preserving the ecosystem was by far the most commonly reported goal. Preservation-oriented goals included increasing biodiversity, protecting the ecosystem from anthropogenic threats, and stabilizing or increasing a population of a sensitive or rare species. For instance, among the goals of the Georgia Mountain Ecosystem Management Project in northeastern Alabama were maintaining and enhancing biodiversity at a variety of scales (e.g., species, community, landscape) as well as protecting rare or unique species and populations. Similarly, the primary goal of the East Fork Management Plan for the Wind River ecosystem in northwestern Wyoming was to perpetuate the region's wildlife by preserving habitat.

The goal of ecosystem preservation is closely linked to ecosystem restoration, which includes restoring ecosystem processes and components and improving water quality. An example is provided by the Indiana Grand Kankakee Marsh Restoration Project. Its ten-year goal is the restoration and protection of 26,500 acres of wetlands and associated uplands in the watershed. Similarly, the goals of the Camp Johnson Sandplain Restoration project in Vermont are to restore and maintain a mosaic of old-growth, presettlement white pine and oak forest and pitch pine and oak heath woodland.

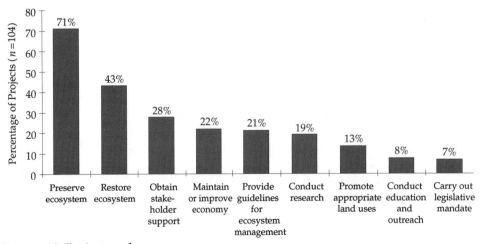

FIGURE 10. Project goals

In some instances, intermediate stages in the process to achieve the first two goals themselves became goals. One of these, obtaining stakeholder support, was reported as an important goal of many projects. Stakeholder support may be obtained through formation of partnerships withall organizations and individuals that have an interest in the ecosystem. Support from stakeholders is especially important if the project area is large and stakeholders are many. A good example is provided by the Blue Mountains Natural Resources Institute in Oregon, which has eighty-three partners to date from such diverse interests as county commissions, chambers of commerce, colleges, private landowners, federal and state agencies, tribal governments, unions, schools, and courts. The partners in this project help determine project direction and assist with funding and project implementation.

Maintaining or improving the local or regional economy was especially important in the Northwest and Southwest, where many economies are resource based. As explained by a respondent from Alaska:

> I think that you have to spend some time in the political arena to cover the social and economic ends of this too. I think we can make a good argument for [the project] biologically, but the reality is that there has to be some social acceptance for it or it won't fly.

Since many of the ecosystem management projects are relatively young, many sought to develop guidelines for planning or implementing projects. Some sites listed development of an ecosystem management plan as a major goal, while others sought to develop a regional framework within which smaller ecosystem management projects could be designed and implemented. For example, the Gulf of Maine Rivers Ecosystem Plan project in several northeastern states provided a regional framework for coordination of local efforts, and the Northern Lower Michigan Ecosystem Management Project was developing a "Resource Conservation Guidelines" document to provide vision and guiding principles for future planning efforts in the region.

Common Strategies for Achieving Goals

To achieve their goals, project managers had to design and implement numerous strategies. The six most commonly reported strategies were research, stakeholder involvement, ecosystem restoration, promotion of compatible human land uses, education and outreach, and land protection through set-asides (Figure 11). The prevalence of some of these strategies is indicative of the young age of many of the projects. Developing an understanding of ecosystem structure and function and fostering cooperation among the multitude of stakeholders in an ecosystem provide a foundation on which ecosystem management projects are built.

Research, the most frequently reported strategy, can be part of all stages of a project. Research helps a natural resource manager determine which issues—biological, social, and economic—he or she faces. In addition, to manage an ecosystem, a manager needs to be aware of the types of ecosystems under his or her management and of the components of these ecosystems. Thus, inventorying was an important component of research, including identifying natural areas and their components and prioritizing these areas for protection. Inventory work was a strategy reported by thirty-eight sites; it included not only ecological factors but social and economic indicators as well. In several cases, inventories were being used as a reliable assessment of an area's problems, to be used by all stakeholders in fashioning

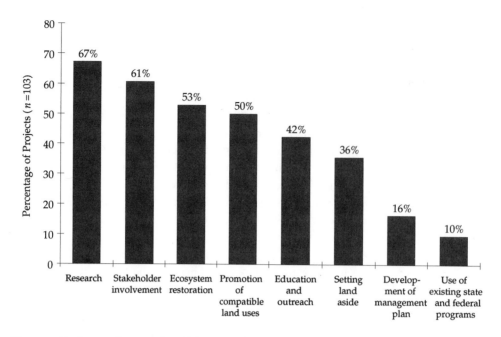

FIGURE 11. Strategies employed by projects

remedial actions. To determine whether management practices were having the de-sired effect on the ecosystem, monitoring was commonly carried out.

Stakeholder involvement was another frequently mentioned strategy. Since eco-systems almost always cross landownership boundaries, a variety of landowners may need to coordinate their land management activities. Even in the case of public lands, where a single agency is the landowner, many people other than the land management agency use the ecosystem for a variety of purposes. Many respondents noted that it was imperative for the success of ecosystem management projects that all stakeholders be involved in development and implementation of project activi-ties. For example, during development of the Ruby Canyon and Black Ridge Ecosys-tem Management Plan in Colorado, the BLM was provided with many valuable in-sights by an ad hoc committee consisting of representatives of user groups; local, state, and federal agencies; grazing interests; and community representatives. Other stakeholder involvement strategies ranged from open houses to federally chartered advisory committees.

Ecosystem restoration strategies involved restoring ecosystem composition and function. Common examples were restoration of fire regime, reforestation, wetland restoration, and control of non-native plants. For example, fire was reintroduced in the Stegall Mountain Natural Area in Missouri in order to restore its igneous glades and oak savannas. A bison herd was reintroduced on the Oklahoma Tallgrass Prairie Preserve in an effort to mimic natural grazing processes, and on the Patrick Marsh Wetland Mitigation Bank Site in Wisconsin, drainage pumps were removed to allow a historic marsh to reclaim the cornfield that had taken its place.

Promotion of human land uses compatible with ecosystem function and structure included the use of innovative agricultural techniques (e.g., conservation tillage and range improvements), erosion reduction strategies, timber management strategies, and management of recreational activities. For instance, conservation tillage is promoted in the Fish Creek Watershed Project in Indiana and Ohio by financially assist-ing farmers in purchasing their first piece of conservation tillage equipment.

Education and outreach strategies are somewhat related to stakeholder involve-ment strategies. To ensure their participation in an ecosystem management project, it may be necessary to instill in stakeholders (including the general public) an aware-ness of and respect for the ecosystem as well as an understanding of the project. Methods employed by project managers included sending newsletters to landown-ers, holding open houses, and having project staff members present their ideas at meetings of stakeholder groups.

Setting aside land for protection consisted primarily of either purchasing land for preserves or securing voluntary easements that restricted harmful land use prac-tices. In a few cases, cooperative land management agreements were arranged, in which a public or private conservation organization oversaw or cooperatively man-aged another landowner's property. For example, Kansas State University's De-partment of Biology manages TNC-owned Konza Prairie Research Natural Area. Similarly, in the Guadalupe-Nipomo Dunes Preserve area of central California, TNC holds a twenty-five-year lease on lands owned by the Santa Barbara County Parks Department and has an agreement with the California State Parks Department to manage state-owned property as a natural area.

WHY WERE THE PROJECTS STARTED?

Ecosystem management projects were initiated for a variety of reasons, and respondents identified more than one reason for most projects (Figure 12). While preexisting agency programs clearly provided an opportunity for interested parties to move forward, many stakeholders were motivated by concern about degradation of the character of unique ecological areas. A sense of crisis was not always present at these sites; nevertheless, a strong shared sense of concern often translated into public pressure for managers to act in new ways.

Agency Policies and Programs Provide Incentive

Fifty-two percent of the respondents reported that their projects were motivated by agency policies and programs. Not surprisingly, agency programs were much more important for state and federal agencies than for nonprofit organizations. In some instances, ecosystem management was adopted by agencies as an overarching management paradigm to be incorporated into most agency activities. In other cases, agency policy required the development of specific ecosystem management plans. Respondents indicated that the Forest Service's New Perspectives Program, the National Estuary Program administered by the EPA, and the North American Waterfowl Management Program administered by the FWS provided incentives for initiation of ecosystem management projects.

The Uniqueness of Ecosystems at Project Areas

Throughout the country, people have come to recognize the uniqueness of many ecosystems and the need for their protection. Uniqueness of the project area motivated the start-up of 47 percent of the ecosystem management projects. Examples of such projects include the Lajas Valley Lagoon System in Puerto Rico and the Albany Pine Bush in New York as well as virtually all projects initiated by TNC. Sometimes

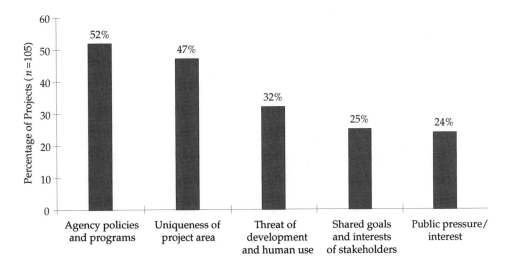

FIGURE 12. Factors motivating project initiation

the presence of threatened or endangered species contributed to the uniqueness of a project area. Indeed, in some cases the project area contained the primary remaining population of a species, as in the case of the Catalina Island mountain-mahogany, a plant proposed for federal listing as an endangered species, and found only on the Santa Catalina Island Ecological Restoration Program site in California.

To Mitigate the Effects of On-Site Stresses and Respond to Threats of Further Damage

Many ecosystems are threatened by human activities, such as urban and industrial development, resource extraction (including timber management, mining, and grazing), and recreation. One-third of the projects in the inventory were motivated by such threats. For instance, a proposed airport was the impetus for the eventual completion of a comprehensive management plan for New Jersey's Pinelands. Increasing recreational pressures of many kinds (ranging from hikers and mountain bikers to river floaters and offroad vehicle users) were important factors in the development of the Ruby Canyon and Black Ridge Ecosystem Management Plan in Colorado.

Shared Goals and Interest of Stakeholders

One-quarter of the projects were started because of shared goals and interests of stakeholders. In some projects, stakeholders recognized opportunities for all to gain through joint action. In others, they understood that they faced common problems. An increased perception that the groups share common ground motivated efforts at coordination and collaboration. A good example is the Northern Lower Michigan Ecosystem Management Project, where the Michigan Department of Natural Resources (MDNR) and the Forest Service recognized that lands under their jurisdiction often share administrative and ecological boundaries. As one respondent from the MDNR explained:

> We were aware that some of the same ecosystems that the Forest Service is involved with are the same ones we're involved with. Sometimes our land shares mutual boundaries, and [in other cases] we're very close to each other. It would certainly be an obvious kind of conclusion that one would make, I would think, that maybe we should work together in our planning.

Public Pressure or Interest

Public pressure or interest was responsible for the initiation of one-fourth of the projects. Public concern for protection of ecosystems, as well as for protection of natural resource–based economies, may result in citizen initiatives or in citizen pressure on agencies to take action. For instance, in northeastern Oregon and southeastern Washington, citizens were concerned that the Forest Service might not be adequately researching natural resource problems relevant to local needs and concerns. These citizens subsequently were instrumental in the formation of the Blue Mountains Natural Resources Institute. Another example from Oregon is the Applegate Partnership, initiated by two community residents in response to shared concern about the degrading resource base in the Applegate River watershed. Since the communities in southwestern Oregon are resource dependent, long-term health of forest resources was seen to be in the interest of many different groups in the region.

WHAT HAVE THE PROJECTS PRODUCED?

Rather than impose a specific definition of success for ecosystem management projects, the research team asked participants to describe the success of the projects in their own terms. In response, virtually every respondent cited specific positive outcomes (Figure 13). These outcomes are presented here as proxies for evaluating success, since they incorporate several possible measuring sticks for success: realization of goals, effective implementation of strategies, establishment of activities that are likely to lead to successful ecosystem management in the future, and other desirable results, whether anticipated or not.

Much Early Success Measured in Process Terms

While one-third of the projects reported specific ecological results, the five outcomes cited most frequently can be viewed as procedural in nature. Achieving better communication, developing a management plan and/or new approaches to management, creating new decision-making structures, and initiating restoration activities do not themselves immediately improve the ecological situation on the

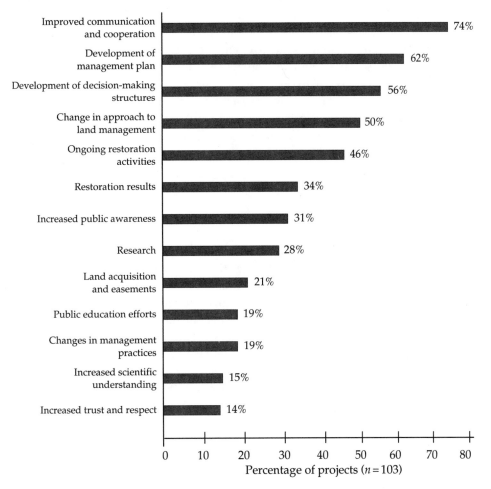

FIGURE 13. Project outcomes

ground, though they do yield other important benefits at project areas. Rather, they establish relationships and management approaches that, it is hoped, will lead to successful ecological results sometime in the future. This finding is partly a result of the relative youth of many of the projects and the fact that ecological results often take years to be observed. It appears that small successes at improving the process of management will motivate larger successes that can be measured in ecological terms in the future.

At the same time, these process improvements are important in themselves. Many respondents seemed exhilarated by successful formation of partnerships and glimpses of a style of management that can garner public enthusiasm and support. Ecosystem management is fundamentally a long-term process of human interaction and decision making that yields improved ecosystem health along the way. The data from this assessment of early experience with ecosystem-based approaches to land management suggests that important process changes are under way.

Increased Communication and Cooperation

Three-fourths of the project respondents cited increased and improved communication and cooperation as an important outcome of their projects. A large portion of respondents cited improvements in interagency relations at all levels of government, including improved relations between two or more federal agencies, between state and federal agencies, or between local and state governments. Another large segment reported improved relations between public agencies and the private sector, including nongovernmental organizations, private landowners, the business community, and the general public. Improvements in communication among stakeholders in the private sector were also evident as landowners and resource users within a project area began to talk to one another in new ways.

A number of projects also resulted in improved communication and cooperation within agencies, in part due to efforts to educate employees about the activities of different offices within an agency. For example, a FWS employee involved in the Gulf of Maine Rivers Ecosystem Plan noted:

> An immediate success is that [FWS] offices are communicating and there is a greater understanding of each other. This will reduce the duplication of effort and will help with resource sharing.

Development of Management Plans

The development of management plans, enunciating the goals and strategies of ecosystem management and often providing guidance to a variety of participating agencies and groups as they implemented their own projects, was the second most commonly reported outcome. A number of these management plans, including draft and final versions, were developed by multiple stakeholders, often through open public forums using a consensus-based approach. Management plans at twenty-two sites were described as internal documents in that they were developed by one institution with minimal or no outside input. In many cases, the plans were meant simply as internal guiding documents rather than part of a broader public process. One FWS respondent explained that his agency needed to look inward before embarking on ecosystem management:

> We are trying to decide how to run our own ship before we bring in others. In some ways, this is internal FWS decision making about our own priorities, and we don't need to involve the other parties initially.

Some project areas had multiple management plans, with each institution developing a plan for lands under its jurisdiction. In the absence of a single, overarching document for the project area, it was unclear whether these plans were unified in their approach, although they were generally described by respondents as having an ecosystem focus.

Development of Decision Making Structures

The creation of partnerships, management committees, and task forces was the third most frequently cited outcome, with 56 percent of the sites reporting the establishment of some type of coordinating body. These decision-making structures were seen as central to the ecosystem management effort and were used to reach consensus among stakeholders, develop management plans, identify tasks, and serve as communication tools, among other functions. They ranged from formal, federally chartered advisory committees to unofficial, loose-knit partnerships among stakeholders.

The vast majority of these decision-making bodies appeared to cross agency and institution boundaries, being far more inclusive of multiple stakeholders in the decision-making process than has been the case traditionally. Only a small portion appeared to be purely internal management teams. In these cases, stakeholder and public involvement may be limited to public comment and response procedures or simply informal consultations.

Several projects had multiple committees addressing different functions such as research, outreach, and funding. For example, National Estuary Program (NEP) committees were structured around groups of stakeholders: The Corpus Christi Bay NEP in Texas has policy, management, citizen, local government, and scientific/technical advisory committees that carry out the work of the project.

Generally, the larger the project area or number of stakeholders, the greater the likelihood of finding partnerships or other decision-making bodies. Only one-third of projects under 50,000 acres in size had multiparty decision-making bodies in place, while three-fourths of those over 1 million acres in size had such structures. In the case of the Indiana Grand Kankakee Marsh Restoration Project, it was recognized that no one agency could singlehandedly accomplish all the project's goals. At the same time, it was understood that potential cooperators had their own agendas that needed to be addressed in a collective effort. In this instance, the partnering concept was used to gain more cooperators, drawing strength from the activities of multiple groups that were involved primarily to pursue their own interests:

> [Our] approach taps into the interests of large numbers of people and lots of groups. The groups themselves . . . may focus on only one particular aspect. They may be a water quality organization. They may be interested in endangered species, or protecting rare habitats. But . . . focusing in on the ecosystem itself and restoring the historic habitats of the ecosystem . . . tends to bring those diverse partnerships together. And they all see something in it for themselves, for their organizations, or for them individually.

Changes in Approaches to Land Management

Fifty percent of the respondents reported changes in management philosophy, including a shift in emphasis from single to multiple species or outputs, an emphasis on landscape-level designs, incorporation of holistic approaches, or a general focus on ecosystem considerations in management decisions. For example, the Plainfield Project in Massachusetts sought to incorporate more landscape-level knowledge rather than knowledge restricted by artificial administrative boundaries. As noted by one respondent: "The purpose is to not do things in a vacuum. [We are looking] at the bigger picture in order to optimize the ecological and economic value."

At the time of this writing, some of these changes were still at the conceptual stage while others had led to changes in specific new management activities. What is important is that these changes in management approach stand in contrast to previous or traditional approaches. For example, a shift away from a traditional buy-and-protect strategy for land preservation is emerging as an important consideration for land managers. Several TNC respondents stated that their organization could no longer achieve conservation goals by simply purchasing a piece of land and putting a fence around it. In the case of the Virginia Coast Reserve, TNC's approach is:

> [to look] beyond simply identifying, buying and protecting a parcel to a much broader ecosystem thinking. ... You need to get out into the community, listen to them and find common goals rather than focusing on the differences. ... Otherwise, you end up with little preserves with who knows what surrounding them. ... You are setting yourself up for disaster. ... Everything that goes on in the buffer has an impact.

A Wide Range of Restoration Activities, Some Showing Ecological Results

Almost half of the respondents cited ongoing ecosystem restoration activities as important outcomes of their projects, and ecological results had been achieved at one-third of the project sites. Restoration of fire regime, reintroduction of native plants and animals, and restoration of hydrology are a few examples of activities reported by the respondents. Other restoration efforts, such as removal of exotic species or control of water pollution, were designed to control anthropogenic stresses. In most situations, restoration was limited to selected portions of a project area. For example, restoration of the fire regime on the Albany Pine Bush project area in upstate New York was quite limited for a variety of ecological and social reasons.

What Has Helped the Projects Move Forward?

The 105 sites in the inventory differ from one another in many ways, and the projects on these sites have progressed largely by building on the supportive aspects of their local settings. Each area has unique resources to draw on as well as unique problems to deal with; the most successful project managers found ways to mobilize those resources in response to site-level problems. In aggregate, though, a set of common factors appears to facilitate success of the projects (Figure 14). The support of stakeholders, public agencies, and the general public often allowed initiating organizations to proceed and contributed to the willingness of many groups to collaborate on project activities. Having resources available to support the projects and having individuals dedicated to carrying out the projects were also critical. At the same time, the character of certain areas and projects also affected the ability of involved agencies and groups to carry projects out.

Collaboration

It is clear that the collaborative efforts of people coming together to create new solutions for managing resources were a vital component of many of the projects. Indeed, collaboration, more than any other variable, was cited by respondents as critical to their projects' progress. As one respondent from the Indiana Grand Kankakee Marsh Restoration Project explained, "The strength of the many is greater than the

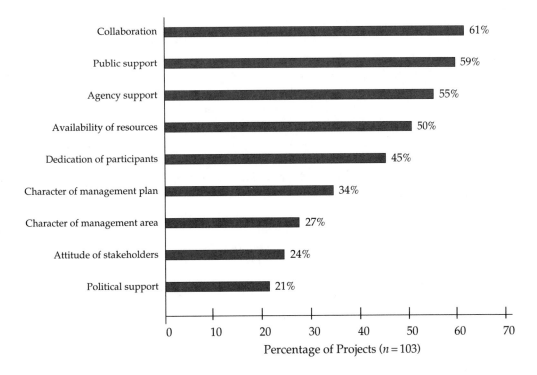

FIGURE 14. Factors facilitating progress

strength of the individual organization." Collaboration within and among public agencies, private landowners, nongovernmental organizations, businesses, and the general public was an important mechanism for increasing cooperation and communication, fostering trust and understanding among participants, and allowing a greater set of interests to be met. In some cases, such efforts have contributed greatly to changing public opinion about government agencies and developing positive relationships among stakeholders. Increases in interagency cooperation were notable, with several respondents reporting a decline in turf battles between agencies. In many cases, agencies displayed greater flexibility and made impressive strides toward coordinating planning and management with other agencies and stakeholders.

Collaborative approaches produced benefits through several mechanisms. Many project managers used consensus-based processes that allowed participants to have an equal voice in decision making. Such processes fostered greater ownership of decisions and allowed varying interests to find common ground. In addition, the ability of participants to share funding responsibilities and pool resources reduced duplication of efforts, created more funding opportunities, and promoted more efficient use of limited resources. One respondent stated that "a recognition . . . among government agencies [that] no single agency can accomplish the task [alone]" was vital to efforts to protect the last remaining wetlands in their region.

Public Support Developed Through Stakeholder Involvement and Community-Based Initiatives

Gaining public support was an important factor for 59 percent of the ecosystem management projects. Many respondents emphasized the importance of identifying all stakeholder groups and involving them in planning and management decisions from a project's inception. Others explained how more conventional efforts had been obstructed by the opposition of stakeholders who, having been excluded from the decision-making process, believed that their interests were not well represented in management plans.

In regions where a large proportion of the land was privately owned, ecosystem restoration or protection efforts progressed only with the support and participation of private landowners. One such example was provided by the Wildlife Habitat Improvement Group in southern Vermont, which relies on a voluntary consortium of private landowners to coordinate woodlands management across ownership boundaries in order to enhance wildlife habitat. Landowner support for this effort derives largely from stability in landownership over many generations. Because of the area's rural character, landowners already have a tradition of communicating with their neighbors and thus know their land and their neighbors well.

Projects rooted in the local community were better received than those perceived as top-down agency directives or outsider initiatives. Gaining support from local community leaders and hiring project personnel from within the community were both cited repeatedly as important. One respondent from a rural project area in the South explained: "The fact that I was a local boy, grew up here, knew lots of folks, and the fact that I didn't have a government uniform on, made all the difference in the world." A respondent from the Molokai Preserves in Hawaii described the importance of being accepted by the community:

> If you want [support] from the community, you have to hire people from the community, someone who knows the community and thinks like the community. . . . I was born and raised here, and this helped [me] get a foothold here, because people are cautious when new groups come in.

Agency Support of Many Types

Respondents from 55 percent of the sites reported the support of agencies from many levels of government as important to their progress. Development of federal policies and programs supportive of ecosystem management has enabled managers to implement ecosystem-level projects. As discussed in the previous section about factors promoting the initiation of projects, FWS, Forest Service, and EPA programs in particular provided the authorization, flexibility, funding, and technical assistance necessary to support nontraditional resource management activities.

State programs also provided assistance to projects. For instance, the Hawaii Department of Land and Natural Resources' Division of Forestry and Wildlife has established a Natural Areas Partnership program to provide matching funds for management of private lands that are permanently dedicated to conservation. Two projects in this study benefit from that program: Molokai Preserves, coordinated by TNC, and the Pu'u Kukui Watershed Management Area, involving lands owned by the Maui Land and Pineapple Company.

While federal and state programs facilitated on-the-ground progress, it is also important that support for ecosystem management initiatives be infused *throughout* agencies. Many respondents noted that the concurrence and enthusiasm of higher-level administrators were instrumental to the progress of their projects. At times, this support resulted from the creation of new agency policies that were supported by top administrators. In cases in which no agencywide ecosystem management policy existed, pilot projects received internal support that allowed project leaders greater latitude in designing innovative programs.

Availability of Resources

Resources of many different kinds were needed to mount the projects, including human resources, funding, time, and equipment. Fifty percent of the respondents credited availability of resources as critical to the success of their projects, and funding was cited as the most important resources-related factor facilitating progress. It was the initial funding of many of these projects that got them off the ground by enabling managers to hire personnel, acquire lands, collect data, or invest in technology for ecosystem management activities. Several sites reported benefiting from an adequate number of staff members, the expertise and technical capability of field personnel, and the presence of full-time researchers and project managers whose responsibilities were limited to the project. In several cases, especially the National Estuary Programs supported by the EPA in cooperation with state agencies, separate offices were established for the projects, which allowed staff, equipment, and expertise to be dedicated exclusively to them.

Physical resources such as equipment were also reported as important. Specifically, several respondents reported that the availability of geographic information system (GIS) technology greatly expanded their management capabilities. In some cases, GIS allowed managers to undertake more comprehensive ecosystem-based analyses and create integrated management designs. In others, it was beneficial simply by promoting standardization of data and allowing more rapid exchange of information among groups.

Dedicated, Energetic, and Capable Participants

Forty-five percent of the respondents noted the importance of having motivated, dedicated individuals to creating and maintaining momentum on ecosystem management projects. Project leaders, community leaders, agency field staff members, natural resource managers, landowners, and elected officials all played this role in various projects, often keeping ecosystem management projects alive despite a lack of resources, political or public support, or agency direction. These individuals served as a source of motivation for change and fostered stakeholders' trust in and support for the goals of projects.

What characterized these individuals? They understood the benefits that innovation provided to the projects and the importance of garnering broad stakeholder support. At the same time, they often tried to downplay their own contributions so as not to dominate the process and demotivate other partners. Often they were willing to take risks and engage in entrepreneurial behavior within their organizations. They were not "superhuman." Rather, they put a lot of energy into moving the projects forward. One respondent in the Northwest summarized the opinions of a number of respondents: "It always boils down to key talented people [who] are willing to invest themselves over and beyond the call of duty."

WHAT OBSTACLES HAVE THE PROJECTS FACED?

Natural resource and environmental management is intrinsically difficult because it involves multiple interests and values, often intangible benefits, and long time scales. Ecosystem-based approaches to land management exacerbate some of these problems by requiring the active involvement of numerous stakeholders, an enhanced focus on ecosystem integrity, and a robust understanding of ecosystem components and processes. Ecosystem management is viewed by many agencies and groups as a new and uncertain approach to management, which can be perceived as threatening to long-standing agency norms and stakeholder interests. At the same time, ecosystem management faces the same set of problems encountered by any project, including limited time, funds, and expertise. Respondents highlighted five common obstacles: opposition from both public and private sources, constrained resources, problematic agency attitudes and procedures, scientific uncertainty, and inadequate or ineffective stakeholder involvement (Figure 15). Less frequently cited obstacles included ecological problems (i.e., problems too great to overcome), continued development pressures, and problems associated with fragmented landownership.

Opposition from the General Public, Political Circles, and Private Interests

Opposition to the concept of ecosystem management, to an effort at a particular site, or to government in general was the most frequently reported obstacle to a project's progress. Resistance from the general public was common, resulting from misperceptions about what ecosystem management entails or simple skepticism in the face

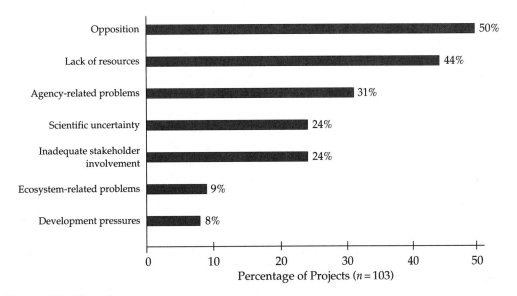

FIGURE 15. Obstacles to progress

of significant changes in land management. As one respondent from the Northeast Chichagof Island effort in Alaska noted, "There's a lot of misconception out there over what [ecosystem management] is, what it means, what it can do." At some sites, such generalized fears about ecosystem management translated into project-level opposition. Commenting on the Block Island Refuge project in Maine, one respondent noted that there was "skepticism by some local citizens on the effect of land protection on local tax rolls. There is a belief that it hurts the local economy." Unfavorable past experience with government regulations or programs helped fuel these negative attitudes on the part of the general public. According to one respondent: "The public feels they've been duped a number of times [by the Forest Service], and that's why this ecosystem management effort that we're into right now has got such a challenge socially."

Political opposition, including outright opposition by politicians at all levels of government, also figured prominently among site-level problems. Several practitioners commented that the political climate created by the 104th Congress generated an overarching sense of uncertainty about the future of ecosystem-based approaches to resource management. Many expressed concern that existing programs and continued funding were no longer as "safe" as they had been and that hopes for securing additional authority or appropriations under the heading of a new and often controversial management paradigm would be slim at best.

Opposition by landowners, industry, property rights advocates, and environmentalists was also highlighted as an obstacle to ecosystem management projects. Landowner opposition was manifested as a generalized distrust of government as well as a cynicism about agencies' motivations for undertaking ecosystem management. Landowners often feared that government participation in a project would lead to additional government regulations or infringement on perceived private property rights. Although landowner participation in most projects is strictly voluntary, it can be difficult to convince the public that state or federal agencies are not involved for regulatory purposes, as a respondent from a project in Colorado explained:

> [There is a] fear that the feds have an ulterior motive . . . to discover threatened and endangered species to shut down operations. . . . People outside the project —area landowners, local commissioners, etc.—still are wary that ecosystem management [might be] an attempt by government agencies to control private lands.

Limited Resources

A shortage or absence of adequate resources—human resources, funding, equipment, time—was reported by 44 percent of project respondents, with the vast majority pointing to funding shortages as a problem for both public and private efforts. Funding shortages were seen as barriers to many site-level activities, ranging from inventory work and ecological research to restoration and land acquisition. Since ecosystem management often achieves ecological results through efforts sustained over many years, respondents also feared that a lack of long-term funding would hamper their ability to achieve results.

There are several reasons for difficulty in obtaining funds. A number of projects reported problems in planning for long-term land management such as ecosystem management when funding processes operate on short-term, yearly cycles. In addition, ecosystem management does not fit into the appropriations structures of most governments. Legislatures at all levels traditionally allocate funding based on line items or program boundaries, or wholly within a specific agency, whereas ecosystem management cuts across line items and administrative boundaries. In general, this management-funding conflict was not resolved and remained frustrating for many project coordinators, as the following respondent from a federally led project indicated:

> You go into the budget exercise in Washington asking for money on a program basis but you're asking managers of the ecosystems to manage everything as "ecosystem management." It is very confusing right now. I think they are doing that because the political arena won't allow funding without programs. ... I think if they went in [asking for] ecosystems appropriations, they would be worse off.

In addition, one respondent noted that being innovative is a two-edged sword when it comes to obtaining financial support from agencies:

> One of the frustrating things was [that] we were light years ahead of other districts, other forests ... when we submitted our request for funding for activities. These are "go" projects. They had already been through the analysis steps, the community supported them, and they were good activities, good actions, good things that needed to happen out there. We were competing with proposed actions by other districts, by other forests, with more traditional projects. As a result, we often lost.

A lack of personnel and time was specified by another large segment of respondents. Several reported that their project duties were only one set of responsibilities competing for their time. Several efforts, often defined as pilot projects, were being led by a single individual whose high motivation nevertheless could not compensate for an overwhelming workload.

Agency-Related Problems

Problems associated with agencies were reported by 31 percent of the respondents, with the largest portion describing institutional obstacles including a lack of interagency coordination and cooperation and administrative red tape. In many cases, these issues were unavoidable given the need to involve all affected stakeholders and the resulting complexities of multiple decision making layers. In other cases, difficulties in communicating or coordinating with personnel from other agencies were due to an absence of interagency relationships. As the respondent from the Elliott State Forest Management Plan in Oregon explained, "For the first six to nine months of the development of the plan, it was mainly ... working our way through figuring out how to work together."

In several instances, federal policies constrained project activities. For example, the Federal Advisory Committee Act (FACA) was seen as a significant barrier to communication and coordination between federal agencies and other stakeholders. According to one respondent from Utah:

> FACA is a big stumbling block affecting things all over the country. ... You will find groups all over the country who are trying to get around it, trying to get into compliance or restricting themselves to discussion only.

Other problems were rooted in the agencies' values and historical behaviors. For example, jurisdictional conflicts between agencies were the source of several problems, despite the good intentions of individuals within those agencies or of project coordinators. "Maintenance of turf wars that have been in existence for thirty years" was the way one respondent described a pattern of interagency behavior that was a source of problems evident at many sites. At other times, conflicts between agency programs led to problems, such as when one agency's cropland incentive program conflicted with another agency's conservation program.

Many respondents cited problems associated with staff members and their values on a more personal level. Commonly cited were conflicting value systems in which an individual's (and an organization's) professional philosophy was in conflict with new land management approaches associated with ecosystem management. Agency staff members often are not accustomed to increased levels of public involvement, not trained in organizational management, and not in possession of the scientific knowledge required to understand and manage ecosystems. In other cases, staff members from different agencies and organizations had vastly different management philosophies, which often were difficult to merge into one project's goals and strategies.

Scientific Uncertainty

Twenty-four percent of the respondents noted problems relating to scientific uncertainty. Some of these problems were the result of conceptual difficulties with ecosystem management. For example, many respondents explained how the lack of a single, agreed-upon definition of ecosystem management led to difficulties in communicating the concept or a particular project to the public. An unclear definition can also lead to confusion among partners and can make funding requests difficult to justify. In addition, it can allow reticent agencies to use the existence of uncertainty as an excuse for not taking action. Insufficient scientific information about a particular project area was reported as a problem by many respondents. The absence of baseline data and an insufficient monitoring program were common.

Inadequate Stakeholder Participation

Various problems relating to stakeholder involvement were reported by approximately one-fourth of the respondents. Fragmented landscapes owned by multiple public and private parties made it difficult to work effectively at the ecosystem level. Project staff members also had difficulty identifying affected landowners and getting them involved in the effort. Distrust among stakeholders and the inherent difficulties of consensus building across a diversity of interests were also seen as problematic in helping a project progress.

WHAT DO THESE EXPERIENCES SUGGEST FOR FUTURE ECOSYSTEM MANAGEMENT PROJECTS?

Respondents had numerous pieces of advice for others contemplating ecosystem management projects (Figure 16). Most advice focused on the project planning process. Many respondents highlighted the need to involve stakeholders and provided specific recommendations for maximizing their involvement. Ensuring adequate resources and obtaining agency support were also viewed as important. Many highlighted the need to broaden the philosophy and scope of management strategies and expand the level of scientific information available to planners. Finally, respondents suggested that ecosystem management projects be developed within a framework of mutual understanding between project staff and local communities.

Involve All Stakeholders Early

Involving stakeholders in project planning and decision making was recommended by 41 percent of the respondents. The largest portion of this group emphasized that all stakeholders, or as many as possible, should be brought into the process. Citing

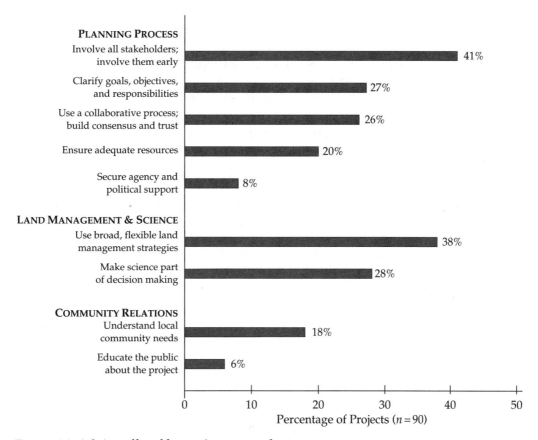

FIGURE 16. Advice offered by project respondents

the importance of proactively seeking representation of all interests, the respondent from the Owl Mountain Partnership in Colorado suggested simply that project managers "beat the brush to find out who has an interest in the project and get them involved."

Site contacts also advised that stakeholders be brought into project planning during the early stages. In their view, stakeholders who were involved from the beginning would help shape the final management plan so that it dealt with the appropriate set of problems and needs. They would also develop a sense of ownership of the efforts. One respondent from the San Pedro River project in southeastern Arizona commented, "The more people who feel ownership in the process and have bought into the project, the longer the project will last." Stakeholders brought in after planning is under way may not understand or agree with the ideas that have been developed. For many respondents, the critical issue was not how much input a project should receive but rather *when* that input should be received. As a district ranger from the Trail Creek Ecosystem Analysis effort in Idaho explained, "The bottom line is you're going to get [input] one way or another anyway. I'd much rather have it up front to begin with, as opposed to having [people later] question the process that you used."

Use an Open, Clear, and Collaborative Process.

Nearly half of the respondents offered a variety of suggestions on *how* to effectively involve stakeholders. The largest portion stressed the importance of communicating regularly and openly with stakeholders, having clear and well-defined goals, and making decisions by consensus. Respondents felt that openness and honesty in the planning process would ensure that all stakeholders' core interests were understood and would build trust among stakeholders—important for projects that depend on voluntary cooperation. Moreover, respondents indicated that a project planning process without these characteristics would lead to the obstacles discussed earlier. Their overall message was "Be open; don't have a hidden agenda."

Repeatedly, respondents stated that a project should have a clear direction, especially in terms of goals, objectives, and desired outcomes. Clear objectives and approaches were seen as particularly important given the complexity of ecosystem management projects. Respondents felt that poorly defined goals and objectives tended to generate public skepticism of ecosystem management approaches and made cooperation among diverse interests more difficult. To ensure clarity, some respondents believed that it was important to have formal mission statements or plans. Others believed that informal attempts to clarify project objectives were adequate. Overall, respondents agreed with the site contact from the Colorado State Forest Ecosystem Planning Project: "Maintain a strong and clear perspective of what you're trying to achieve."

Several respondents suggested that working by consensus and forming partnerships were ways to ensure constructive stakeholder involvement. A few comments addressed the approach a project manager should take with regard to working with others in a collaborative fashion: "Leave your ego at the door," "Be persistent," and "Be prepared to spend lots of time with lots of people." One series of comments pointed specifically to the need to set up decision-making bodies such as committees, working groups, or "partnerships, partnerships, partnerships," as one respondent from Virginia stated emphatically. Some believed that organizing stakeholders and interests into a cohesive group was a means of ensuring consistent representation and participation. Partnerships were recommended as a way of sharing resources and gaining greater feedback and more diversity of opinion.

Secure Adequate Resources and Agency Support

Advice relating to resources and internal support was offered by approximately one-fifth of the site contacts. Most offered comments about the need to line up adequate funding and personnel. Some comments about funding centered on project managers producing long-term funding plans to ensure the sustainability of multiyear projects. Others commented on the importance of having funding mechanisms organized around ecosystems rather than traditional program planning categories such as recreation, timber, and wildlife enhancement.

Respondents also emphasized the need to have consistent, well-trained staff members involved in ecosystem management efforts. One individual stated: "The process of ecosystem planning is a full-time, 100 percent process that requires a full-time ecosystem planning staff. Otherwise it won't work." Some respondents noted the importance of having staff members with particular skills. A respondent from the Rainwater Basin Joint Venture in Nebraska argued that it was important to hire a coordinator who is skilled at working with people. A few site contacts specifically advised hiring personnel who were trained in mapping and database technologies such as geographic information systems.

The support of supervisors and upper-level managers was also viewed as crucial for the success of ecosystem management projects. Many project managers had to "go out on a limb" to promote their projects and expressed the hope that future efforts would be supported more broadly by agency management. Upper-level managers and agency directors must support ecosystem management if a natural resource manager is to receive the resources, political support, and authority to undertake projects that may be viewed as unusual or risky.

Use Broad, Flexible, Science-Based Land Management Approaches

Respondents indicated that the geographic and temporal scales of land management activities are important in ecosystem management projects. Many respondents called for taking a regional perspective or taking into account ecological links to lands outside the project area. Several suggested considering the hierarchical nature of ecosystems when planning project activities. A resource manager from the Upper Huerfano Ecosystem project in Colorado explained: "You need to look at the landscape at different scales that are appropriate in answering different management questions. You need to look at the scale that is appropriate for that question. Boundaries can be limiting." Other managers similarly noted that projects need to incorporate long-term concerns, since changes to ecological systems can take decades. For example, a resource manager from the San Pedro River project in southern Arizona stated that "long-term commitment is required in projects like these where issues are complex and continually changing."

Respondents also indicated that flexible, adaptive management approaches should be practiced because of the scientific uncertainty surrounding some ecosystem management efforts. One respondent from the Verde River Greenway project in Arizona claimed, "You have to be ready to work with a multitude of issues and know that it is an ongoing process that changes." Rather than wait for complete scientific information before moving ahead with ecosystem management projects, project managers should begin activities based on the information at hand and be ready to shift management approaches and undertake different activities as new information becomes available. As a respondent from the Elliott State Forest Management Plan in Oregon indicated:

> Our approach to it is that we've built in some mechanisms for coping with at least small to moderate change. That's where we tried to structure the plan in an adaptive management context—building in mechanisms for revising the plan as needed. [The plan] points the direction that we want to go, but we realize that the direction may change a few degrees as we're on our way. We're going to need to be flexible enough to change our management as our knowledge changes.

Twenty-eight percent of the respondents emphasized the important role of scientific information in ecosystem management efforts. Most of these individuals stressed the importance of having adequate amounts of high-quality scientific information and finding ways to integrate that information into the decision-making process. Developing baseline data, instituting monitoring programs, and involving experts were all seen as important activities. Indeed, reliable data was of utmost importance to many project managers, and worth going to considerable lengths to acquire. As one responded advised, "Use the best science available, and if it is not available, go out and get it, pay for it, because your plan has to make sense ecologically."

Ensure Communities and Agencies Understand Each Other's Concerns

More than one-fourth of the respondents indicated a need for greater understanding at the local community level. Understanding the community's needs is important to the long-term viability of a project, as a participant in the Verde River Greenway project in Arizona observed:

> You have to incorporate the wants and needs of the people in the community. You can't just come in as an outsider and impose your own ideas and values on the area and the people. You have to try to work with the people, because they are the ones who will ultimately protect the resource.

Similarly, the respondent from the Owl Mountain Partnership in Colorado noted:

> You can't look at ecosystem management only in terms of what it can do for native plant and animal species. From the standpoint of sustainability, people have to be strongly involved.

Educating the public and building public awareness also were viewed as important. Not only must project staff members be aware of local concerns, but local residents also must be aware of a project's purpose, ongoing progress, and results. The need to develop a two-way learning process was cited by many respondents. In their view, enhanced understanding could lead to public support for ecosystem management projects, especially in areas where stakeholders are wary of public agencies or nonprofit groups.

CONCLUSIONS

Many experiments in ecosystem management are under way across the United States. Some employ strategies that are new and different, while others employ "off-the-shelf" ideas that have not been possible to implement in the past. Progress in ecosystem management is not confined to the activities of the federal government or any one agency. Instead, many different groups are playing important roles in ecosystem management, including numerous federal and state agencies, and nongovernmental groups of all kinds. While many of these efforts are quite young and have years to go before they can be deemed true success stories, they provide a rich database for learning about ecosystem management as it unfolds on the ground.

The image these efforts provide of the experience to date is hopeful. Clearly, implementing ecosystem management projects is challenging, since it involves collaboration, complex systems, problematic policies, and a skeptical public. But people are dealing with these challenges effectively. Their successes can be enhanced by increasing the understanding and skills of the people involved in ecosystem management projects, renewing policies and programs that influence the environment in which projects take place, and recognizing that ecosystem management is both a set of goals and a long term process of change. Nevertheless, participants in these projects are excited about the promise of these approaches to achieving sustainable resource management in the future.

Participation and Enthusiasm Are Widespread

If anything stands out from the inventory and assessment, it is the finding that many efforts to define and implement ecosystem management are under way in all regions of the United States, and the early experience with these efforts is very positive. While the catalog includes 105 sites, several times that number easily could have been included from the larger set of sites identified in the course of the study. Most of these sites were not textbook cases of ecosystem management in that their projects rarely succeeded at managing on an ecosystem scale and did not include all elements of an ecosystem management process. Rather, most were moving in that direction by looking outward to the regional scale, expanding the scope of issues and interests involved in making decisions, and grounding decision making in a more complex level of understanding about landscape components and processes.

Considering the comments received during the inventory from individuals in all sectors of society, it is hard not to be optimistic about the future of ecosystem-based approaches to land management. People are excited about what is happening "on the ground"; some are exhilarated by their successes. There is a shared sense of promise about these approaches and an energy that comes from being creative and acting in ways that are different from those in the past. As a fuzzy symbol of change more than an exact set of standards and guidelines, ecosystem management is helping to break the inertia evident in the past fifty years of resource management in the United States, and it is allowing individuals and groups to try out strategies not possible in the recent past. Many of these strategies are not new. Rather, the seeds of good ideas that have been with us for some time are finally being allowed to bear fruit.

Often the successes are quite small, but they are important. A rancher and an environmentalist talking to each other for the first time or two agencies sharing information may seem like a small step forward, but in many places such events are important steps that confound conventional wisdom. And some of these small steps promise larger successes in the future. Establishing relationships and trust, building understanding, and generating information can provide a foundation for real progress in building communities that are sustainable both ecologically and economically.

Public and Private Sectors Are Involved at All Levels

The high level of variation evident in the inventory of current experience bodes well for the future of ecosystem management. Just as a monoculture can be problematic ecologically, so can a monoculture of ideas. Fortunately, several federal agencies, numerous state agencies, and several nongovernmental organizations are active in the development of ecosystem management approaches. Different styles are evident in the ways these institutions approach work on the ground, and as a result, a rich storehouse of experience is developing from which future efforts can draw. In addition, since ecosystem management is not the sole turf of any one agency or group, it is more likely to survive top-down changes in the direction of an agency.

Since much of the rhetoric associated with ecosystem management in recent years has come from federal officials, it was surprising to discover how much activity is under way at the state level. State agencies are being innovative. Projects like the Northern Delaware Wetlands Rehabilitation Program and the Colorado State Forest Ecosystem Planning Project evidence a great deal of creativity in which state agencies play lead roles in restoring and protecting critical areas. Several states are attempting to use ecosystem management as a philosophy underlying their overall approach to resource management. For example, in New Hampshire, new forest plans are being drafted that incorporate ecological ideas rather than viewing forest resources largely as a source of commodity production. In Missouri, two state departments are leading a statewide coordinated resource management planning effort that involves several state and federal agencies, and in Washington, the Department of Fish and Wildlife is developing new management plans for all of its lands.

It is reassuring to know that ecosystem management is under development across different levels of government. Even though the term "ecosystem management" has been deemed "dead on arrival" on Capitol Hill,[3] it is alive and well in many other places in the country. A shift in power from federal to state governments may undercut some of the incentives and resources promoting ecosystem-based approaches, but it will not destroy the considerable interest in these approaches on the part of many state agencies and nongovernmental organizations.

People generally are trying out ecosystem-based approaches to land management not because of a top-down mandate or because it is trendy. Rather, they are trying to deal with real problems at the ground level—problems whose resolution requires more information and collaboration and a greater consideration of larger spatial and temporal scales. These problems will not go away by themselves, and the ideas underlying ecosystem management—knowing the land well and knowing and working with one's neighbors—make a lot of sense as ways to solve these problems. In addition, the opportunities that ecosystem management approaches provide to invest resources as efficiently as possible, by cutting down duplication and fostering resource-sharing arrangements, make them very appropriate for tight fiscal times. As a result, ecosystem management will be a continuing theme underlying land

---

3. John H. Cushman Jr., "Timber! A New Idea Is Crashing," *New York Times*, January 22, 1995, p. E5.

management in this country regardless of temporary shifts in direction emanating from the U.S. Congress or other centers of power.

The power of individual achievements evident in these projects is also reassuring. Dedicated, energetic individuals have accomplished a great deal, often in spite of institutional structures that have hampered their efforts. To individual managers, the message is clear: Innovative, creative approaches can bear fruit regardless of how difficult they may appear. To the rest of us, the meaning is equally clear: The future of ecosystem management lies in the collected efforts of many individuals on the ground, and these efforts can succeed in spite of considerable institutional resistance.

Implementation Involves Unique Challenges

The overall sense of optimism about the future of ecosystem management in the United States does not diminish the reality that accomplishing effective ecosystem-scale management is difficult for many reasons. By requiring more collaborative work from diverse stakeholders with often conflicting interests, ecosystem-based approaches to land management run head-on into the problems commonly associated with human relations and group decision making. Stakeholders who view each other with hostility are not easily molded into a cohesive decision-making group. In addition, collaborative approaches require agency personnel to set aside some traditional notions, including a sense that they are the all-knowing technical experts whose job is to come up with the right answers. Collaborative approaches are often time-consuming, involving numerous meetings that seemingly produce benefits only in the long term.

The increased level of understanding needed to implement ecosystem management approaches is also challenging. Many respondents noted the importance of getting "good science." The nascent level of understanding of many ecosystem processes and stresses means that most projects need to start with an investment in research and inventory work. Many project managers will need to organize data through geographic information systems and other data systems. Simply combining existing information sources requires standardizing data sets that were gathered through varying methods and with different purposes in mind. All this takes time, money, and effort, often resulting in a perceived delay in necessary protection and restoration work. In addition, generating the information required for ecosystem management requires scientists to share ideas with one another and requires scientists and managers to collaborate, both of which have been problematic in the past.

Many current policies make it difficult to practice effective ecosystem management. Participation in numerous land management planning processes initiated by different federal and state agencies consumes the limited time and resources available to nongovernmental stakeholders. Multiple, independent planning processes often make it difficult to take a larger-scale perspective. Government budgeting processes that can ensure funding only for one- or two-year project cycles (and that require results in the short term to justify continued funding) make it difficult to implement programs that will yield tangible benefits over long periods. Policies governing agencies' administrative procedures have also been problematic. One example is the Forest Service's practice of transferring personnel from one location to another for career development. Moving staff members disrupts the ground-level relationships needed for effective, collaborative management. The Federal Advisory

Committee Act (FACA), which structures relations between federal employees and nongovernmental groups, makes the formation of collaborative groups burdensome.

Finally, many sectors of the public are anxious about ecosystem management and its effect on private property rights. This reaction comes partly from a lack of understanding of ecosystems and partly from a fear that management strategies will unduly constrain human uses. It also results from a perception, described repeatedly by the respondents in this study, that government agencies have been heavy-handed in the past and hence are not to be trusted in the future. Public anxiety has also been fanned by a political atmosphere in which government has been set up as a scapegoat for many ills. Project managers expressed the concern that public misperceptions of ecosystem management and local hostility to it might worsen in the short term. They also greatly feared cuts in federal programs that provide some of the resources and tools employed in these efforts. Programs such as the U.S.D.A. Conservation Reserve and other Farm Bill programs and the EPA's Clean Lakes Project assisted a number of the efforts examined in this assessment.

People Are Succeeding in Spite of Problems

Many projects are moving forward in spite of the foregoing concerns. Indeed, many respondents wished to emphasize to others that it is important not to be stopped by a perception that such issues are overwhelming. Projects have succeeded through the loose-knit decision-making structures commonly associated with collaborative efforts. Agency officials have had to take care to avoid acting and being seen as arms of the heavy-handed government. One respondent noted the importance of "not wearing a government uniform" to meetings. Everyone has had to be careful to listen to and acknowledge the legitimate arguments of various stakeholders. Participants have also had to share credit for successes and ensure that each involved group maintains some level of ownership of the projects.

Groups have succeeded in dealing with complex ecological and social systems in part by focusing on small pieces of the larger puzzles and by managing adaptively—that is, by undertaking activities experimentally while investing in information that will determine whether the strategies are effective over time. Many respondents noted the usefulness of geographic information system technology as a means of combining and organizing information. Creation of collaborative management groups has also allowed for sharing of information among scientists and between scientists and managers in ways that have been highly productive.

Individuals have succeeded in finding their way around problematic government policies. Often this has required participants to act as entrepreneurs, pushing projects forward on the strength of their effort and personality while finding ways to cut through red tape and get around constraints caused by traditional ways of doing things. Project managers have leveraged funding and staffing in unique and creative ways. Many have simply ignored policies such as FACA, hoping that the obvious benefits of their approaches would protect them from repercussions.

At the site level, public fears and hostility have not been the overwhelming barrier many project managers anticipated. Indeed, a number of respondents expressed relief that their fears of outright opposition to their efforts from local property rights groups were not realized. Although there may be an initial sense of mistrust among groups, it does not take much to break down stereotypes. When government officials stop acting as an outside authority out to guide wayward children and instead play the role of facilitator and technical advisor, the public appears willing to put its mistrust aside. Many respondents noted that their fears about what might happen were much worse than what actually happened. Indeed, in many cases participants were surprised and elated by their successes:

> I didn't think there was any way in the world we'd ever get anything done [in the Steering Committee] without killing somebody. And we never even had a good fist fight. We had some shouting matches and a little bit of arguing. But it seems like the process brought everyone to the table with a little more open mind. We managed to work around problems.

Enhanced Skills and Supportive Policies and Programs Are Needed

Acknowledging the importance of individual accomplishments does not diminish the need for institutional structures and policies that support individual efforts. Indeed, constantly asking people to "swim upstream" will lead to their exhaustion. There is a clear need for personnel training and public education and for programs that provide resources, information, and incentives to on-the-ground managers.

Training and Education. Ecosystem management clearly asks more of both agency staff members and the public than is needed in traditional management approaches. Understanding the complexity of ecological and social systems requires an expanded level of knowledge of both of these systems. Agency staff members need to receive continuing education as the science of ecosystems and ecosystem management develops. Techniques such as gap analysis and geographic information systems need to be absorbed, along with approaches to economic development and social impact assessment. Specialists are needed, but they must be educated about the broader context of their expertise: how their knowledge draws on and relates to that of others. Intermediaries are also needed who can provide the critical interface between the host of disciplines and groups that need to participate in ecosystem-based approaches. Educational outreach is also needed to inform and motivate groups in the general public. All involved parties need to cultivate good "people" skills. As decision-making and information networks become more collaborative, the ability to communicate, listen, respect others' opinions, and learn from past failures becomes increasingly important.

Policies and Programs. There is also a critical need to reinvest in a set of federal and state policies and programs that provide resources, information, and incentives to managers and other stakeholders involved in ecosystem management efforts. The presence of preexisting policies and programs was an important source of motivation for starting projects and a highly ranked factor contributing to their success. It would be wrong to assume that these programs can succeed on the force of personality alone. The important public benefits that can result from ecosystem management activities justify spending scarce public resources on these efforts and maintaining the incentives for participation provided by other regulatory programs.

Most of the projects in the inventory benefited from public investment of funds or personnel in one way or another. Even projects with no federal or state partner often benefited from tax credits realized through donations of land or easements. Federal support—in the form of dollars spent either on project activities or on staff members—has served as an important catalyst in many efforts. Government has an important role to play as a source of information and technical assistance. It represents a set of larger-scale concerns (both geographic and temporal) that are important to effective decision making. Government programs have also been a significant incentive to stakeholders to come to the ecosystem management table. Having the regulatory provisions of the Endangered Species Act and the National Forest Management Act in the background has encouraged key stakeholders to get involved and their initial anxieties and hostilities diminished as their involvement increased. There is an important role for authorities to play in identifying long-term societal goals, in giving diverse public groups an incentive to pursue these goals through collaborative mechanisms, and in monitoring projects' performance over time. In most of the cases studied in this inventory, these roles have been played by government institutions.

Ecosystem Management is Both a Process and a Set of Goals

The image of ecosystem management that emerges from this inventory is a process-oriented one. Ecosystem management is a long-term process of understanding and decision making that requires the involvement of multiple sources of expertise and numerous stakeholders. It is also a process of organizational change in which agencies and groups must act in ways that have not been traditional. Finally, it is a process of ecological restoration in which actions accumulate over long periods of time to recreate critical natural system components and processes.

Based on this assessment of the experiences at 105 sites, the overall message to practitioners about the process of ecosystem management is to "know your land, and know your neighbors." At one level, this simple message is time-honored: the New England farmer and the Colorado rancher have always had to understand the character of their lands and cooperate with adjacent landowners. But the meaning of the message and what it takes to implement it are more complex today. The knowledge required to truly understand the landscape and the interests that are affected by the outcomes of land management are much greater than in the early days of American development history.

Some of the smaller messages to practitioners of ecosystem management read like the litany of advice that grandmothers have given since the beginning of human civilization: Don't be arrogant; be proud of accomplishments, but share credit for them; learn from mistakes and change direction accordingly; be honest; listen; be patient and persistent; have respect for all opinions; and know that what you do affects others. These are fundamental rules of interpersonal behavior, and they underscore the sense of those involved in these projects that ecosystem management is fundamentally a process of knowing, acting, and changing.

At the same time, ecosystem management is not management toward any end. Rather, it seeks to protect and restore the ecological integrity of landscapes while building sustainable economies and effective organizational and decision-making structures. Ecosystem management projects need to maintain a process orientation while keeping these overall goals in sight. From this assessment of early experience with ecosystem-based approaches to land management, it appears that managers and stakeholders are succeeding at walking this line, sometimes against great odds. Even if the term "ecosystem management" goes out of favor as political officials and agencies change, the innovative approaches to resource management that have developed under this label are likely to persist. Collaboration across landownerships and interests and shifts in management approaches are occurring primarily because they make sense to the individuals involved. Learning from their experiences can help us move steadily toward a sustainable future. The foregoing assessment can provide a baseline for this future, and the catalog of sites that follows can provide practitioners with images to emulate.

PART TWO

CATALOG
OF
ECOSYSTEM MANAGEMENT
PROJECTS

SUMMARY INFORMATION
ON
105 SELECTED PROJECTS

LOCATION OF 105 SELECTED PROJECTS FEATURED IN CATALOG

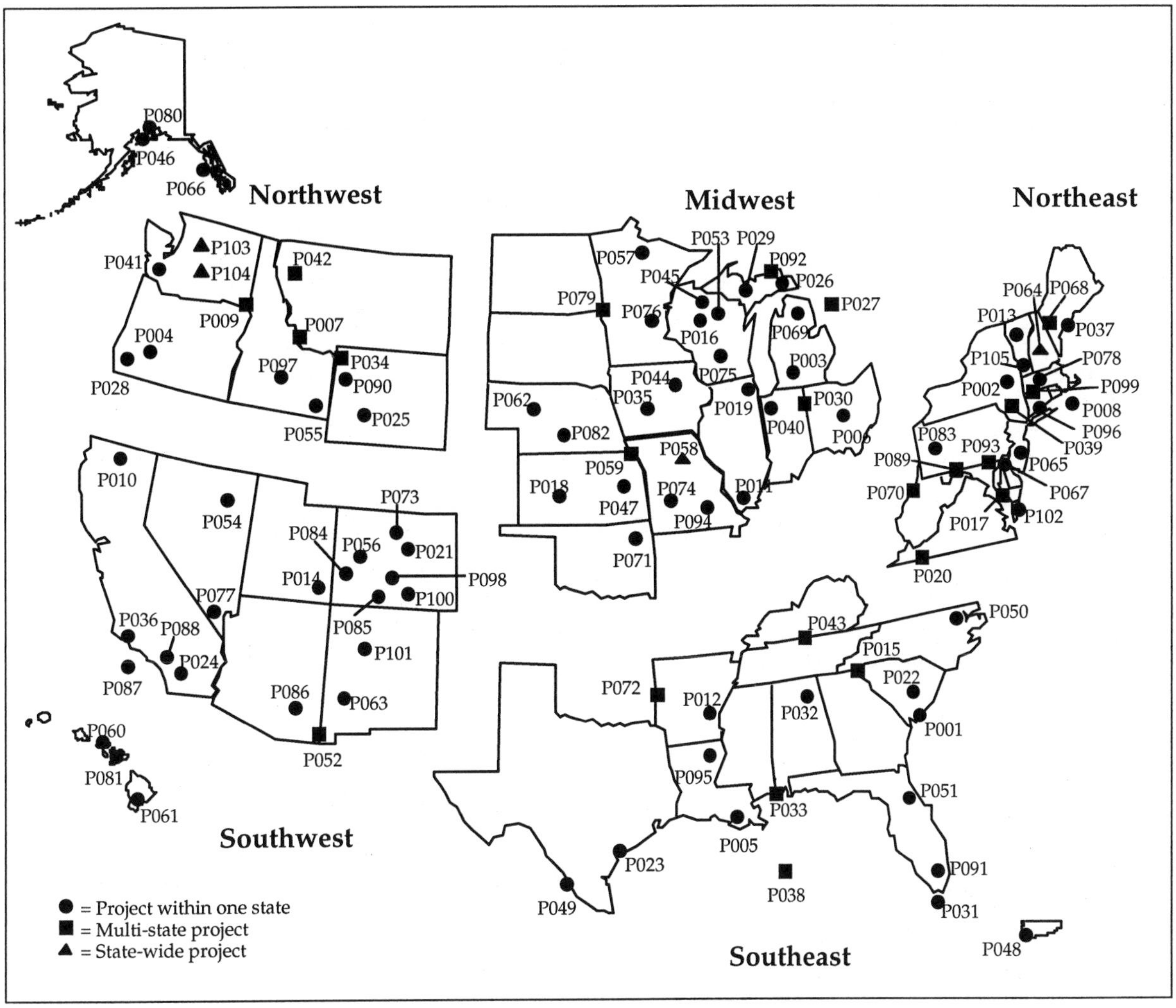

FIGURE 17. Location of projects (*Note:* locations are approximate and do not reflect scale of project area)

Project description page numbers, and key to ID numbers on map

PAGE	ID	PROJECT
81	P001	ACE Basin
83	P002	Albany Pine Bush
85	P003	Allegan State Game Area
87	P004	Applegate Partnership
89	P005	Barataria-Terrebonne National Estuary Program
91	P006	Big Darby Creek Partnership
93	P007	Bitteroot Ecosystem Management Research Project
95	P008	Block Island Refuge
97	P009	Blue Mountains Natural Resources Institute
99	P010	Butte Valley Basin
101	P011	Cache River Wetlands
103	P012	Cache/Lower White Rivers Ecosystem Management Plan
105	P013	Camp Johnson Sandplain Restoration
107	P014	Canyon Country Partnership
109	P015	Chattooga River Project
111	P016	Chequamegon National Forest Landscape Analysis and Design

PAGE	ID	PROJECT
113	P017	Chesapeake Bay Program
115	P018	Cheyenne Bottoms Wildlife Area
117	P019	Chicago Wilderness
119	P020	Clinch Valley Bioreserve
121	P021	Colorado State Forest Ecosystem Planning Project
123	P022	Congaree River Corridor Water Quality Planning Assessment
125	P023	Corpus Christi Bay National Estuary Program
127	P024	Dos Palmas Oasis
129	P025	East Fork Management Plan
131	P026	Eastern Upper Peninsula Ecosystem Management Consortium
133	P027	Ecosystem Charter for the Great Lakes–St. Lawrence Basin
135	P028	Elliott State Forest Management Plan
137	P029	Escanaba River State Forest
139	P030	Fish Creek Watershed Project
141	P031	Florida Bay Ecosystem Management Area

REGIONAL DISTRIBUTION OF ECOSYSTEM MANAGEMENT PROJECTS

NORTHEAST (ME, NH, VT, MA, RI, CT, NY, NJ, PA, DE, MD, DC, VA, WV)

Albany Pine Bush (NY)

Block Island Refuge (RI)

Camp Johnson Sandplain Restoration (VT)

Chesapeake Bay Program (MD, VA, DC, DE, WV, PA, NY)

Clinch Valley Bioreserve (VA, NC)

Gulf of Maine Rivers Ecosystem Plan (ME, NH, MA)

Hudson River/New York Bight Ecosystem (NY, NJ)

New Hampshire Forest Resources Plan (NH)

New Jersey Pinelands (NJ)

Northern Delaware Wetlands Rehabilitation Program (DE)

Northern Forest Lands Council (NY, VT, NH, ME)

Plainfield Project (MA)

Robbie Run Study Area (PA)

Sideling Hill Creek Bioreserve (MD, PA)

State Line Serpentine Barrens (PA, MD)

Tidelands of the Connecticut River (CT)

Upper Farmington River Management Plan (CT, MA)

Virginia Coast Reserve (VA)

Wildlife Habitat Improvement Group (VT)

SOUTHEAST (NC, SC, GA, KY, TN, FL, AL, MS, LA, TX, AR, PR)

ACE Basin (SC)

Barataria-Terrebonne National Estuary Program (LA)

Cache/Lower White Rivers Ecosystem Management Plan (LA)

Chattooga River Project (GA, SC, TN)

Congaree River Corridor Water Quality Planning Assessment (SC)

Corpus Christi Bay National Estuary Program (TX)

Florida Bay Ecosystem Management Area (FL)

Georgia Mountain Ecosystem Management Project (AL)

Grand Bay Savanna (MS, AL)

Gulf of Mexico Program (TX, LA, MS, AL, FL)

Interior Low Plateau (KY, TN, AL)

Lajas Valley Lagoon System (PR)

Lower Rio Grande Ecosystem Plan (TX)

Lower Roanoke River Bioreserve (NC)

Lower St. Johns River Ecosystem Management Area (FL)

Ouachita National Forest (AR, OK)

South Florida/Everglades Ecosystem Restoration Initiative (FL)

Tensas River Basin Initiative (LA)

MIDWEST (OH, MI, IN, IL, WI, MN, ND, SD, IA, NE, MO, OK)

Allegan State Game Area (MI)

Big Darby Creek Partnership (OH)

Cache River Wetlands (IL)

Chequamegon National Forest Landscape Analysis and Design (WI)

Cheyenne Bottoms Wildlife Area (KS)

Chicago Wilderness (IL)

Eastern Upper Peninsula Ecosystem Management Consortium (MI)

Ecosystem Charter for the Great Lakes–St. Lawrence Basin (MN, WI, IL, IN, MI, OH, PA, NY, Ontario & Quebec)

Escanaba River State Forest (MI)

Fish Creek Watershed Project (IN, OH)

Green Valley State Park Ecosystem Management Plan (IA)

Indiana Grand Kankakee Marsh Restoration Project (IN)

Iowa River Corridor Project (IA)

Karner Blue Butterfly Habitat Conservation Plan (WI)

Konza Prairie Research Natural Area (KS)

Marathon County Forest (WI)

Minnesota Peatlands (MN)

Missouri Coordinated Resource Management (MO)

Missouri River Mitigation Project (KS, NE, IA, MO)

Nebraska Sandhills Ecosystem (NE)

Northern Lower Michigan Ecosystem Management Project (MI)

Ohio River Valley Ecosystem (IL, IN, OH, PA, NY, WV, KY, TN, VA, MD,)

Oklahoma Tallgrass Prairie Preserve (OK)

Partners for Prairie Wildlife (MO)

Patrick Marsh Wetland Mitigation Bank Site (WI)

Phalen Chain of Lakes Watershed Project (MN)

Prairie Pothole Joint Venture (IA, MN, MT, ND, SD)

Rainwater Basin Joint Venture (NE)

St. Marys River Remedial Action Plan (MI, Ontario)

Stegall Mountain Natural Area (MO)

SOUTHWEST (NM, AZ, CA, NV, UT, CO, HI)

Butte Valley Basin (CA)

Canyon Country Partnership (UT)

Colorado State Forest Ecosystem Planning Project (CO)

Dos Palmas Oasis (CA)

Guadalupe-Nipomo Dunes Preserve (CA)

Malpai Borderlands Initiative (AZ, NM)

Marys River Riparian/Aquatic Restoration Project (NV)

Mesa Creek Coordinated Resource Management Plan (CO)

Molokai Preserves (HI)

Natural Resource Roundtable (HI)

Negrito Project (NM)

Owl Mountain Partnership (CO)

Piute/El Dorado Desert Wildlife Management Area (NV)

Pu'u Kukui Watershed Management Area (HI)

Ruby Canyon and Black Ridge Ecosystem Management Plan (CO)

San Luis Valley Comprehensive Ecosystem Management Plan (CO)

San Pedro River (AZ)

Santa Catalina Island Ecological Restoration Program (CA)

Santa Margarita River (CA)

Trout Mountain Roadless Area (CO)

Upper Huerfano Ecosystem (CO)

Verde River Greenway (NM)

NORTHWEST (MT, WY, ID, OR, WA, AK)

Applegate Partnership (OR)

Bitteroot Ecosystem Management Research Project (ID, MT)

Blue Mountains Natural Resources Institute (OR, WA)

East Fork Management Plan (WY)

Elliott State Forest Management Plan (OR)

Greater Yellowstone Ecosystem (WY, MT, ID)

Integrated Landscape Management for Fish and Wildlife (WA)

Interior Columbia Basin Ecosystem Management Project (WA, OR, ID, MT, WY, NV, UT)

Kenai River Watershed Project (AK)

McPherson Ecosystem Enhancement Project (ID)

Northeast Chichagof Island (AK)

Prince William Sound–Copper River Ecosystem Initiative (AK)

Snake River Corridor Project (WY)

Trail Creek Ecosystem Analysis (ID)

Wild Stock Initiative (WA)

Wildlife Area Planning (WA)

STATE-BY-STATE LISTING OF 105 SELECTED PROJECTS

(*Note:* projects in multiple states are listed under each state)

ID	PROJECT

ALABAMA

P032	Georgia Mountain Ecosystem Management Project
P033	Grand Bay Savanna
P038	Gulf of Mexico Program

ALASKA

P046	Kenai River Watershed Project
P066	Northeast Chichagof Island
P080	Prince William Sound–Copper River Ecosystem Initiative

ARIZONA

P052	Malpai Borderlands Initiative
P086	San Pedro River
P101	Verde River Greenway

ARKANSAS

P012	Cache/Lower White Rivers Ecosystem Management Plan
P072	Ouachita National Forest

CALIFORNIA

P010	Butte Valley Basin
P024	Dos Palmas Oasis
P036	Guadalupe-Nipomo Dunes Preserve
P087	Santa Catalina Island Ecological Restoration Program
P088	Santa Margarita River

COLORADO

P021	Colorado State Forest Ecosystem Planning Project
P056	Mesa Creek Coordinated Resource Management Plan
P073	Owl Mountain Partnership
P084	Ruby Canyon and Black Ridge Ecosystem Management Plan
P085	San Luis Valley Comprehensive Ecosystem Mgmt. Plan
P098	Trout Mountain Roadless Area
P100	Upper Huerfano Ecosystem

CONNECTICUT

P096	Tidelands of the Connecticut River
P099	Upper Farmington River Management Plan

DELAWARE

P017	Chesapeake Bay Program
P067	Northern Delaware Wetlands Rehabilitation Program

DISTRICT OF COLUMBIA

P017	Chesapeake Bay Program

FLORIDA

P031	Florida Bay Ecosystem Management Area
P038	Gulf of Mexico Program
P051	Lower St. Johns River Ecosystem Management Area
P091	South Florida/Everglades Ecosystem Restoration Initiative

GEORGIA

P015	Chattooga River Project

HAWAII

P060	Molokai Preserves
P061	Natural Resource Roundtable
P081	Pu'u Kukui Watershed Management Area

IDAHO

P034	Greater Yellowstone Ecosystem
P042	Interior Columbia Basin Ecosystem Management Project
P055	McPherson Ecosystem Enhancement Project
P097	Trail Creek Ecosystem Analysis

ILLINOIS

P011	Cache River Wetlands
P019	Chicago Wilderness
P027	Ecosystem Charter for the Great Lakes–St. Lawrence Basin
P070	Ohio River Valley Ecosystem

INDIANA

P027	Ecosystem Charter for the Great Lakes–St. Lawrence Basin
P030	Fish Creek Watershed Project
P040	Indiana Grand Kankakee Marsh Restoration Project
P070	Ohio River Valley Ecosystem

IOWA

P035	Green Valley State Park Ecosystem Management Plan
P044	Iowa River Corridor Project
P059	Missouri River Mitigation Project
P079	Prairie Pothole Joint Venture

KANSAS

P018	Cheyenne Bottoms Wildlife Area
P047	Konza Prairie Research Natural Area
P059	Missouri River Mitigation Project

KENTUCKY

P043	Interior Low Plateau
P070	Ohio River Valley Ecosystem

LOUISIANA

P005	Barataria-Terrebonne National Estuary Program
P038	Gulf of Mexico Program
P095	Tensas River Basin Initiative

MAINE

P037	Gulf of Maine Rivers Ecosystem Plan
P068	Northern Forest Lands Council

MARYLAND

P017	Chesapeake Bay Program
P070	Ohio River Valley Ecosystem
P089	Sideling Hill Creek Bioreserve
P093	State Line Serpentine Barrens

MASSACHUSETTS

P037	Gulf of Maine Rivers Ecosystem Plan
P078	Plainfield Project
P099	Upper Farmington River Management Plan

MICHIGAN

P003	Allegan State Game Area
P026	Eastern Upper Peninsula Ecosystem Management Consortium
P027	Ecosystem Charter for the Great Lakes–St. Lawrence Basin
P029	Escanaba River State Forest
P069	Northern Lower Michigan Ecosystem Management Project
P092	St. Marys River Remedial Action Plan

MINNESOTA

P057	Minnesota Peatlands
P027	Ecosystem Charter for the Great Lakes–St. Lawrence Basin
P076	Phalen Chain of Lakes Watershed Project
P079	Prairie Pothole Joint Venture

MISSISSIPPI

P033	Grand Bay Savanna
P038	Gulf of Mexico Program

ID	PROJECT

MISSOURI

P058	Missouri Coordinated Resource Management
P059	Missouri River Mitigation Project
P074	Partners for Prairie Wildlife
P094	Stegall Mountain Natural Area

MONTANA

P007	Bitteroot Ecosystem Management Research Project
P034	Greater Yellowstone Ecosystem
P042	Interior Columbia Basin Ecosystem Management Project
P079	Prairie Pothole Joint Venture

NEBRASKA

P059	Missouri River Mitigation Project
P062	Nebraska Sandhills Ecosystem
P082	Rainwater Basin Joint Venture

NEW HAMPSHIRE

P037	Gulf of Maine Rivers Ecosystem Plan
P064	New Hampshire Forest Resources Plan
P068	Northern Forest Lands Council

NEW JERSEY

P039	Hudson River/New York Bight Ecosystem
P065	New Jersey Pinelands

NEW MEXICO

P063	Negrito Project
P052	Malpai Borderlands Initiative

NEW YORK

P002	Albany Pine Bush
P017	Chesapeake Bay Program
P027	Ecosystem Charter for the Great Lakes–St. Lawrence Basin
P039	Hudson River/New York Bight Ecosystem
P068	Northern Forest Lands Council
P070	Ohio River Valley Ecosystem

NEVADA

P042	Interior Columbia Basin Ecosystem Management Project
P054	Marys River Riparian/Aquatic Restoration Project
P077	Piute/El Dorado Desert Wildlife Management Area

NORTH CAROLINA

P015	Chattooga River Project
P050	Lower Roanoke River Bioreserve

NORTH DAKOTA

P079	Prairie Pothole Joint Venture

OHIO

P006	Big Darby Creek Partnership
P027	Ecosystem Charter for the Great Lakes–St. Lawrence Basin
P030	Fish Creek Watershed Project
P070	Ohio River Valley Ecosystem

OKLAHOMA

P071	Oklahoma Tallgrass Prairie Preserve

OREGON

P004	Applegate Partnership
P009	Blue Mountains Natural Resources Institute
P028	Elliott State Forest Management Plan
P042	Interior Columbia Basin Ecosystem Management Project

PENNSYLVANIA

P017	Chesapeake Bay Program
P027	Ecosystem Charter for the Great Lakes–St. Lawrence Basin
P070	Ohio River Valley Ecosystem

ID	PROJECT

PENNSYLVANIA -- continued

P083	Robbie Run Study Area
P089	Sideling Hill Creek Bioreserve
P093	State Line Serpentine Barrens

PUERTO RICO

P048	Lajas Valley Lagoon System

RHODE ISLAND

P008	Block Island Refuge

SOUTH CAROLINA

P001	ACE Basin
P015	Chattooga River Project
P022	Congaree River Corridor Water Quality Planning Assessment

SOUTH DAKOTA

P079	Prairie Pothole Joint Venture

TENNESSEE

P020	Clinch Valley Bioreserve
P070	Ohio River Valley Ecosystem

TEXAS

P023	Corpus Christi Bay National Estuary Program
P038	Gulf of Mexico Program
P049	Lower Rio Grande Ecosystem Plan

UTAH

P014	Canyon Country Partnership

VERMONT

P013	Camp Johnson Sandplain Restoration
P068	Northern Forest Lands Council
P104	Wildlife Habitat Improvement Group

VIRGINIA

P017	Chesapeake Bay Program
P020	Clinch Valley Bioreserve
P070	Ohio River Valley Ecosystem
P102	Virginia Coast Reserve

WASHINGTON

P009	Blue Mountains Natural Resources Institute
P041	Integrated Landscape Management for Fish and Wildlife
P042	Interior Columbia Basin Ecosystem Management Project
P105	Wild Stock Initiative
P103	Wildlife Area Planning

WEST VIRGINIA

P017	Chesapeake Bay Program
P070	Ohio River Valley Ecosystem

WISCONSIN

P016	Chequamegon National Forest Landscape Analysis and Design
P027	Ecosystem Charter for the Great Lakes–St. Lawrence Basin
P045	Karner Blue Butterfly Habitat Conservation Plan
P053	Marathon County Forests
P075	Patrick Marsh Wetland Mitigation Bank Site

WYOMING

P025	East Fork Management Plan
P034	Greater Yellowstone Ecosystem
P042	Interior Columbia Basin Ecosystem Management Project
P090	Snake River Corridor Project

AGENCIES, ORGANIZATIONS, AND INDIVIDUALS INITIATING PROJECTS

PROJECT	Federal agencies [1]					State Agncs [2]	Other Gov't [3]	NGOs [4]		Other [5]
	USFS	FWS	BLM	EPA	Misc			TNC	Other	
ACE Basin						◆		◆	DU	
Albany Pine Bush									Local	
Allegan State Game Area						◆				
Applegate Partnership										Cit
Barataria-Terrebonne National Estuary Program				◆		◆				
Big Darby Creek Partnership										X
Bitteroot Ecosystem Management Research Project	◆									
Block Island Refuge								◆	Local	
Blue Mountains Natural Resources Institute	◆									Cit
Butte Valley Basin	◆					◆				
Cache River Wetlands		◆				◆		◆	DU	
Cache/Lower White Rivers Ecosystem Management Plan		◆								
Camp Johnson Sandplain Restoration						◆				
Canyon Country Partnership	◆		◆		NPS	◆	◆			
Chattooga River Project	◆									
Chequamegon National Forest Landscape Analysis & Design	◆									
Chesapeake Bay Program				◆		◆				
Cheyenne Bottoms Wildlife Area						◆				
Chicago Wilderness		◆					◆	◆		FMNH
Clinch Valley Bioreserve								◆		
Colorado State Forest Ecosystem Planning Project						◆				
Congaree River Corridor Water Quality Planning Assessment					NPS, USGS, NBS					
Corpus Christi Bay National Estuary Program				◆		◆				Cit
Dos Palmas Oasis			◆					◆		
East Fork Management Plan						◆				
Eastern Upper Peninsula Ecosystem Management Consortium						◆				
Ecosystem Charter for the Great Lakes–St. Lawrence Basin									Reg'l	
Elliott State Forest Management Plan						◆				
Escanaba River State Forest						◆				
Fish Creek Watershed Project		◆				◆		◆		
Florida Bay Ecosystem Management Area						◆				
Georgia Mountain Ecosystem Management Project					TVA					
Grand Bay Savanna		◆						◆		

1. Federal agencies:
    Department of Agriculture: USFS= Forest Service; NRCS=Natural Resources Conservation Service
    Department of Interior: FWS=Fish and Wildlife Service; BLM=Bureau of Land Management; NPS=National Park Service; NBS=National Biological Service; USGS=Geological Survey; DOI=Department of Interior (projects initiated by department, not individual agency)
    Misc: EPA=US Environmental Protection Agency; ACOE=Army Corps of Engineers; TVA=Tennessee Valley Authority (semi-independent)

2. State agencies: Include, for example, state departments of natural resources, environment, environmental protection, environmental quality, planning, transportation.

3. Other government: Local=local, county, municipal, parish government; Trib=native American tribal government or corporations.

4. Non-governmental organizations (NGOs): TNC=The Nature Conservancy; DU=Ducks Unlimited; Reg'l=Regional NGO; Local=Local NGO.

5. Other: Ind=Industry or other for-profit organization; Cit=Private citizen(s); Lndw=Private landowner(s); Univ=University; IJC=International Joint Commission; Ont=Ontario Ministry of Environment and Energy; X=No clear initiator; FMNH=Field Museum of Natural History; Pres=President Bill Clinton; Cong=U.S. Congress.

PROJECT	Federal agencies [1]					State Agncs [2]	Other Gov't [3]	NGOs [4]		Other [5]
	USFS	FWS	BLM	EPA	Misc			TNC	Other	
Greater Yellowstone Ecosystem	◆				NPS					
Green Valley State Park Ecosystem Management Plan						◆				
Guadalupe-Nipomo Dunes Preserve								◆	Reg'l	
Gulf of Maine Rivers Ecosystem Plan		◆								
Gulf of Mexico Program				◆						
Hudson River/New York Bight Ecosystem		◆								
Indiana Grand Kankakee Marsh Restoration Project		◆				◆				
Integrated Landscape Management for Fish and Wildlife						◆				
Interior Columbia Basin Ecosystem Management Project										Pres
Interior Low Plateau						◆			Reg'l	
Iowa River Corridor Project					NRCS					
Karner Blue Butterfly Habitat Conservation Plan						◆				
Kenai River Watershed Project				◆		◆		◆		
Konza Prairie Research Natural Area										Univ
Lajas Valley Lagoon System		◆								
Lower Rio Grande Ecosystem Plan		◆								
Lower Roanoke River Bioreserve								◆		
Lower St. Johns River Ecosystem Management Area						◆				
Malpai Borderlands Initiative										Ldwn
Marathon County Forests							◆			
Marys River Riparian/Aquatic Restoration Project			◆			◆				
McPherson Ecosystem Enhancement Project	◆									
Mesa Creek Coordinated Resource Management Plan			◆							Ind
Minnesota Peatlands						◆				
Missouri Coordinated Resource Management						◆				
Missouri River Mitigation Project		◆			ACOE	◆				
Molokai Preserves								◆		
Natural Resource Roundtable						◆				
Nebraska Sandhills Ecosystem		◆								
Negrito Project	◆									Cit
New Hampshire Forest Resources Plan						◆				
New Jersey Pinelands									Local	Cit
Northeast Chichagof Island	◆					◆				
Northern Delaware Wetlands Rehabilitation Program						◆				
Northern Forest Lands Council										X
Northern Lower Michigan Ecosystem Management Project	◆					◆				

1. Federal agencies:
   Department of Agriculture: USFS= Forest Service; NRCS=Natural Resources Conservation Service
   Department of Interior : FWS=Fish and Wildlife Service; BLM=Bureau of Land Management; NPS=National Park Service; NBS=National Biological Service; USGS=Geological Survey; DOI=Department of Interior (projects initiated by department, not individual agency)
   Misc: EPA=US Environmental Protection Agency; ACOE=Army Corps of Engineers; TVA=Tennessee Valley Authority (semi-independent)

2. State agencies: Include, for example, state departments of natural resources, environment, environmental protection, environmental quality, planning, transportation.

3. Other government: Local=local, county, municipal, parish government; Trib=native American tribal government or corporations.

4. Non-governmental organizations (NGOs): TNC=The Nature Conservancy; DU=Ducks Unlimited; Reg'l=Regional NGO; Local=Local NGO.

5. Other: Ind=Industry or other for-profit organization; Cit=Private citizen(s); Lndw=Private landowner(s); Univ=University; IJC=International Joint Commission; Ont=Ontario Ministry of Environment and Energy; X=No clear initiator; FMNH=Field Museum of Natural History; Pres=President Bill Clinton; Cong=U.S. Congress.

PROJECT	Federal agencies [1]					State Agncs [2]	Other Gov't [3]	NGOs [4]		Other [5]
	USFS	FWS	BLM	EPA	Misc			TNC	Other	
Ohio River Valley Ecosystem		◆								
Oklahoma Tallgrass Prairie Preserve								◆		
Ouachita National Forest	◆									
Owl Mountain Partnership										Cit, Lndw
Partners for Prairie Wildlife						◆				
Patrick Marsh Wetland Mitigation Bank Site						◆				
Phalen Chain of Lakes Watershed Project						◆	◆			
Piute/El Dorado Desert Wildlife Management Area							◆			
Plainfield Project	◆									
Prairie Pothole Joint Venture		◆								
Prince William Sound–Copper River Ecosystem Initiative					DOI, NBS					
Pu'u Kukui Watershed Management Area										Ind
Rainwater Basin Joint Venture		◆				◆			DU	
Robbie Run Study Area	◆									
Ruby Canyon and Black Ridge Ecosystem Management Plan			◆							
San Luis Valley Comprehensive Ecosystem Management Plan		◆								
San Pedro River			◆					◆	Local	
Santa Catalina Island Ecological Restoration Prog.									Local	
Santa Margarita River								◆		
Sideling Hill Creek Bioreserve								◆		
Snake River Corridor Project							◆			
South Florida/Everglades Ecos. Restoration Init.					DOI					
St. Marys River Remedial Action Plan				◆		◆				IJC, Ont
State Line Serpentine Barrens								◆		
Stegall Mountain Natural Area						◆				
Tensas River Basin Initiative									Local	
Tidelands of the Connecticut River								◆		
Trail Creek Ecosystem Analysis	◆									
Trout Mountain Roadless Area	◆									
Upper Farmington River Management Plan										Cit, Cong
Upper Huerfano Ecosystem						◆				
Verde River Greenway						◆				
Virginia Coast Reserve								◆		
Wild Stock Initiative						◆	Trib			
Wildlife Area Planning						◆				
Wildlife Habitat Improvement Group										Lndw

(1) <u>Federal agencies</u>:
    Department of Agriculture: USFS= Forest Service; NRCS=Natural Resources Conservation Service
    Department of Interior: FWS=Fish and Wildlife Service; BLM=Bureau of Land Management; NPS=National Park Service; NBS=National
      Biological Service; USGS=Geological Survey; DOI=Department of Interior (projects initiated by department, not individual agency)
    Misc: EPA=US Environmental Protection Agency; ACOE=Army Corps of Engineers; TVA=Tennessee Valley Authority (semi-independent)

(2) <u>State agencies</u>: Include, for example, state departments of natural resources, environment, environmental protection, environmental quality, planning, transportation.

(3) <u>Other government</u>: Local=local, county, municipal, parish government; Trib=native American tribal government or corporations.

(4) <u>Non-governmental organizations (NGOs)</u>: TNC=The Nature Conservancy; DU=Ducks Unlimited; Reg'l=Regional NGO; Local=Local NGO.

(5) <u>Other</u>: Ind=Industry or other for-profit organization; Cit=Private citizen(s); Lndw=Private landowner(s); Univ=University; IJC=International Joint Commission; Ont=Ontario Ministry of Environment and Energy; X=No clear initiator; FMNH=Field Museum of Natural History; Pres=President Bill Clinton; Cong=U.S. Congress.

AGE OF ECOSYSTEM MANAGEMENT PROJECTS

1 TO 2 YEARS OLD	Year Initiated
Congaree River Corridor Water Quality Plan'g Assessment	1995
Marathon County Forests	1995
Bitteroot Ecosystem Management Research Project	1994
Chicago Wilderness	1994
Florida Bay Ecosystem Management Area	1994
Georgia Mountain Ecosystem Management Project	1994
Green Valley State Park Ecosystem Management Plan	1994
Gulf of Maine Rivers Ecosystem Plan	1994
Hudson River/New York Bight Ecosystem	1994
Interior Columbia Basin Ecosystem Management Project	1994
Interior Low Plateau	1994
Karner Blue Butterfly Habitat Conservation Plan	1994
Lajas Valley Lagoon System	1994
Lower Rio Grande Ecosystem Plan	1994
Lower Roanoke River Bioreserve	1994
Lower St. Johns River Ecosystem Management Area	1994
New Hampshire Forest Resources Plan	1994
Northern Lower Michigan Ecosystem Management Project	1994
Ohio River Valley Ecosystem	1994
Plainfield Project	1994
Robbie Run Study Area	1994
Ruby Canyon and Black Ridge Ecosystem Mgmt. Plan	1994
Santa Margarita River	1994

3 TO 5 YEARS OLD	
Canyon Country Partnership	1993
Colorado State Forest Ecosystem Planning Project	1993
Indiana Grand Kankakee Marsh Restoration Project	1993
Iowa River Corridor Project	1993
Kenai River Watershed Project	1993
McPherson Ecosystem Enhancement Project	1993
Missouri Coordinated Resource Management	1993
Northeast Chichagof Island	1993
Owl Mountain Partnership	1993
Phalen Chain of Lakes Watershed Project	1993
Prince William Sound–Copper River Ecosystem Initiative	1993
Snake River Corridor Project	1993
South Florida/Everglades Ecosystem Restoration Initiative	1993
Cache/Lower White Rivers Ecosystem Management Plan	1992
Chattooga River Project	1992
Chequamegon Nat'l Forest Landscape Analysis & Design	1992
Corpus Christi Bay National Estuary Program	1992
Eastern Upper Peninsula Ecosystem Mngt. Consortium	1992
Ecosystem Charter for the Great Lakes–St. Lawrence Basin	1992
Elliott State Forest Management Plan	1992
Mesa Creek Coordinated Resource Management Plan	1992
Natural Resource Roundtable	1992
Negrito Project	1992
Northern Delaware Wetlands Rehabilitation Program	1992
Partners for Prairie Wildlife	1992
Stegall Mountain Natural Area	1992
Tensas River Basin Initiative	1992
Wild Stock Initiative	1992
Wildlife Area Planning	1992
Applegate Partnership	1991
Barataria-Terrebonne National Estuary Program	1991
Cache River Wetlands	1991

3-5 years old--*continued*	Year Initiated
Patrick Marsh Wetland Mitigation Bank Site	1991
Sideling Hill Creek Bioreserve	1991
State Line Serpentine Barrens	1991
Upper Huerfano Ecosystem	1991
Albany Pine Bush	1990
Camp Johnson Sandplain Restoration	1990
Grand Bay Savanna	1990
Malpai Borderlands Initiative	1990
Nebraska Sandhills Ecosystem	1990
Piute/El Dorado Desert Wildlife Management Area	1990
San Luis Valley Comprehensive Ecosystem Mgmt. Plan	1990
Tidelands of the Connecticut River	1990
San Pedro River	early 1990s

6 TO 10 YEARS OLD	
Blue Mountains Natural Resources Institute	1989
Cheyenne Bottoms Wildlife Area	1989
Dos Palmas Oasis	1989
Integrated Landscape Management for Fish and Wildlife	1989
Ouachita National Forest	1989
Rainwater Basin Joint Venture	1989
Trail Creek Ecosystem Analysis	1989
Fish Creek Watershed Project	1988
Guadalupe-Nipomo Dunes Preserve	1988
Gulf of Mexico Program	1988
Northern Forest Lands Council	1988
Oklahoma Tallgrass Prairie Preserve	1988
Santa Catalina Island Ecological Restoration Program	1988
Virginia Coast Reserve	1988
ACE Basin	1987
Prairie Pothole Joint Venture	1987
Pu'u Kukui Watershed Management Area	1987
Butte Valley Basin	1986
Greater Yellowstone Ecosystem	1986
Upper Farmington River Management Plan	1986
Verde River Greenway	1986
Escanaba River State Forest	1985
Marys River Riparian/Aquatic Restoration Project	1985
St. Marys River Remedial Action Plan	1985
Trout Mountain Roadless Area	1985
Wildlife Habitat Improvement Group	1985
Clinch Valley Bioreserve	1984
East Fork Management Plan	mid-1980s

OVER 10 YEARS OLD	
Minnesota Peatlands	1984
Molokai Preserves	1983
Allegan State Game Area	1982
Block Island Refuge	1980s
Chesapeake Bay Program	1980
Missouri River Mitigation Project	circa 1975
New Jersey Pinelands	1970s
Konza Prairie Research Natural Area	1956
South Florida Ecosystem Restoration Initiative	1993

NOT SPECIFIED
Big Darby Creek Partnership

LANDOWNERSHIP PATTERNS AT PROJECT AREAS

ENTIRELY PUBLIC LAND

Allegan State Game Area
Bitteroot Ecosystem Management Research Project
Camp Johnson Sandplain Restoration
Chequamegon National Forest Landscape Analysis & Design
Colorado State Forest Ecosystem Planning Project
Congaree River Corridor Water Quality Planning Assessment
East Fork Management Plan
Elliott State Forest Management Plan
Escanaba River State Forest
Konza Prairie Research Natural Area
Marathon County Forests
McPherson Ecosystem Enhancement Project
Minnesota Peatlands
Negrito Project
Patrick Marsh Wetland Mitigation Bank Site
Robbie Run Study Area
Trail Creek Ecosystem Analysis
Trout Mountain Roadless Area

PREDOMINANTLY PUBLIC LAND

Butte Valley Basin
Canyon Country Partnership
Chattooga River Project
Florida Bay Ecosystem Management Area
Greater Yellowstone Ecosystem
Green Valley State Park Ecosystem Management Plan
Mesa Creek Coordinated Resource Management Plan
Northeast Chichagof Island
Ouachita National Forest
Piute/El Dorado Desert Wildlife Management Area
Prince William Sound–Copper River Ecosystem Initiative
Ruby Canyon and Black Ridge Ecosystem Management Plan
Stegall Mountain Natural Area
Verde River Greenway

MIXED PUBLIC AND PRIVATE LANDS

Albany Pine Bush
Applegate Partnership
Blue Mountains Natural Resources Institute
Chesapeake Bay Program
Cheyenne Bottoms Wildlife Area
Clinch Valley Bioreserve
Dos Palmas Oasis
Eastern Upper Peninsula Ecosystem Management Consortium
Ecosystem Charter for the Great Lakes–St. Lawrence Basin
Guadalupe-Nipomo Dunes Preserve
Integrated Landscape Management for Fish and Wildlife
Interior Columbia Basin Ecosystem Management Project
Karner Blue Butterfly Habitat Conservation Plan
Kenai River Watershed Project
Malpai Borderlands Initiative
Molokai Preserves
New Hampshire Forest Resources Plan
New Jersey Pinelands
Northern Delaware Wetlands Rehabilitation Program
Northern Lower Michigan Ecosystem Management Project
Ohio River Valley Ecosystem

Owl Mountain Partnership
San Luis Valley Comprehensive Ecosystem Mgmt. Plan
San Pedro River
Santa Margarita River
Snake River Corridor Project
South Florida/Everglades Ecosystem Restoration Initiative
Upper Huerfano Ecosystem
Wild Stock Initiative

PREDOMINANTLY PRIVATE LAND

ACE Basin
Barataria-Terrebonne National Estuary Program
Big Darby Creek Partnership
Block Island Refuge
Cache River Wetlands
Cache/Lower White Rivers Ecosystem Management Plan
Chicago Wilderness
Corpus Christi Bay National Estuary Program
Georgia Mountain Ecosystem Management Project
Grand Bay Savanna
Gulf of Maine Rivers Ecosystem Plan
Gulf of Mexico Program
Hudson River/New York Bight Ecosystem
Indiana Grand Kankakee Marsh Restoration Project
Interior Low Plateau
Iowa River Corridor Project
Lajas Valley Lagoon System
Lower Rio Grande Ecosystem Plan
Lower Roanoke River Bioreserve
Lower St. Johns River Ecosystem Management Area
Marys River Riparian/Aquatic Restoration Project
Missouri Coordinated Resource Management
Missouri River Mitigation Project
Natural Resource Roundtable
Nebraska Sandhills Ecosystem
Northern Forest Lands Council
Partners for Prairie Wildlife
Phalen Chain of Lakes Watershed Project
Plainfield Project
Prairie Pothole Joint Venture
Sideling Hill Creek Bioreserve
St. Marys River Remedial Action Plan
State Line Serpentine Barrens
Tensas River Basin Initiative
Tidelands of the Connecticut River
Upper Farmington River Management Plan
Wildlife Area Planning

ENTIRELY PRIVATE LAND

Fish Creek Watershed Project
Oklahoma Tallgrass Prairie Preserve
Pu'u Kukui Watershed Management Area
Rainwater Basin Joint Venture
Santa Catalina Island Ecological Restoration Program
Virginia Coast Reserve
Wildlife Habitat Improvement Group

LAND USES IN PROJECT AREAS

PROJECT \ LAND USE	Development--urban, rural, industrial, other	Agriculture--crops or unspecified	Agriculture--grazing and ranching	Forest or timber management	Wildlife management	Hydrologic uses--diversions, impoundments, navigation	Mining--hard rock or unspecified	Mining--oil and gas	Fishing--commercial	Recreation--motorized or unspecified	Recreation--non-motorized	Preserves or open space	Hunting--subsistence	Other
ACE Basin		3		2					2		2	2		
Albany Pine Bush	2	2									2	2		
Allegan State Game Area				2	2						2	2		
Applegate Partnership	2	2	2	2	2	2	2	2		2	2	2		2
Barataria-Terrebonne National Estuary Program		1	3					3				3		
Big Darby Creek Partnership	3	1										3		
Bitteroot Ecosystem Management Research Project			2	2		2					2	2		
Block Island Refuge	2	3										2		
Blue Mountains Natural Resources Institute		2	2	2							2	2		2
Butte Valley Basin		2	2		2						2	2		
Cache River Wetlands	3	2		2										
Cache/Lower White Rivers Ecosystem Mgmt. Plan		1							3		3	2		
Camp Johnson Sandplain Restoration														1
Canyon Country Partnership	2	3	2				3	2			1			
Chattooga River Project	3	3	3	1		3	3				2	2		
Chequamegon N.F. Landscape Analysis & Design				2			2				2	2		
Chesapeake Bay Program	2	2	2	2		2			2	2	2	2		
Cheyenne Bottoms Wildlife Area		2			2							2		
Chicago Wilderness	1	3										3		
Clinch Valley Bioreserve	2		2				3							
Colorado State Forest Ecosystem Planning Project			3	2								2		
Congaree River Corridor Water Quality Plng. Asmt.											2	2		
Corpus Christi Bay National Estuary Program	2	2	2			2			2	2	2	2		
Dos Palmas Oasis		2										2		
East Fork Management Plan											2	2		
Eastern Upper Penin. Ecosystem Mgmt. Consortium	3	2		2	2						2	2	2	
Ecosystem Charter for Great Lakes-St. Lawrence Basin	2	2	2	2	2	2	2	2	2	2	2	2		
Elliott State Forest Management Plan			3	1	2							2		
Escanaba River State Forest				2	2		3				2	2	3	
Fish Creek Watershed Project		1		3										
Florida Bay	2	2							3		2	1		
Georgia Mountain	2	2				2					2			
Grand Bay Savanna	3	3		3					2		2	1		

Explanation of symbols: 1–primary use; 2–one of several major uses; 3–minor use

PROJECT	Development--urban, rural, industrial, other	Agriculture--crops or unspecified	Agriculture--grazing and ranching	Forest or timber management	Wildlife management	Hydrologic uses--diversions, impoundments, navigation	Mining--hard rock or unspecified	Mining--oil and gas	Fishing--commercial	Recreation--motorized or unspecified	Recreation--non-motorized	Preserves or open space	Hunting--subsistence	Other
Greater Yellowstone Ecosystem	2		2	2	2	2	2	2		2	2	2		
Green Valley State Park Ecosystem Management Plan		3	3							2	2			
Guadalupe-Nipomo Dunes Preserve		3	3			3	3	3		2	2			
Gulf of Maine Rivers Ecosystem Plan	2	2										3		
Gulf of Mexico Program	2	2	2	2				2	2	2	2			
Hudson River/New York Bight Ecosystem	1	2		2		3						3		
Indiana Grand Kankakee Marsh Restoration Project	3	1		3										
Integrated Landscape Mgmt. for Fish and Wildlife				1								2		
Interior Columbia Basin Ecosystem Mgmt. Project		2	2	2			2			2	2	2		
Interior Low Plateau	2	2	2	2		2	2	2		2	2		2	
Iowa River Corridor Project		1		3						2				
Karner Blue Butterfly Habitat Conservation Plan	2	2	2	2	2									
Kenai River Watershed Project	2			2		3	3		2	2	2	2	2	
Konza Prairie Research Natural Area														1
Lajas Valley Lagoon System	3	2	2	3								3		
Lower Rio Grande Ecosystem Plan	2	2	2								3	2		
Lower Roanoke River Bioreserve	2	2		2					2		2	2		
Lower St. Johns River Ecosystem Management Area	2	2		2			3					3		
Malpai Borderlands Initiative			1	3						3	3			
Marathon County Forest				1						2	2			
Marys River Riparian/Aquatic Restoration Project		2										1		
McPherson Ecosystem Enhancement Project		2	2			2				3	3			
Mesa Creek Coordinated Resource Management Plan			1											
Minnesota Peatlands				3							2	1		
Missouri Coordinated Resource Management	3	2	2	2			3							
Missouri River Mitigation Project	2	2				2	3		3	2	2			
Molokai Preserves		2	2			2				3	3	2	3	
Natural Resource Roundtable	2		2	2										
Nebraska Sandhills Ecosystem		3	1											
Negrito Project			2	2	2					3	3		2	
New Hampshire Forest Resources Plan	3	2		2						2	2	2		
New Jersey Pinelands	2	2				2	2			2	2			
Northeast Chichagof Island	3	3		2								2	2	
Northern Delaware Wetlands Rehabilitation Program	1	3												
Northern Forest Lands Council	2	3	3	2						3	3	3		
Northern Lower Michigan Ecosystem Mgmt. Project	2	2		2	2			2		2	2			

Explanation of symbols: 1–primary use; 2–one of several major uses; 3–minor use

PROJECT	Development--urban, rural, industrial, other	Agriculture--crops or unspecified	Agriculture--grazing and ranching	Forest or timber management	Wildlife management	Hydrologic uses--diversions, impoundments, navigation	Mining--hard rock or unspecified	Mining--oil and gas	Fishing--commercial	Recreation--motorized or unspecified	Recreation--non-motorized	Preserves or open space	Hunting--subsistence	Other
Ohio River Valley Ecosystem	2	3		2		2	2	2						
Oklahoma Tallgrass Prairie Preserve			2					3			2	1		
Ouachita National Forest	3	2	3	2			3			2	2	2		
Owl Mountain Partnership		3	2	2		2				2				
Partners for Prairie Wildlife		2	2									2		—
Patrick Marsh Wetland Mitigation Bank Site											2	1		
Phalen Chain of Lakes Watershed Project	2										2	3		
Piute/El Dorado Desert Wildlife Management Area	2					2	2			2	2	2		
Plainfield Project	2	3		2								2		
Prairie Pothole Joint Venture		1	2											
Prince Wm. Sound–Copper River Ecosystem Initiative	2			2		2			2	2	2	2	2	
Pu'u Kukui Watershed Management Area												1		
Rainwater Basin Joint Venture	3	1												
Robbie Run Study Area				2	2									
Ruby Canyon and Black Ridge Ecosystem Mgmt. Plan		1								2	2			3
San Luis Valley Comprehensive Ecostm. Mgmt. Plan	2	1	2	3		3	2				3			
San Pedro River	2		2				2							
Santa Catalina Island Ecological Restoration Program	3										2	2		
Santa Margarita River	1					2						2		
Sideling Hill Creek Bioreserve	3	3										1		
Snake River Corridor Project	2		2			2					2	2		
South Florida/Everglades Ecos. Restoration Initiative														
St. Marys River Remedial Action Plan	3		3			3						1		
State Line Serpentine Barrens	2	2										1		
Stegall Mountain Natural Area											2	2		
Tensas River Basin Initiative		1										2		
Tidelands of the Connecticut River	1	3												2
Trail Creek Ecosystem Analysis			3	3						2	2			3
Trout Mountain Roadless Area				2							2			
Upper Farmington River Management Plan	1	3				2					2			
Upper Huerfano Ecosystem	3	2	2	2		2				2	2	2		
Verde River Greenway	2	2	2								2	2		
Virginia Coast Reserve		2							2	3	3	2		
Wild Stock Initiative	2	2	2	2		2								
Wildlife Area Planning				2	2					2	2			
Wildlife Habitat Improvement Group	1													

Explanation of symbols: 1–primary use; 2–one of several major uses; 3–minor use

SIZE OF PROJECT AREAS

50,000 ACRES OR LESS	Size (acres)
Robbie Run Study Area	60
Camp Johnson Sandplain Restoration	200
Patrick Marsh Wetland Mitigation Bank Site	270
Verde River Greenway	410
Georgia Mountain Ecosystem Management Project	1,250
State Line Serpentine Barrens	2,100
Green Valley State Park Ecosystem Management Plan	2,300
Tidelands of the Connecticut River	2,500
Guadalupe-Nipomo Dunes Preserve	3,897
Albany Pine Bush	4,000
Stegall Mountain Natural Area	5,387
Wildlife Habitat Improvement Group	5,600
Block Island Refuge	6,400
Pu'u Kukui Watershed Management Area	8,600
Konza Prairie Research Natural Area	8,616
Northern Delaware Wetlands Rehabilitation Program	10,000
McPherson Ecosystem Enhancement Project	11,000
Plainfield Project	13,632
Dos Palmas Oasis	14,000
Phalen Chain of Lakes Watershed Project	14,790
Trail Creek Ecosystem Analysis	20,000
Molokai Preserves	22,000
Congaree River Corridor Water Quality Planning Assessment	22,200
Indiana Grand Kankakee Marsh Restoration Project	26,500
Marathon County Forests	27,000
Trout Mountain Roadless Area	32,000
Butte Valley Basin	32,625
Oklahoma Tallgrass Prairie Preserve	37,000
Bitteroot Ecosystem Management Research Project	40,000
Cheyenne Bottoms Wildlife Area	41,000
Santa Catalina Island Ecological Restoration Program	42,135
Virginia Coast Reserve	45,000
Lajas Valley Lagoon System	48,000
Partners for Prairie Wildlife	49,920
Allegan State Game Area	50,000
Iowa River Corridor Project	50,000
Missouri River Mitigation Project	50,000

50,001 TO 250,000 ACRES	
East Fork Management Plan	54,000
Grand Bay Savanna	60,000
Cache River Wetlands	60,000
Sideling Hill Creek Bioreserve	66,000
Fish Creek Watershed Project	70,400
Colorado State Forest Ecosystem Planning Project	70,768
Elliott State Forest Management Plan	93,000
Mesa Creek Coordinated Resource Management Plan	100,000
Ruby Canyon and Black Ridge Ecosystem Management Plan	118,700

50-001 to 250,000 acres--continued	Size (acres)
Chattooga River Project	120,000
Negrito Project	120,000
Natural Resource Roundtable	122,000
Minnesota Peatlands	146,224
Chicago Wilderness	200,000
Owl Mountain Partnership	246,000
Upper Huerfano Ecosystem	250,000

250,001 TO 1,000,000 ACRES	
Northeast Chichagof Island	275,000
Marys River Riparian/Aquatic Restoration Project	332,800
ACE Basin	350,000
Big Darby Creek Partnership	371,000
Escanaba River State Forest	420,400
Applegate Partnership	500,000
San Pedro River	512,000
Piute/El Dorado Desert Wildlife Management Area	531,000
Florida Bay Ecosystem Management Area	640,000
Tensas River Basin Initiative	750,000
Malpai Borderlands Initiative	802,000
Integrated Landscape Management for Fish and Wildlife	839,000
Wildlife Area Planning	840,000
Chequamegon National Forest Landscape Analysis and Design	850,000
Santa Margarita River	1,000,000

OVER 1,000,000 ACRES	
New Jersey Pinelands	1,100,000
Kenai River Watershed Project	1,400,000
Clinch Valley Bioreserve	1,408,000
Ouachita National Forest	1,600,000
Lower St. Johns River Ecosystem Management Area	1,777,000
San Luis Valley Comprehensive Ecosystem Management Plan	2,560,000
Rainwater Basin Joint Venture	2,700,000
Cache/Lower White Rivers Ecosystem Management Plan	2,800,000
Eastern Upper Peninsula Ecosystem Management Consortium	3,880,000
Barataria-Terrebonne National Estuary Program	4,000,000
New Hampshire Forest Resources Plan	5,000,000
Corpus Christi Bay National Estuary Program	7,735,680
Karner Blue Butterfly Habitat Conservation Plan	9,000,000
Hudson River/New York Bight Ecosystem	11,500,000
South Florida/Everglades Ecosystem Restoration Initiative	11,500,000+
Nebraska Sandhills Ecosystem	12,500,000
Canyon Country Partnership	15,000,000+
Blue Mountains Natural Resources Institute	19,000,000
Greater Yellowstone Ecosystem	19,000,000
Northern Lower Michigan Ecosystem Management Project	19,000,000
Northern Forest Lands Council	26,000,000

Over 1,000,000 acres--continued	Size (acres)
Interior Low Plateau	30,000,000
Prince William Sound–Copper River Ecosystem Initiative	30,000,000
Lower Rio Grande Ecosystem Plan	38,400,000
Chesapeake Bay Program	41,000,000
Wild Stock Initiative	42,000,000
Gulf of Maine Rivers Ecosystem Plan	44,233,600
Missouri Coordinated Resource Management	45,000,000
Prairie Pothole Joint Venture	64,000,000
Ohio River Valley Ecosystem	92,160,000
Interior Columbia Basin Ecosystem Management Project	144,000,000
Ecosystem Charter for the Great Lakes–St. Lawrence Basin	189,000,000
Gulf of Mexico Program	410,000,000

OTHER

Upper Farmington River Management Plan	14 river miles
Snake River Corridor Project	69 river miles
St. Marys River Remedial Action Plan	75 river miles
Lower Roanoke River Bioreserve	137 river miles

ANTHROPOGENIC ECOSYSTEM STRESSES ON PROJECT AREAS

PROJECT \ STRESS	Agricultural practices	Disruption of fire regime	Exotic species	Grazing & range management	Hydrologic alteration	Land conversion–agriculture	Land conversion–urban	Mining	Overfishing, hunting, collecting	Non-point source pollution	Point source pollution	Recreation	Roads or other infrastructure	Timber & forest management	Other	No significant stresses
ACE Basin																♦
Albany Pine Bush		♦	♦				♦			♦			♦			
Allegan State Game Area															♦	
Applegate Partnership		♦	♦		♦		♦	♦								
Barataria-Terrebonne National Estuary Program					♦	♦				♦	♦					
Big Darby Creek Partnership	♦					♦	♦			♦	♦					
Bitteroot Ecosystem Management Research Project		♦	♦				♦							♦		
Block Island Refuge							♦								♦	
Blue Mountains Natural Resources Institute		♦		♦	♦								♦	♦	♦	
Butte Valley Basin	♦		♦	♦	♦	♦										
Cache River Wetlands					♦										♦	
Cache/Lower White Rivers Ecosystem Management Plan	♦				♦	♦				♦						
Camp Johnson Sandplain Restoration		♦					♦						♦			
Canyon Country Partnership				♦			♦			♦		♦	♦			
Chattooga River Project		♦					♦			♦				♦		
Chequamegon National Forest Landscape Analysis & Design		♦				♦	♦					♦		♦		
Chesapeake Bay Program	♦				♦	♦	♦			♦	♦		♦			
Cheyenne Bottoms Wildlife Area	♦	♦			♦											
Chicago Wilderness		♦	♦												♦	
Clinch Valley Bioreserve		♦		♦				♦			♦					
Colorado State Forest Ecosystem Planning Project		♦		♦									♦	♦	♦	
Congaree River Corridor Water Quality Planning Assessment			♦													
Corpus Christi Bay National Estuary Program				♦	♦				♦	♦	♦					
Dos Palmas Oasis	♦		♦													
East Fork Management Plan							♦						♦	♦		
Eastern Upper Peninsula Ecosystem Management Consortium		♦												♦		
Ecosystem Charter for the Great Lakes–St. Lawrence Basin	♦		♦		♦	♦	♦			♦	♦	♦				
Elliott State Forest Management Plan																
Escanaba River State Forest	♦	♦								♦		♦	♦		♦	
Fish Creek Watershed Project										♦						
Florida Bay Ecosystem Management Area	♦				♦	♦	♦						♦		♦	
Georgia Mountain Ecosystem Management Project		♦			♦	♦						♦		♦		
Grand Bay Savanna		♦			♦		♦			♦	♦					
Greater Yellowstone Ecosystem		♦		♦	♦		♦	♦		♦	♦	♦		♦		
Green Valley State Park Ecosystem Management Plan	♦															
Guadalupe-Nipomo Dunes Preserve			♦					♦			♦	♦				

PROJECT \ STRESS	Agricultural practices	Disruption of fire regime	Exotic species	Grazing & range management	Hydrologic alteration	Land conversion–agriculture	Land conversion–urban	Mining	Overfishing, hunting, collecting	Non-point source pollution	Point source pollution	Recreation	Roads or other infrastructure	Timber & forest management	Other	No significant stresses
Gulf of Maine Rivers Ecosystem Plan					♦		♦				♦					
Gulf of Mexico Program	♦									♦					♦	
Hudson River/New York Bight Ecosystem			♦				♦			♦	♦		♦		♦	
Indiana Grand Kankakee Marsh Restoration Project					♦	♦										
Integrated Landscape Management for Fish and Wildlife					♦					♦				♦		
Interior Columbia Basin Ecosystem Management Project	♦	♦	♦	♦	♦					♦			♦	♦		
Interior Low Plateau						♦	♦			♦	♦					
Iowa River Corridor Project	♦						♦									
Karner Blue Butterfly Habitat Conservation Plan	♦	♦				♦	♦						♦	♦	♦	
Kenai River Watershed Project					♦		♦		♦			♦	♦	♦		
Konza Prairie Research Natural Area																♦
Lajas Valley Lagoon System	♦		♦	♦	♦	♦				♦						
Lower Rio Grande Ecosystem Plan	♦	♦		♦	♦	♦	♦	♦	♦	♦	♦	♦	♦			
Lower Roanoke River Bioreserve	♦				♦				♦	♦	♦	♦	♦			
Lower St. Johns River Ecosystem Management Area	♦						♦			♦	♦		♦			
Malpai Borderlands Initiative		♦					♦									
Marathon County Forests			♦				♦					♦				
Marys River Riparian/Aquatic Restoration Project	♦			♦	♦	♦										
McPherson Ecosystem Enhancement Project		♦	♦		♦									♦		
Mesa Creek Coordinated Resource Management Plan		♦	♦	♦	♦					♦						
Minnesota Peatlands					♦											
Missouri Coordinated Resource Management	♦		♦	♦	♦	♦	♦						♦			
Missouri River Mitigation Project					♦	♦				♦	♦					
Molokai Preserves	♦	♦	♦	♦	♦	♦	♦	♦	♦				♦	♦		
Natural Resource Roundtable		♦			♦		♦			♦				♦		
Nebraska Sandhills Ecosystem	♦				♦					♦						
Negrito Project		♦		♦								♦		♦		
New Hampshire Forest Resources Plan							♦									
New Jersey Pinelands	♦	♦								♦						
Northeast Chichagof Island					♦				♦					♦	♦	
Northern Delaware Wetlands Rehabilitation Program			♦		♦		♦			♦	♦	♦	♦			
Northern Forest Lands Council												♦				
Northern Lower Michigan Ecosystem Management Project						♦	♦									
Ohio River Valley Ecosystem	♦			♦	♦			♦		♦	♦	♦		♦		
Oklahoma Tallgrass Prairie Preserve																
Ouachita National Forest		♦		♦		♦				♦			♦	♦		
Owl Mountain Partnership													♦	♦		
Partners for Prairie Wildlife	♦	♦	♦	♦			♦									
Patrick Marsh Wetland Mitigation Bank Site							♦			♦						
Phalen Chain of Lakes Watershed Project		♦	♦		♦		♦			♦			♦	♦		

70

PROJECT	Agricultural practices	Disruption of fire regime	Exotic species	Grazing & range management	Hydrologic alteration	Land conversion--agriculture	Land conversion--urban	Mining	Overfishing, hunting, collecting	Non-point source pollution	Point source pollution	Recreation	Roads or other infrastructure	Timber & forest management	Other	No significant stresses
Piute/El Dorado Desert Wildlife Management Area	♦			♦			♦	♦	♦			♦	♦			
Plainfield Project								♦						♦		
Prairie Pothole Joint Venture	♦			♦	♦	♦										
Prince William Sound–Copper River Ecosystem Initiative			♦						♦		♦	♦		♦		
Pu'u Kukui Watershed Management Area			♦		♦											
Rainwater Basin Joint Venture	♦		♦		♦	♦						♦				
Robbie Run Study Area			♦												♦	
Ruby Canyon and Black Ridge Ecosystem Management Plan		♦	♦									♦				
San Luis Valley Comprehensive Ecosystem Management Plan		♦	♦		♦		♦					♦				
San Pedro River	♦	♦	♦		♦		♦	♦			♦					
Santa Catalina Island Ecological Restoration Program		♦	♦											♦		
Santa Margarita River		♦			♦	♦										
Sideling Hill Creek Bioreserve			♦							♦				♦		
Snake River Corridor Project				♦			♦	♦						♦		
South Florida/Everglades Ecosystem Restoration Initiative	♦		♦		♦		♦			♦						
St. Marys River Remedial Action Plan			♦		♦		♦			♦	♦			♦		
State Line Serpentine Barrens		♦	♦				♦	♦	♦					♦		
Stegall Mountain Natural Area		♦						♦								
Tensas River Basin Initiative	♦				♦	♦				♦						
Tidelands of the Connecticut River			♦				♦									
Trail Creek Ecosystem Analysis							♦			♦		♦	♦			
Trout Mountain Roadless Area									♦				♦	♦		
Upper Farmington River Management Plan																♦
Upper Huerfano Ecosystem		♦		♦								♦				
Verde River Greenway					♦		♦	♦		♦	♦	♦		♦		
Virginia Coast Reserve	♦				♦					♦	♦					
Wild Stock Initiative	♦			♦	♦	♦	♦			♦				♦	♦	
Wildlife Area Planning			♦							♦				♦	♦	
Wildlife Habitat Improvement Group															♦	

MOST SIGNIFICANT OUTCOMES OF PROJECTS

PROJECT \ OUTCOME	Improved communication & cooperation	Development of management plan	Development of decision-making structures	Change in approach to land management	Ongoing restoration activities	Restoration results	Increased public awareness	Research	Land acquisition and easements	Public education efforts	Changes in management practices	Increased scientific understanding	Increased trust and respect	Other
ACE Basin	♦	♦	♦						♦					
Albany Pine Bush		♦			♦	♦	♦		♦					
Allegan State Game Area		♦		♦	♦	♦					♦			
Applegate Partnership	♦			♦	♦		♦			♦	♦		♦	
Barataria-Terrebonne National Estuary Program	♦	♦	♦					♦				♦	♦	
Big Darby Creek Partnership	♦	♦	♦	♦		♦	♦				♦			♦
Bitteroot Ecosystem Management Research Project														
Block Island Refuge		♦		♦	♦	♦	♦		♦					
Blue Mountains Natural Resources Institute	♦			♦	♦		♦	♦		♦			♦	
Butte Valley Basin	♦	♦			♦	♦	♦	♦						
Cache River Wetlands	♦		♦	♦					♦		♦			
Cache/Lower White Rivers Ecosystem Management Plan		♦	♦									♦		
Camp Johnson Sandplain Restoration	♦	♦		♦	♦	♦	♦	♦						
Canyon Country Partnership	♦	♦								♦	♦			
Chattooga River Project	♦		♦		♦			♦						
Chequamegon National Forest Landscape Analysis and Design		♦		♦				♦			♦			
Chesapeake Bay Program	♦		♦	♦	♦	♦	♦	♦						
Cheyenne Bottoms Wildlife Area		♦	♦		♦	♦								
Chicago Wilderness			♦	♦										
Clinch Valley Bioreserve	♦	♦	♦	♦			♦							
Colorado State Forest Ecosystem Planning Project	♦	♦	♦	♦			♦							
Congaree River Corridor Water Quality Planning Assessment	♦		♦											
Corpus Christi Bay National Estuary Program	♦		♦	♦					♦					
Dos Palmas Oasis	♦	♦		♦	♦	♦			♦					
East Fork Management Plan		♦			♦	♦						♦	♦	
Eastern Upper Peninsula Ecosystem Management Consortium	♦		♦	♦			♦	♦	♦				♦	
Ecosystem Charter for the Great Lakes–St. Lawrence Basin	♦	♦						♦						
Elliott State Forest Management Plan	♦	♦		♦										
Escanaba River State Forest	♦	♦	♦	♦			♦	♦			♦			
Fish Creek Watershed Project	♦		♦			♦	♦			♦	♦		♦	
Florida Bay Ecosystem Management Area	♦	♦										♦		
Georgia Mountain Ecosystem Management Project	♦	♦												
Grand Bay Savanna	♦				♦				♦					
Greater Yellowstone Ecosystem														

PROJECT / OUTCOME	Improved communication & cooperation	Development of management plan	Development of decision-making structures	Change in approach to land management	Ongoing restoration activities	Restoration results	Increased public awareness	Research	Land acquisition and easements	Public education efforts	Changes in management practices	Increased scientific understanding	Increased trust and respect	Other
Green Valley State Park Ecosystem Management Plan	♦	♦	♦	♦	♦	♦								
Guadalupe-Nipomo Dunes Preserve	♦			♦	♦	♦			♦	♦	♦			
Gulf of Maine Rivers Ecosystem Plan	♦	♦		♦										
Gulf of Mexico Program	♦		♦	♦	♦	♦	♦	♦		♦				
Hudson River/New York Bight Ecosystem	♦	♦											♦	
Indiana Grand Kankakee Marsh Restoration Project	♦	♦	♦	♦	♦			♦			♦	♦		
Integrated Landscape Management for Fish and Wildlife	♦	♦	♦											
Interior Columbia Basin Ecosystem Management Project	♦					♦								
Interior Low Plateau		♦						♦				♦		
Iowa River Corridor Project		♦		♦				♦						
Karner Blue Butterfly Habitat Conservation Plan	♦													
Kenai River Watershed Project	♦													
Konza Prairie Research Natural Area	♦	♦		♦	♦	♦		♦			♦	♦		
Lajas Valley Lagoon System		♦												
Lower Rio Grande Ecosystem Plan	♦	♦	♦	♦										
Lower Roanoke River Bioreserve		♦	♦				♦	♦	♦	♦				
Lower St. Johns River Ecosystem Management Area	♦	♦	♦											
Malpai Borderlands Initiative	♦		♦	♦	♦				♦					
Marathon County Forests		♦	♦	♦	♦					♦				
Marys River Riparian/Aquatic Restoration Project	♦	♦		♦	♦	♦	♦		♦					
McPherson Ecosystem Enhancement Project														
Mesa Creek Coordinated Resource Management Plan	♦	♦			♦	♦								
Minnesota Peatlands					♦						♦			
Missouri Coordinated Resource Management	♦	♦	♦	♦			♦				♦			
Missouri River Mitigation Project	♦	♦								♦				
Molokai Preserves	♦		♦	♦	♦	♦		♦	♦	♦		♦		
Natural Resource Roundtable	♦	♦	♦										♦	
Nebraska Sandhills Ecosystem	♦	♦	♦		♦	♦				♦			♦	
Negrito Project	♦		♦					♦					♦	
New Hampshire Forest Resources Plan	♦		♦	♦			♦							
New Jersey Pinelands	♦		♦	♦				♦						♦
Northeast Chichagof Island	♦		♦					♦						
Northern Delaware Wetlands Rehabilitation Program	♦	♦	♦	♦	♦	♦	♦				♦			
Northern Forest Lands Council	♦	♦	♦	♦										
Northern Lower Michigan Ecosystem Management Project	♦	♦	♦				♦							
Ohio River Valley Ecosystem	♦	♦		♦				♦						
Oklahoma Tallgrass Prairie Preserve	♦	♦			♦				♦		♦		♦	
Ouachita National Forest	♦	♦	♦	♦	♦			♦			♦			

PROJECT \ OUTCOME	Improved communication & cooperation	Development of management plan	Development of decision-making structures	Change in approach to land management	Ongoing restoration activities	Restoration results	Increased public awareness	Research	Land acquisition and easements	Public education efforts	Changes in management practices	Increased scientific understanding	Increased trust and respect	Other
Owl Mountain Partnership	♦		♦	♦	♦			♦				♦	♦	
Partners for Prairie Wildlife	♦	♦		♦						♦				
Patrick Marsh Wetland Mitigation Bank Site		♦			♦	♦			♦	♦				
Phalen Chain of Lakes Watershed Project	♦	♦	♦	♦	♦		♦			♦	♦			
Piute/El Dorado Desert Wildlife Management Area		♦	♦		♦	♦			♦		♦			
Plainfield Project	♦			♦	♦		♦			♦				
Prairie Pothole Joint Venture	♦	♦	♦		♦	♦								
Prince William Sound–Copper River Ecosystem Initiative	♦		♦											
Pu'u Kukui Watershed Management Area		♦			♦	♦								
Rainwater Basin Joint Venture	♦	♦	♦		♦	♦	♦	♦	♦	♦		♦		
Robbie Run Study Area				♦										
Ruby Canyon and Black Ridge Ecosystem Management Plan	♦	♦	♦	♦			♦				♦			
San Luis Valley Comprehensive Ecosystem Management Plan			♦	♦	♦	♦	♦		♦				♦	
San Pedro River	♦	♦	♦						♦	♦				
Santa Catalina Island Ecological Restoration Program					♦	♦								
Santa Margarita River	♦			♦					♦	♦				
Sideling Hill Creek Bioreserve	♦	♦		♦			♦							
Snake River Corridor Project	♦													
South Florida/Everglades Ecosystem Restoration Initiative	♦		♦		♦									
St. Marys River Remedial Action Plan		♦	♦		♦	♦	♦			♦				
State Line Serpentine Barrens		♦							♦	♦				
Stegall Mountain Natural Area				♦	♦	♦					♦			
Tensas River Basin Initiative	♦	♦	♦						♦			♦	♦	
Tidelands of the Connecticut River		♦	♦	♦						♦				
Trail Creek Ecosystem Analysis	♦	♦		♦	♦	♦			♦			♦		
Trout Mountain Roadless Area			♦											
Upper Farmington River Management Plan	♦	♦	♦											♦
Upper Huerfano Ecosystem	♦		♦	♦			♦							
Verde River Greenway	♦	♦		♦	♦		♦				♦		♦	
Virginia Coast Reserve		♦	♦	♦	♦	♦	♦		♦					
Wild Stock Initiative	♦				♦		♦	♦				♦		
Wildlife Area Planning	♦	♦			♦							♦		
Wildlife Habitat Improvement Group	♦	♦	♦		♦	♦	♦	♦			♦			

75

Descriptions of 105 Selected Ecosystem Management Projects

INDEX OF 105 ECOSYSTEM MANAGEMENT PROJECT DESCRIPTIONS

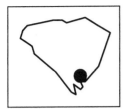

ACE BASIN

Location:
Southeastern South Carolina

Project size:
350,000 acres

Initiators:
South Carolina Department of Natural Resources, The Nature Conservancy, Ducks Unlimited, U.S. Fish and Wildlife Service

PROJECT AREA DESCRIPTION

The lower watersheds of the Ashepoo, Combahee, and Edisto rivers collectively make up the ACE Basin ecosystem, extending roughly from Interstate 95 to the St. Helena Sound estuary and Atlantic coast. This coastal region consists of southern pine and hardwood uplands, bottomland forested wetlands (including tupelo gum, cypress, sweetgum, and maple), and freshwater, saltwater, and brackish marshes. The extended ecosystem includes the estuary, beaches, and barrier islands. Over 130 species of birds and seventeen wading bird rookeries are found in this sparsely-populated, largely undeveloped area. Several federal and state listed threatened and endangered species reside in the basin, including the wood stork, loggerhead turtle, and two plants, pondberry and Canby dropwort.

The ACE Basin was once extensively cleared, first for pine lumber and poles supplied to the port of Charleston for export in the 1600s. Later, tidal forested wetlands were cleared for large rice plantations in the 1700s. Following the collapse of rice farming during the Civil War, many failed plantations were bought by wealthy northern industrialists for hunting reserves, who kept them undivided, maintaining their flooded rice fields for waterfowl and the upland habitat for deer and turkey. Other plantation lands reverted to bottomland hardwoods.

Today, timber, hunting, and commercial and recreational fishing are the dominant human uses of the area. Landownership is primarily private, with several large industrial forest landowners. Many former plantations remain largely intact. Public lands include the ACE Basin National Wildlife Refuge (NWR), two state Wildlife Management Areas (WMA), and the National Estuarine Research Reserve.

ECOSYSTEM STRESSES

Stresses associated with the extensive clearing in the 1600s and 1700s are unknown. However, the area is now ecologically stable, whether in plantations or natural forests. Current uses of the land appear to cause only minimal stress to the system. Resort development, and to a lesser degree roads, pose the greatest potential threats, due to an expanding population base and demand for housing from Charleston only forty miles away.

PROJECT DESCRIPTION

In 1988, an ACE Basin task force was formed through the cooperative efforts of The Nature Conservancy (TNC), Ducks Unlimited/Wetlands America, the South Carolina Department of Natural Resources (SC DNR), the U.S. Fish and Wildlife Service (FWS), and private landowners, who recognized the values in protecting this relatively undeveloped ecosystem. Another catalyst of the effort was the development of the North American Waterfowl Management Plan in 1986, which includes the ACE Basin under the Atlantic Coast Joint Initiative. Westvaco Corporation, the largest of the landowners, supported the task force in July 1991 with a memorandum of understanding outlining forestry practices on their 15,500 acres within the basin. The ACE is part of TNC's "Last Great Places" campaign.

The effort is focussed principally on

safeguarding the ecosystem against urban development. A management plan for the ACE Basin was developed by the task force in 1988. Its goals are to 1) protect 200,000 acres by the end of the century through conservation easements, cooperative management agreements (such as Westvaco's), and outright purchases; 2) assist landowners with resource management; and 3) preserve compatible traditional land management practices which have effectively become part of the ecosystem's integrity. No monitoring is specified in the management plan, although some individual institutions are including monitoring as part of their management efforts.

The task force consists of one representative of each member organization and is chaired by a private landowner representative. Additional staff from the member organizations provide biological, legal, and public relations expertise. The task force is managed on an informal basis with decisions reached through general consensus.

STATUS AND OUTLOOK

The effort has achieved over 50 percent of the initial protection goal (90,000 acres), with 35,000 acres acquired by federal and state agencies, and more than 30,000 acres in conservation easements in addition to Westvaco's lands and previously-protected sites. As a result, the protection goal was increased to 200,000 acres. The federal ACE Basin NWR (11,000 acres to date) and the state's Donnelley WMA (8,000 acres) are newly created; the state has expanded the existing Bear Island WMA to 12,000 acres. In addition, the newly-created National Estuarine Research Reserve includes coastal islands totaling approximately 12,000 acres and is administered by SC DNR with oversight from the National Oceanic and Atmospheric Administration.

In 1995, TNC and several local community institutions initiated a strategic planning process to develop a vision for achieving economic growth compatible with long-term conservation of the basin's natural resources.

Much remains to be done beyond land acquisition, especially with private landowner education and assistance, and improving public awareness in general.

Factors Facilitating Progress

Perhaps the greatest asset to this effort is the basin's undeveloped, relatively stable ecological condition. Broad-scale cooperation, vision, and commitment across public and private institutions is the primary reason this effort has moved ahead. Political support from the state's U.S. Senate delegation has resulted in the availability of federal land acquisition funds.

Obstacles to Progress

Lack of awareness by the local public, who are concerned that land will be "locked up," traditional uses eliminated, and economic opportunities lost, appears to be the greatest obstacle to this effort's progress. The threat of resort development is still significant, as many critical tracts remain unprotected.

Contact information:

Mr. Mike Prevost
Director, ACE Basin Bioreserve
  Project
The Nature Conservancy
P.O. Box 848
8675 Willtown Road
Hollywood, SC 29449
(803) 889-2427
Fax: (803) 889-3282

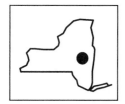

ALBANY PINE BUSH

Location:
Eastern New York

Project Size:
4,000 acres

Initiator:
Friends of the Pine Bush (now known as Save the Pine Bush, Inc.)

PROJECT AREA DESCRIPTION

The Albany Pine Bush (APB) barrens once stretched across 25,000 acres, largely between Schenectady and Albany. Apparently, the historical ecosystem corresponded with the distribution of sandy soils deposited on the floor of glacial Lake Albany. The APB consists of gently-rolling sand dunes running generally southwest-northeast, with dunes ranging from under twenty feet to 100 feet tall, the highest reaching nearly 400 feet. Ravines up to 100 feet deep and containing groundwater runoff cut into parts of the ecosystem. The present-day Pine Bush Preserve encompasses 2,290 acres and is located on the western edge of Albany.

Seven different vegetative community types have been identified in the Pine Bush Preserve, including pitch pine-scrub oak barrens, vernal ponds, successional southern hardwood and northern hardwood forests, Appalachian oak-pine forests, red maple hardwood swamps, and shallow emergent marshes. The area contains one federally listed endangered species, the Karner blue butterfly, and more than two dozen state listed rare plants, reptiles, and amphibians, including the hognose snake and Jefferson/spotted salamander.

Current uses of the ecosystem are many, and include urban, commercial and residential development, agriculture, open space, and public and private preserves.

ECOSYSTEM STRESSES

Because of the preserve's location in a metropolitan area, and its sandy soil texture, considered ideal for construction, the most significant stress to the landscape is land conversion to urban uses. Fire suppression is another significant stress and disrupts natural vegetative processes; at least two of the natural vegetative communities are known to be fire-dependent. Associated roads and other infrastructure have caused severe fragmentation, also making prescribed fire management more difficult. Finally, non-point source pollution from urban nutrient and pesticide runoff and point source pollution from a chemical waste site pose threats to the underlying aquifer.

PROJECT DESCRIPTION

Efforts to protect the APB began in the 1960s with a group of local citizens who formed the Friends of the Pine Bush, now known as Save the Pine Bush, Inc., in the 1907s. With the assistance of The Nature Conservancy (TNC), Friends of the Pine Bush convinced the town of Guilderland and the New York State Department of Environmental Conservation (NYSDEC) to purchase several hundred acres of land for a preserve.

Over the ensuing years, Save the Pine Bush sued developers and the city of Albany on several occasions, halting development on several sites that were then acquired for protection. One case against Albany in 1987 resulted in the city funding the first three years of research and management of the APB Preserve and allowed the first staff person to be hired. In 1988, the state legislature created the Albany Pine Bush Preserve Commission and charged it with overseeing and ensuring protection and management of the preserve. The goals of the APB preserve and commission are to 1) protect the fragile ecology of the APB by protecting appropriate lands, managing the preserve, and allowing appropriate uses; and 2) provide an educational and recreational resource for the public. Specific strategies for achieving these goals include: 1) protecting nearly 4,000 acres in the preserve through land acquisition, easements, and voluntary cooperative management agreements with willing landowners; 2) on-the-ground management (prescribed fire, weedy/exotic plant control through fire, chemical and mechanical means); 3) research and monitoring; 4) public education; 5) trail system development; and 6) decision-maker/public awareness efforts.

As the decision-making body for the APB Preserve, the commission consists of senior figures from six land-owning institutions (TNC, towns of Colonie and Guilderland, city of Albany, NYSDEC, New York State Office of Parks and Recreation), and three private citizens. A technical committee, which assists the commission, consists of representatives from the six organizations, along with occasional, non-permanent ex-officio members. The organizations provide in-kind services to operate the

preserve. For example, the city of Albany provides equipment and personnel to assist with the preparation of firebreaks each year; and NYSDEC and the town of Colonie provide rangers and a conservation officer, respectively, as burn crew members.

STATUS AND OUTLOOK

Since the 1970s, 2,290 acres have been protected through acquisitions and easements, at a cost of over $25 million. Restoration activities began in 1991 with controlled burns on a limited scale. Today, burning is more extensive and herbicide and mechanical control techniques are being used on black locust (non-native) and aspen (native) species that are encroaching in inappropriate places.

In May 1995, the commission released a draft protection plan for the preserve. Following a public comment period, the protection plan was finalized and was to have been voted on for approval by the APB Commission in late March 1996.

In November 1994, an executive director and preserve steward were hired. The steward's primary responsibilities are to develop a trail system, post boundaries, and make the preserve more recognizable.

The public is much more aware of the preserve and its programs, which may benefit efforts to manage recreation on the preserve.

Factors Facilitating Progress

Public support, especially from local municipalities and state agencies, as well as the protection efforts of Save the Pine Bush and TNC, have been credited with helping this effort progress. Funding from the state and TNC has also been instrumental. Finally, the proximity of the APB to universities, schools, and volunteers has benefited the effort.

Obstacles to Progress

A shortage of funding for land acquisition, program operations, research, and monitoring is a continuing problem. The ability to conduct controlled burns is limited by air quality problems prevalent in the northeast U.S. and by citizen complaints.

Contact information:

Ms. Stephanie Gebauer
The Nature Conservancy
Albany Pine Bush
1653 Central Avenue
Albany, NY 12205
(518) 464-6496
Fax: (518) 464-6761

ALLEGAN STATE GAME AREA

Location:
Southwestern Michigan

Project size:
50,000 acres

Initiator:
Michigan Department of Natural Resources

PROJECT AREA DESCRIPTION

Several ecosystems are represented in the Allegan State Game Area (ASGA). Half of the ASGA is characterized by dry-mesic northern forest, dominated by oak and white pine. Another 15,000 acres is dry-mesic southern forest (oak-hickory). The ASGA also supports oak barrens, dry-sand prairie, and small coastal plain marshes. The ASGA is home to ten federally listed threatened and endangered species and sixty-five state listed species. An example of the former is the Karner blue butterfly. Another federally listed species, the peregrine falcon, migrates through the area.

The ASGA is primarily used for wildlife and timber management. Hunting is the most important recreational use. Other uses include fishing, berry picking, mushroom collecting, wildlife viewing, and snowmobiling.

ECOSYSTEM STRESSES

History has left its mark on the ASGA. Logging of white pine and hardwoods around the turn of the century was followed by repeated, uncontrolled fires. Unusually dense stands of white and black oak resulted, persisting through a subsequent period of fire suppression. Fire suppression has also stressed other ecosystems in the ASGA.

During the early logging era, an exhaustive road system was developed in the ASGA. Many of these roads still exist today. The state, however, recently closed several because of their stress to the area.

Another ecosysem stress is the occurrence of exotic pests like gypsy moth, pine shoot beetle, and Japanese beetle. In addition, the large number of non-indigenous species has led to a restructuring of plant communities and to the local disappearance of some prairie species.

PROJECT DESCRIPTION

In order to comply with funding requirements, the Wildlife Division of the Michigan Department of Natural Resources is required to manage the ASGA using a comprehensive management plan. Master plans are reviewed at least once every ten years. Review of the 1978 plan and subsequent planning resulted in the current plan whose implementation started in 1994.

One of the goals of the current plan is to manage wildlife in concert with plant communities and ecosystems. To reach this goal, the following steps were taken. During the planning period, the entire ASGA was inventoried for forest types, unique plant and animal species, unique plant communities, presettlement vegetation types, and historical and archaeological sites. In addition, most plant species, amphibians, reptiles, mammals, and fish were identified. Based on these extensive inventories, the Wildlife Division determined to which communities forest stands and key species belong. Management techniques vary based on the type of plant community. Once every five years, a stand will be reviewed, and if needed, specific management practices applied. Open areas may be maintained using prescribed burns, cuttings, and conventional tilling. Forests may be managed using clear-cutting, shelterwood harvesting, selection harvests, or "doing nothing" in areas set aside as potential old growth.

Many agencies and organizations have been instrumental in the development of the management plan, as well as in its implementation. Although the Wildlife Division manages the ASGA and is thus the lead agency, others such as the Michigan Natural Features Inventory, Michigan State University, U.S.D.A. Forest Service, and various divisions within the Michigan Department of Natural Resources have been involved. Public input was solicited also.

STATUS AND OUTLOOK

According to the five-year review cycle, 20 percent of all stands have been reviewed every year for needed treatments. Prescribed burns have been emphasized in recent years as one of several available management techniques. Several prescribed burns have already been carried out and have led to an increase in the Karner blue butterfly population. The effects of management techniques on other wildlife has been observed already or will be apparent soon.

Factors Facilitating Progress

The long-term planning process and the ability to set long-term goals has been beneficial. The combination of using a "top-down" (plant community or ecosystem) approach to management integrated with a "bottom up" (stand) approach works well. Development of an ecosystem-based plan has taken time. It has been very helpful that supervisors have allowed staff to take that time, rather than pressing for a less comprehensive plan based on outputs.

Obstacles to Progress

The planning process was slowed by many meetings and correspondence, although such communication was required by the process. For instance, politicians needed to be updated continually.

Public education efforts, considered an essential component that must occur before progress can be made in other aspects of the planning efforts, are needed to address what has been a

highly emotional conflict between some stakeholders with opposing views. The public has not always responded favorably to the plan. In some cases, this can be attributed to incorrect perceptions. In other cases, a proposed change in the status quo has been perceived as threatening certain recreational uses of the ASGA.

Outbreaks of pests and diseases, such as gypsy moth and red pine diseases, have required a rethinking of some management techniques.

Contact information:

Mr. John Lerg
Wildlife Biologist
Michigan Department of Natural Resources
Allegan State Game Area
4590 118th Avenue
Allegan, MI 49010
(616) 673-2430

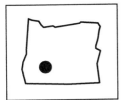

APPLEGATE PARTNERSHIP

Location:
Southern Oregon

Project size:
500,000 acres

Initiators:
Private citizens

PROJECT AREA DESCRIPTION

The Siskiyou Mountains are the dominant landform in the project area. The Siskiyou chain runs east-west across southern Oregon and northern California, connecting the Oregon Coast Range to the Cascades and the Sierras. Elevation ranges from 700 to 7,000 feet. The area's micro-climates include high alpine to nearly desert conditions. Accordingly, rainfall in the Applegate River watershed varies from seventeen to forty inches per year. The Siskiyous represent an important biological bridge in the Pacific Northwest, resulting in one of the most diverse areas in the United States. In addition to the Siskiyou salamander, the Applegate counts one of the largest spotted owl populations in the country among its federally listed threatened and endangered species.

During the Ice Age, the southerly glacial flow stopped at the Applegate River watershed. The watershed's boundaries are also the project's boundaries. Within the project area, grazing and extensive recreational uses occur alongside late successional old-growth reserves. The lowlands are largely dedicated to agricultural production while timber production occurs in the uplands. Nearly 10 percent of the watershed is owned by large industrial forest concerns. The remainder includes 35 percent owned by the U.S.D.A. Forest Service (USFS), 35 percent by the Bureau of Land Management (BLM), and 20 percent by other private landowners. Nearly 12,000 people live in the Applegate. A number of towns and communities, none incorporated, are located there.

ECOSYSTEM STRESSES

Mining has had a dramatic effect on the ecosystem since presettlement times. Although currently there is little mining activity, large portions of the landscape were burnt in the 1800s to facilitate access to minerals. Past logging, road construction, fire suppression, and development have greatly altered this area. However, the northern spotted owl injunction stopped virtually all logging on USFS and BLM lands, and clear-cutting has been replaced by selective thinning. Land conversion to residential use represents a significant threat to the three counties in the Applegate. Surrounding urban areas view the Applegate as a place for upscale residential development and are pressuring local county commissioners to rezone the watershed's lands from agricultural use to rural residential use. Although some residents own land on a speculative basis, most residents in the Applegate are fighting the rezoning efforts.

PROJECT DESCRIPTION

Concerned about preserving the environment and traditional economic lifestyles in the rural community, two individuals, one a resident of the area and an environmentalist, the other the owner of an aerial forestry management company, pulled together a group of sixty residents three years ago. The group was drawn together by a common desire to prevent degradation to the Applegate's natural resource base, which was also the economic base for many of the region's jobs. Sierra Club and Audubon Society members, ranchers, farmers, Farm Bureau representatives, loggers, community groups, and USFS and BLM officials attended the meeting. The group nominated eighteen people to decide which nine members of their group would constitute a new board. Among the criteria for board selection were 'leaving baggage from your organization behind' and committing time to make the project work.

The goal of the partnership is to reestablish the health of the forest and watershed through a cooperative community effort, using natural resource principles that promote ecosystem health and natural diversity. Each meeting of the partners is open to the public, is convened by a facilitator (which is rotated weekly), and often includes fields trips. An overall assessment of the condition of watershed, the ecosystem's components, and natural processes is currently under way. Individual landowners decide how management of their lands fit within the stated goal. Ongoing projects include restoration and recovery of riparian areas, tree plantings, road reconstruction or removal, selective thinning, fuels reduction, reintroduction of fire, and encouraging small landowners and the timber industry to voluntarily adopt practices that promote long-term ecological and economic health.

STATUS AND OUTLOOK

Three years later, the group of concerned residents continues to meet weekly. Although attendance varies, enthusiasm for the project and its goals has not died down. An effort to reduce the meetings to twice a month was made and then discarded after participants missed the weekly dialogue.

As a result of the project, there is less competition for resources and less antagonism in the Applegate. Relationships have developed between previous adversaries and have contributed to resolutions and problem-solving which extend beyond the Applegate. More direct communication on ecological and economic issues occurs between the region's residents and many landowners are integrating their management activities into the Applegate framework.

Factors Facilitating Progress

Face-to-face communication was cited as an important ingredient to progress. The meetings play an important role in fostering community spirit and support. Through the project, residents solve their problems the "old way"—by sitting down, talking about them, and working together to find common solutions. While the meetings are sometimes intense and emotional, the shared commitment to the community's well-being keeps participants returning weekly. Professional facilitation assisted in the beginning, as did conflict resolution training (which is still ongoing). Shared leadership,

open access to information, and field trips also contributed positively.

Obstacles to Progress

The Federal Advisory Committee Act (FACA) presented a significant obstacle. USFS and BLM representatives have made an important contribution to the group. Community access and expanded dialogue with federal personnel is necessary, given the large percentage of the project area's land under federal jurisdiction. Although the representatives still attend the group's regular meetings, exchange information, and provide support for its projects, FACA requires additional sensitivity when discussing projects on federal lands. (For instance, federal officials share information and discuss projects, but leave the meetings at the time the partnership wants to develop a formal recommendation about a federal action.) During the first year start-up phase, extensive education within the community was needed to get all group members to stay on board.

Contact information:

Mr. Jack Shipley
Board Member
Applegate Partnership
1340 Missouri Flat Road
Grants Pass, OR 97527
(541) 846-6917

BARATARIA-TERREBONNE NATIONAL ESTUARY PROGRAM

Location:
South-central Louisiana

Project size:
4 million acres

Initiators:
U.S. Environmental Protection Agency, State of Louisiana

PROJECT AREA DESCRIPTION

Consisting of two adjacent river basins, the project area is generally bounded on the north and east by the Mississippi River, on the west by the Atchafalaya Basin floodway levee, and the south by the Gulf of Mexico. The basins are divided by Bayou Lafourche, a major transportation artery that once served as a main channel of the Mississippi River. The entire project area lies within the Mississippi deltaic plain and is characterized by very little topographical relief. The basins consist of large tracts of freshwater, brackish, and saltwater marshes and shallow open bays. Southern Louisiana contains 40 percent of the coastal wetlands in the lower forty-eight states.

Originally, much of the area consisted of forested wetlands. Currently, agriculture (sugar cane primarily) is the predominant land use, with limited cattle grazing and oil and gas development. Urban expansion has occurred primarily on old abandoned distributary ridges of the Mississippi, the only areas that do not flood frequently.

ECOSYSTEM STRESSES

Hydrologic disruption is the primary stress to the ecosystem, especially since the great Mississippi flood of 1927, after which extensive levees were built along the river. These levees have led to a loss of sediment, nutrients, and freshwater that annually nourished the bottomland forests and marshes. The region has experienced an 80 percent loss of its original wetlands, which is continuing at the rate of approximately twenty-one square miles per year. The loss in habitat has significantly altered or eliminated living resources in the basins.

Numerous canals for oil and gas well access, transportation, and drainage have reduced sheet flow. Finally, water quality degradation—eutrophication to toxics pollution—has resulted from agricultural, urban, oil and gas, and sewage runoff.

PROJECT DESCRIPTION

This effort is one of twenty-one National Estuary Programs (NEPs), a nationwide program authorized by the Clean Water Act. Because of its important aquatic biodiversity, the Louisiana Governor submitted a nomination to the U.S. Environmental Protection Agency (EPA) in 1989 for the "BT" to be an NEP site. Following its acceptance, the program officially began in late 1991. It will last five years. Its goals (as for all NEPs) are to 1) identify ecosystem problems; 2) characterize the problems by identifying data gaps, research needs, and statuses and trends; and 3) produce a management plan. Implementation will then follow for twenty to forty years, to be assumed by the state. The effort is coordinated by the Louisiana Department of Environmental Quality, with EPA providing 75 percent of the funds and the state providing the remainder.

In order to include as many stakeholders and planning aspects as possible, five committees have been established. Two committees—policy and management—have primary control of the process. The other three committees—science and technology, citizen advisory, and local government—make recommendations to the management committee. The latter three committees serve a two-way function, educating their constituents on the need and progress of this effort and representing constituent interests to the agencies. In all, over 100 participants are included on the committees, including government officials, from parish-level to agency heads; academicians; landowners; sugar and cattle growers; and representatives from levee boards, planning committees, environmental organizations, and the oil and gas industry.

As required by EPA, a Comprehensive Conservation Management Plan (CCMP) is being prepared, to be divided into four sections: 1) the Estuary Compact is a forty-page executive summary of the effort, to be distributed to the general public; 2) a technical supplement provides detailed information on each of the plan's action steps; 3) a shorter, ten-page executive summary will be developed for distribution to legislators; and 4) a second, lengthier technical document will present monitoring strategies and information on how the BTNEP will continue to conform with other federal and state management programs.

STATUS AND OUTLOOK

Drafts of the first two sections of the CCMP were completed in December 1995 and are being circulated for public comment. Drafts of the second two sections were due to be completed by June 1996.

Factors Facilitating Progress

The committee structure has been a significant benefit; com-

mittees and their members have worked well together. The committee function has extended beyond simply serving as sounding boards for conflicting interests. Being inclusive of all interests from the program's inception has helped foster respect among stakeholders, with the effort benefiting from significant stakeholder support as a result.

Another benefit has been the small size of the coordinating office—nine staff—allowing for reduced "red tape" and greater flexibility and assistance among personnel in carrying out their responsibilities. Finally, the program's coordinators consider this effort to be far more open than other state or federal projects, with greater representation from local government and citizens, both of whom traditionally have not been included in planning processes.

Obstacles to Progress

While essential to the effort's progress, the five committees present another layer of bureaucracy in a planning process, for instance, leading to difficulties with coordination between committees and EPA. Identifying the program's needs has taken longer than anticipated. Consequently, some deadlines approved by EPA at the project's inception have been difficult to meet and may no longer be appropriate.

In the future, funding will be a limiting factor, primarily because there is no federal funding guar-

anteed by EPA beyond the planning phases. Implementation and securing funding invariably must be assumed by the state. Whereas specific project funds are potentially available under different statutes (e.g., water treatment facilities under the Clean Water Act), securing those funds is a long process that does not necessarily consider broader project needs. To a great degree, the program's continuation will depend on the state's legislature, which may only consider the program on a year-to-year basis.

Conflict between stakeholders may arise with the development of specific management strategies, when all stakeholders may have to make compromises. Finally, conflicting agency goals and mandates have been cited as another obstacle.

Contact information:

Mr. Richard DeMay
Science and Technology
  Coordinator
Barataria Terrebonne National
  Estuary Program
P.O. Box 2663
Thibodaux, LA 70310
(800) 259-0869; (504) 447-0868
Fax: (504) 447-0870
E-mail: BTEP-RD@nich-
nsunet.nich.edu

BIG DARBY CREEK PARTNERSHIP

Location:
West-central Ohio

Project size:
371,000 acres

PROJECT AREA DESCRIPTION

Big Darby Creek, an eighty-two mile-long stream, is the most diverse aquatic system of its size in the Midwest. Eighty percent of its 580-square-mile watershed consists of farmland. However, portions of the riparian corridor are covered by hardwoods such as buckeye, sycamore, silver maple, and box-elder. On slopes and bluffs along the stream, remnant prairie species can be found. The creek is home to eighty-six species of fish and forty species of mollusks, including the federally listed endangered Scioto madtom (a fish), northern riffle shell and northern club shell, and several state listed species. Eighty-two miles of Big Darby Creek and its major tributary, Little Darby Creek, have been designated as both National and State Scenic Rivers.

ECOSYSTEM STRESSES

Due to the rural character of the Big Darby Creek watershed, the creek has not been subjected to large amounts of industrial or municipal waste, and therefore has retained a natural balance of aquatic species. Nevertheless, a major stress on the ecosystem has been non-point source pollution resulting from traditional farming practices. A decrease in water quality has resulted from sediment originating from agricultural fields, and deforestation. Decreasing water quality has posed a threat to the inhabitants of the stream. The continuous westward expansion of the city of Columbus and its suburban communities poses an additional, growing threat, as it leads to increasing erosion levels from construction sites and increased storm water runoff.

PROJECT DESCRIPTION

The uniqueness of Big Darby Creek has been recognized by many individuals and organizations in the region. Around 1989, their joint concerns and activities led to the formation of the Big Darby Creek Partnership, which consists of more than forty private and public organizations. Partners include the Ohio Department of Natural Resources (ODNR), The Nature Conservancy (TNC), Operation Future Association, U.S. Environmental Protection Agency, Ohio State University Extension, and many others.

Although the partnership is a loose-knit organization without an official mission document, the partners recognize that the overarching goal of their efforts is to ensure the long-term ecological health of the Big Darby Creek watershed. Each partnering agency or organization maintains its own specific goals and strategies. For example, strategies employed by the Scenic River Division in the ODNR and the Operation Future Association include reforestation of the riparian corridor, innovative zoning schemes, encouragement of alternative agricultural practices, and conservation easements through purchases and donations. TNC also uses conservation techniques such as conservation easements and land acquisition in addition to voluntary landowner agreements. These strategies are not only intended to preserve this unique riverine habitat, but they also seek to accommodate human uses within this ecosystem.

The Scenic River Division, the Ohio Environmental Protection Agency, the local Soil and Water Conservation District, and the city of Columbus monitor the watershed using fish and macroinvertebrate surveys, water quality, change in farming practices, among others, to evaluate the effectiveness of management strategies.

STATUS AND OUTLOOK

Although the partnership is relatively young, it has succeeded in bringing together a large number of organizations for the sake of preserving the watershed. It is still too early in the project to know if populations of sensitive species are increasing in Big Darby Creek, but an endangered species is now found in a tributary (Little Darby) where it was never found before. Also, the use of no-till agricultural techniques has increased by more than 50 percent in the watershed, thereby decreasing sedimentation into the creek.

Factors Facilitating Progress

Some of the project's success may be attributed to the fact that Big Darby Creek was well-studied years before it was designated as a scenic river and became the focus of this partnership. The availability of this information and the interests of private citizens and organizations have contributed greatly to this project. In addition, the partners share their views with and learn from each other, improving the understanding and cooperation that benefits Big Darby Creek.

Obstacles to Progress

Most of the watershed is held in private ownership and some townships have not passed protective zoning. However, with the continuing education of landowners in particular and the public in general, there has been an increasing understanding and acceptance of efforts aimed at protecting Big Darby Creek.

Contact information:

Mr. Stuart Lewis
Assistant Chief
Ohio Department of Natural Resources
1889 Fountain Square
Columbus, OH 43224
(614) 265-6453

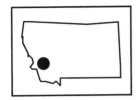

BITTERROOT ECOSYSTEM MANAGEMENT RESEARCH PROJECT

Location:
Eastern Idaho, western Montana

Project size:
40,000 acres initially; to be expanded to 200,000 acres

Initiator:
U.S.D.A. Forest Service

PROJECT AREA DESCRIPTION

Steep east-west canyons carve this area along the Idaho-Montana border. Douglas-fir and western larch grow along the flatiron faces of the Bitterroot Range that look east over Missoula. However, most of the forest cover consists of lodgepole pine, Engelmann spruce, Douglas-fir, and grand fir. Alpine larch is present at higher elevations of the north facing slopes. The area is not believed to host any federal or state listed endangered species.

The initial 40,000-acre project site is defined by the borders of the Stevensville West-Central Integrated Resource Analysis Unit. The area is bounded by the ridge between Silverthorn and Sherritt Creek, extending south to encompass the entire Gash Creek watershed. All of the land is owned by the U.S.D.A. Forest Service (USFS). More than half is designated as wilderness.

ECOSYSTEM STRESSES

The fifteen-year fire regime which favored the growth of ponderosa pine and western larch has been actively suppressed throughout the nearly 100 years of USFS ownership. The logging of most of the high value pine and larch timber has led to further changes in stand structure and species composition. As a result, thick stands of Douglas-fir and grand fir now stress the ecosystem for nutrients and water. The current composition is especially susceptible to fire, insects, and disease. West winds create a high potential for fire to come ripping out of the wilderness area in the mid to late summer and threaten recently constructed subdivisions nestled near the national forest. Finally, eastern brook trout have been introduced and breed with native bull trout. Bull trout are difficult to find in areas where they used to be abundant.

PROJECT DESCRIPTION

The Forestry Science Laboratory in Missoula started this project in early 1994 in order to assist the USFS in its goal of restoring forest health. The research team is seeking to restore the dominance of ponderosa pine and western larch which would be less susceptible to fire, insect, and disease; these would also be highly conducive to wildlife populations and an important contribution to the region's economic base.

The five-year research project employs several strategies, including applying models of vegetation management that plot succession, comparing riparian areas in different stages of succession to determine their influence on bird nesting patterns, and replanting ponderosa pine and western larch. Also, the Forest Service will conduct prescribed burns in an attempt to mimic the natural fire regime to support the regeneration of pine and larch stands. The specific factors to be monitored and evaluated have yet to be established.

STATUS AND OUTLOOK

Several cooperative agreements focused on scientific research have been signed with University of Montana scientists. Project decisions rest with the USFS district ranger, yet many environmental groups and "wise use" groups routinely participate in public involvement processes. The Forest Service will also use focus group analysis to better understand local sentiments. The focus group analysis involves Forest Service staff and cooperators attending the regular meetings of stakeholder groups in and around the Bitterroot Valley.

Factors Facilitating Progress

Agency support for the application of ecosystem-based approaches is credited with allowing the project to move ahead. The research team's progress has been enhanced by applying technology to management activities. Scientific information concerning forest health and resources had been fragmented and dispersed throughout several Forest Service offices. The information, now centralized in the laboratory's geographic information systems (GIS) database, is being used in successful modeling efforts. The assembled scientific team is confident in its ability to convert the forest. Key personnel on the science and management team are likely to stay in place throughout important phases of the project, allowing for progress to continue on schedule.

Obstacles to Progress

The complex social systems in the Bitterroot Valley provide a challenging backdrop for the USFS's research activities. Resentment toward USFS lingers in the valley from the controversies over clear-cutting in the 1960s and 1970s. Furthermore, a strong presence of home rule advocates could inhibit the work of the research scientists. Within the forest products industry, there is concern that ecosystem management will lead to a significant drop in available timber. There was concern in early 1995 that Congress would rescind several million dollars in USFS research funds, thus jeopardizing the ability of this project to continue.

Contact information:

Dr. Clint Carlson
Team Leader
USDA Forest Service
Forestry Science Laboratory
PO Box 8089
Missoula, MT 59807
(406) 329-3485
Fax: (406) 543-2663

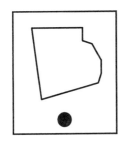

BLOCK ISLAND REFUGE

Location:
Southern Rhode Island

Project size:
6,400 acres

Initiators:
Block Island Conservancy, The Nature Conservancy

PROJECT AREA DESCRIPTION

This project area encompasses the entire land area of Block Island. The island consists of glacial till and is morainal and undulating in topography. This relatively small island contains a great deal of habitat diversity including beaches, dunes, bluffs, wetlands, marshes and grasslands, a large salt water pond at the center of the island, and perched wetlands and kettle hole ponds interspersed throughout. The vegetative cover includes maritime shrublands and grasslands. Federally listed threatened and endangered species include the American burying beetle; Block Island is one of only three known sites for this species. Currently, half of the island is undeveloped, a quarter is in conservation status, and a quarter has already been developed.

ECOSYSTEM STRESSES

The biggest threat to the ecosystem is residential development and the conversion of grasslands to lawns with little ecological diversity. A second threat is the seasonal human disturbance to beach communities.

PROJECT DESCRIPTION

Island residents have a history of conservation ethics. The project is rooted in the 1970s when a cherished piece of land was threatened by development. A community group, the Block Island Conservancy, was created to raise money to buy this parcel. Subsequently, The Nature Conservancy (TNC) and the State Department of Environmental Management became involved. In the 1980s, the Block Island Land Trust was formed by residents to counter development, which was outpacing preservation during those years. This trust

lobbied the state legislature to fund the group through a tax on real estate.

The goals of the project are preservation and enhancement of the ecological values of the island. The strategies to meet these goals focus on restoration of grasslands and aggressive land acquisition. Emphasis is also placed on general environmental education and targeted education to help landowners understand tax laws so they will place conservation easements on their land. A longer-term goal is the reintroduction of native species that have been extirpated from the island. A second future goal is to make the economy of the island less dependent on development and more dependent on sustainable activities.

STATUS AND OUTLOOK

A proactive land acquisition program has been adopted by TNC. The U.S. Fish and Wildlife Service (FWS) has formed a partnership with TNC to aid in preservation efforts, although FWS's land acquisition capabilities have been severely reduced due to recent budget cuts.

While it is too early to gauge the results of the outreach and education programs, restoration efforts have allowed several species to recolonize the grasslands. In order to measure progress in this area, the burying beetle, breeding birds, and rare plants are monitored. The land acquisition program has also been very successful with 300-400 additional acres preserved over the last four years.

Factors Facilitating Progress

The degree of cooperation and support at all levels of the community has been extraordinarily helpful. The tangibility of the threat of development has been a motivating factor for people to become involved. This was especially true during the 1980s when the pace of development on the island was extremely high. Also, a recent comprehensive town plan indicated that the number of new houses allowed under current zoning could double the number of existing houses. This threat of development has stimulated greater concern and support among the island's inhabitants for conservation efforts. Finally, a January 1996 oil spill off the coast, with the resulting death of numerous seabirds, has led to even greater awareness of the island's resources.

Obstacles to Progress

Ensuring future funding to protect additional habitats is critical to continued progress. Because land preservation is tied to local economies, the politics involved with preservation can be a disincentive for community involvement.

Finally, creating innovative strategies to deal with the seasonal disturbances is another challenge.

Contact information:

Mr. Chris Littlefield
Bioreserve Manager
The Nature Conservancy
Block Island Refuge
P.O. Box 1287
Block Island, RI 02807
(401) 466-2129

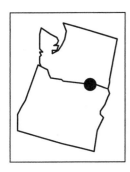

BLUE MOUNTAINS NATURAL RESOURCES INSTITUTE

Location:
Northeastern Oregon, southeastern Washington

Project size:
19 million acres

Initiators:
U.S.D.A. Forest Service, private citizens

PROJECT AREA DESCRIPTION

The Blue Mountains ecoprovince is generally mountainous with fairly large, flat valleys. In presettlement times, fire-adapted species such as ponderosa pine, and to some extent lodgepole pine and western larch, dominated the forests in this ecoprovince. As a result of fire suppression and selective logging, large numbers of Douglas-fir and white fir have become established in relatively dry areas to which they are not well adapted. The project area also includes range and crop land. The Blue Mountains provide vital habitat for many wildlife and fish species, including the Snake River chinook salmon, which is federally listed as threatened.

More than 40 percent of the land is federally owned, 10 percent of which is designated as Wilderness or National Recreation Area. The remainder of the land is mostly privately owned. The most prevalent land uses are timber management, outdoor recreation (including hunting), cattle grazing, and farming.

ECOSYSTEM STRESSES

Forests in the area have recently suffered from an insect and pest epidemic, due to the susceptibility of Douglas-fir and white fir to insects and disease. Fire suppression has led to unprecedented levels of fuel build-up, which could potentially lead to a catastrophic forest fire. Dams on the Columbia and Snake rivers have severely affected salmon populations. Salmon populations are also negatively affected by temperature increases in some streams from a reduction of shade and other factors. Grazing, logging, and roads contribute to erosion and stream sedimentation.

PROJECT DESCRIPTION

In 1989, several citizens voiced concerns that the U.S.D.A. Forest Service (USFS) was not researching local problems, and that the best scientific information was not being used. These citizens were concerned that disputes over forest management would develop into region-wide conflicts. In addition, the forests were plagued by insect and disease problems. In response to these concerns, the Blue Mountains Natural Resources Institute (BMNRI) was formed by USFS in a partnership with a wide range of stakeholders. The goal of the BMNRI is to enhance the long-term economic and social benefits derived from the area's natural resources in a sustainable and ecologically-sensitive manner. In order to reach this goal, the BMNRI sponsors, facilitates, and encourages research on ecological, economic, and social issues. It also promotes outreach, including technology transfer, education, demonstration of innovative management strategies, and facilitation of cooperation of the various stakeholders in the area.

BMNRI's policies and direction are guided by the recommendations of a board of directors, which is composed of individuals representing federal, state, local and tribal agencies, as well as private landowners, industry, environmental interests, and civic groups. The board is a federally chartered advisory committee. In addition, BMNRI has eighty-three partners. A partner is an organization that agrees with BMNRI's mission, is willing to lend support to accomplishing the goals that the partner has in common with BMNRI, and agrees that cooperation is key to resolving critical resource issues.

STATUS AND OUTLOOK

The increased level of communication and cooperation among diverse interests, and an increased level of integrated research projects have been some of the important results of this project. A level of trust has developed among the people who live in the region, which is not found elsewhere in Oregon and Washington.

Factors Facilitating Progress

The local initiation of the project and the will to succeed have helped the project proceed. The willingness of people to work together to solve issues has been very helpful. In addition, BMNRI maintains credibility by basing its positions on science. The flexibility of BMNRI's programs and the quality of its products and projects are other positive factors.

Obstacles to Progress

The process for the board of directors to obtain Federal Advisory Committee status took four years, a lengthy process that was very discouraging and distracting to the board members. Initially, some administrative problems existed, but those have been overcome. Inadequate staff and funding are other obstacles. Implementation of projects on federal lands have been delayed because of a lengthy approval process.

Contact information:

Ms. Lynn Starr
Blue Mountains Natural
*   Resources Institute*
1401 Gekeler Lane
La Grande, OR 97850
(503) 962-6529
Fax: (503) 962-6504

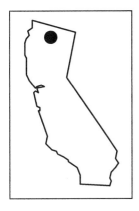

BUTTE VALLEY BASIN

Location:
Northern California

Project size:
32,625 acres

Initiators:
*U.S.D.A. Forest
Service,
California
Department of Fish
and Game*

PROJECT AREA DESCRIPTION

Just south of the Oregon border, the Butte Valley basin lies on an ancient lake bed, whose sandy terraces, dunes, and lakeshore support communities of salt shrub, perennial bunchgrasses, intermountain sage steppe, and juniper woodlands. The basin ecosystem provides habitat for the bald eagle, Swainson's hawk (a state listed threatened species), and burrowing owl (a state listed sensitive species). The basin also supports many waterfowl species. Since the early 1900s, this basin was severely de-watered for agricultural purposes. After the drought of the 1930s, the federal government bought the basin's submarginal farmlands in order to stabilize them. Currently, the most prevalent land uses include livestock grazing, wildlife viewing, waterfowl hunting, and agriculture. Landownership is mixed federal, state, and private.

ECOSYSTEM STRESSES

Early in the century, dikes and ditches were built to drain wetlands and irrigate farmland. In 1968, structures were built to pump flood water to the Klamath River. These water projects lowered the water table considerably, causing the loss of wetland and riparian habitats as well as native grasses. Crested wheatgrass, introduced in the 1950s to stabilize the soil, has thrived in these droughty soils where native grasses can no longer grow. Fire suppression has been a problem for the regeneration of the sage steppe and juniper woodland communities. Finally, overgrazing throughout the century has proven to be a stress to the basin ecosystem.

PROJECT DESCRIPTION

In 1991, the federally owned lands were designated as the Butte Valley National Grassland. Five years earlier, the U.S.D.A. Forest Service (USFS) had begun a coordinated resource management planning process to restore traditional waterfowl and wetland habitats in the area. Recognizing their common interest in wetland restoration, the Klamath National Forest and the California Department of Fish and Game have combined their planning efforts in the Butte Valley. Also involved in the planning process are Ducks Unlimited, the Butte Valley Resource Conservation District, and participating private landowners and grazing permittees. Implementation of the wetland restoration projects has been under way since 1992. Future project efforts will focus on development of a comprehensive restoration and management plan for the grasslands.

The current goals of the project are 1) to restore and maintain Butte Valley's native grassland and wetland ecosystems; 2) to restore and maintain the productivity of native rangeland vegetation for livestock and wildlife; 3) to optimize storage and use of water for wetland restoration and increased groundwater levels by confining excess water to the basin rather than pumping it to the Klamath River; and 4) to provide an ecological approach to multiple use management.

STATUS AND OUTLOOK

Many of the project goals outlined above have already been met. By 1993, water control structures designed by the project team had provided enough water for the results of wetland restoration efforts to be observed. By the end of 1995, all wetland projects, including structural improvements and a second flooding episode, had been completed as planned. Monitoring plans and monitoring points have been established to measure changes in water quality and bird and wildlife use resulting from the restoration efforts.

More work is being undertaken for the grasslands. Reintroducing fire and limiting grazing, and testing alternative grazing treatments, will be needed to restore upland vegetation. Experimental seeding programs have attempted to reestablish native grasses, including cooperative projects with the County Extension and local schools. Prescribed burning has been applied in some of the sage communities. Hopefully, smaller projects such as these will soon be part of a larger comprehensive plan for the grasslands.

Factors Facilitating Progress
The collection of recent, accurate baseline data on the biological and physical components of the basin ecosystem, incorporated into a comprehensive, integrated geographic information systems (GIS) database, has allowed the planning team to use a landscape-based analysis and design process. Furthermore, adequate monitoring programs have been in place to evaluate current restoration projects. Adequate funding provided equally by the state of California, Ducks Unlimited, and the Forest Service has been available. Finally, there has been strong public participation and input in determining the desired future conditions for Butte Valley that the project hopes to achieve.

Obstacles to Progress
Despite this public participation, there have been misconceptions about what the project is trying to accomplish. Furthermore, the pervading local attitude maintains that groundwater is for growing crops and that providing for waterfowl is an inappropriate use of water. Efforts to educate the public on the project have relieved some of this opposition, but public skepticism remains.

Contact information:

Mr. Jim Stout
Resource Officer
USDA Forest Service
Klamath National Forest
Goosenest Ranger District
37805 Highway 97
Macdoel, CA 96058
(916) 398-4391
Fax: (916) 398-4599

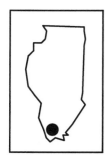

CACHE RIVER WETLANDS

Location:
Southern Illinois

Project size:
60,000 acres

Initiators:
*U.S. Fish and
Wildlife Service,
Illinois Department
of Conservation,
The Nature
Conservancy,
Ducks Unlimited*

PROJECT AREA DESCRIPTION

The Cache River is a 110-mile long river in southern Illinois, which drains an area of 470,000 acres. The Cache River Wetlands project focuses on the wetlands corridor along the lower fifty miles of the river. The wetlands are characterized by bottomland hardwood forests (oak-hickory) and bald cypress-tupelo swamps, and are interspersed with low, oak-dominated hills.

The area supports a remarkable diversity of plant and animal species. Federally listed threatened and endangered species that can be found in the project area include the gray bat, Indiana bat, interior least tern, and bald eagle. In addition, more than sixty state listed species occur in the area. The wetlands are a major stop-over for migratory neotropical songbirds as well as for migratory waterfowl. The area is also graced by the presence of many raptors, a heron rookery, bobcats and river otters. Farming and timber management are important land uses in the surrounding watershed.

ECOSYSTEM STRESSES

As late as the turn of the century, the area remained almost entirely wooded. However, logging began as the area was settled, followed by agriculture. With the onset of agriculture, drainage projects began and were continued until the 1970s. Today, only approximately 20 percent of the original Cache River wetlands remain. The extensive drainage network and the lack of forest cover has led to rapid runoff of rain water, and has resulted in major stream bank erosion. The eroded materials subsequently settle in the wetlands, reducing water quality and smothering swamps.

Another stress is the fragmentation of the remaining forest. Animal populations which cannot travel between forest patches may become genetically isolated. In addition, fragmented forests have too many edges, leading to parasitism and predation of the nests of migratory neotropical birds.

PROJECT DESCRIPTION

Recognizing the natural values of the Cache River Wetlands twenty years ago, The Nature Conservancy (TNC) and the Illinois Department of Conservation (IDOC) started buying land that is now known as the Cache River State Natural Area. After the passage of the North American Wetlands Conservation Act, the Cypress Creek National Wildlife Refuge was established in 1991. Shortly thereafter, the Cache River Wetlands Joint Venture Partnership was formed to manage the wetlands. This partnership consists of TNC, IDOC, the U.S. Fish and Wildlife Service (FWS), and Ducks Unlimited. Together, the partners have acquired more than half of the 60,000 acres targeted by the project. In 1993, the Cache River Consortium was established, which is a larger advisory body that comprises the four partners and several state and federal agencies.

In order to "knit the ecosystem back together," the partnership uses several strategies which focus on restoration and water resource planning. To restore the bottomland hardwood forests, TNC uses a mechanized tree-planting operation. Five tons of the native hardwood nuts and acorns that have been planted were collected by local Boy Scouts. FWS has employed the services of prison crews to plant tree seedlings.

Hydrologic restoration is carried out through plugging ditches, building dams, and other methods to flood old farmlands.

The U.S. Army Corps of Engineers recently began a three-year, $1.3 million feasibility study to evaluate measures planned to mimic natural hydrologic processes. In addition to restorative efforts, the U.S. Environmental Protection Agency has funded a Water Resource Planning Initiative. Its aim is to identify areas where erosion is a problem and to develop solutions to mitigate this problem. The initiative is made up of twenty-five landowners throughout the watershed, and is cosponsored by the U.S.D.A. Natural Resources Conservation Service and TNC.

STATUS AND OUTLOOK

The Corps of Engineers has completed the first year of its study. The water resource plan has also been completed, and the project is now securing funding and moving into implementation. Thirty-eight thousand acres have been acquired, of which 4,000 have been reforested. Finally, hydrologic restoration and a water quality monitoring program have begun.

Factors Facilitating Progress

Factors contributing to the project's progress are the significance of the natural area, good planning, strong Congressional support, a dedicated local support group and the strong interest and commitment of all involved agencies and organizations.

Obstacles to Progress

Land is only acquired from willing sellers. As a result, the partnership has not been able to obtain some large key parcels. In addition, there is great concern about the availability of Land and Water Conservation Fund appropriations from Congress. These funds are needed for continued acquisition of lands by the FWS.

Contact information:

Mr. John Penberthy
Project Manager
The Nature Conservancy
Cache River Office
Route 1, Box 53E
Ullin, IL 62992
(618) 634-2524
Fax: (618) 634-9656
E-mail: cachebio@aol.com

CACHE/LOWER WHITE RIVERS ECOSYSTEM MANAGEMENT PLAN

Location:
Eastern Arkansas

Project size:
2.8 million acres

Initiator:
U.S. Fish and Wildlife Service

PROJECT AREA DESCRIPTION

Comprising 12 percent of the Mississippi Alluvial Valley (MAV), the Lower White River watershed, including the Cache River, a tributary, was once part of a vast hardwood-forested wetland complex. While the ecosystem is very flat, its topography is complemented by a complex and dynamic hydromorphologic system of rivers and bayous, braided stream terraces, meander belts, back swamps, and elevated terraces. The area is vital to waterfowl migration, breeding, and wintering, and boasts one of the largest populations of black bears in the South.

The wetlands that remain are nationally and internationally recognized for their value, not only to commercial and recreational fisheries, but to the region's 52 mammal species, 232 birds, 95 fish, and 48 reptiles and amphibians. The region includes 5 federally listed threatened or endangered species: interior least tern, bald eagle, pallid sturgeon, and the fat pocketbook and pink mucket mussels.

Today, most of the area is in row crop agriculture. The watershed contains the largest remaining hardwood-forested wetland complex on any Mississippi River tributary in the MAV: 350,000 mostly contiguous acres, of which 229,000 acres are in public ownership (two National Wildlife Refuges, four State Management Areas).

ECOSYSTEM STRESSES

Large-scale habitat destruction is the primary stress, with over 85 percent of the ecosystem having been converted to agriculture. Associated stresses include disruption of hydrological and sedimentation processes, water quality degradation, and biodiversity loss, although the latter is difficult to estimate due to a lack of baseline data as a result of the extensive habitat conversion. Several species are known to have been extirpated from the region, including Bachman's warbler, Carolina parakeet, ivory-billed woodpecker, mountain lion, and red wolf.

PROJECT DESCRIPTION

The Arkansas-Idaho Land Exchange Act of 1992 mandated that a Comprehensive Management Plan (CMP) be drawn up by the U.S. Fish and Wildlife Service (FWS) for the newly-expanded White and Cache Rivers National Wildlife Refuges. Subsequently, the CMP effort was expanded to include an ecosystem management framework for the watershed, drawing in various other stakeholders, including several conservation and land preservation organizations and various state agencies.

A planning review team was formed to draw up the CMP, which contains three goals: 1) restoration and conservation of wetland ecosystem functions in the remaining forested-wetland complex; 2) partial restoration of ecosystem function on private lands, toward sustainable use; and 3) provision and encouragement, where appropriate, public use opportunities which are compatible with restoration and conservation efforts of ecosystem functions. Fourteen strategies have been drawn up to fulfill these goals, in areas ranging from hydrologic function to cultural resources.

STATUS AND OUTLOOK

The CMP was finalized in early 1995. Formalizing the unofficial interagency team is another goal. While some pilot projects have been identified, drawing up specific plans to accomplish the goals and objectives is the largest and most important next step. No coordinated implementation has occurred at this point.

Factors Facilitating Progress

Interagency cooperation, even though only at a relatively informal level thus far, has been the primary benefit *to* this effort and a result *from* this effort. Despite the need for further education on ecosystem management, the interagency team participants were nevertheless familiar with the ecosystem management approach. Their familiarity covered a broad range of concepts, but it was nevertheless sufficient to allow the CMP development to progress. Finally, strong leadership from the project leader, and support from conservation organizations, the Congressional delegation, and federal and state agencies, have been cited as positive influences.

Obstacles to Progress

Stakeholder involvement has been incomplete, particularly from the farming community, but also from the state Soil and Water Conservation Commission. FWS expects to make a greater effort to identify community leaders among the largely unorganized and unrepresented farmers for future activities on the effort.

A newer ongoing effort to develop a statewide wetland conservation plan is being conducted by the Arkansas Wetlands and Water Resources Task Force. Appointed by the governor, this task force has representatives from nearly every sector, including the agricultural community, and is co-chaired by the commission's executive director and the director of the state TNC chapter. This effort is having positive effects on the Cache/Lower White rivers plan in terms of increasing stakeholder involvement.

Federal reorganization and downsizing, and conflicts between traditional FWS goals and ecosystem management goals, are additional concerns.

The lack of any major university in the Cache and Lower White rivers region has resulted in a lack of biological data on the region. Finally, funding continuity is a concern, due to state and federal cutbacks, and difficulties in raising funds for the effort when so little data is available on the ecosystem.

Education of project partners, including personnel from natural resource management agencies, has been identified as an important need.

Contact information:

Dr. Scott Yaich
Wildlife Management Biologist
U.S. Fish & Wildlife Service
Wildlife & Habitat Management
  Office
P.O. Box 396
St. Charles, AR 72140
(501) 282-3213
Fax: (501) 282-3391

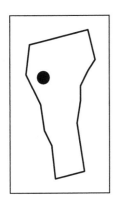

CAMP JOHNSON SANDPLAIN RESTORATION

Location:
Northwestern Vermont

Project size:
200 acres

Initiator:
Vermont Non-Game and Natural Heritage Program

PROJECT AREA DESCRIPTION

The Camp Johnson Sandplain consists of mostly level delta terraces dissected by erosional streams. These terraces are composed of deltaic sand deposits of the glacially-swollen Winooski River delta near Lake Champlain. Vegetation is characteristic of a pine-oak-heath sandplain community, including black, white, and red oak, red maple, white pine, pitch pine, huckleberry, teaberry, sheep laurel, beaked hazelnut, and wild sarsaparilla. In presettlement times, this area was subject to periodic fires. The area is home to several state listed threatened species: mountain rice-grass, large whorled pogonia, blunt-leaved milkweed, low bindweed, and harsh sunflower.

The entire restoration area is on military land owned by the U.S. Air Force and managed by the Vermont Military Department. The forested area is used for limited infantry training and orienteering. Land adjacent to the military base is either town park land or privately owned.

ECOSYSTEM STRESSES

Fire suppression has interfered with the regeneration of some tree species in the community, especially pitch pine, although controlled burning has been initiated. An additional stress is the development of private land adjacent to the base, which is further fragmenting the ecosystem. An existing unimproved road is a stress to a minor degree, since it permits military tanks to access the forest, causing damage.

PROJECT DESCRIPTION

This project was initiated in 1990 by Vermont's Nongame and Natural Heritage Program (NNHP). NNHP realized that in the state, the threatened pine-oak-heath sandplain community was best represented at Camp Johnson. The Vermont Military Department was willing to work with NNHP, following in the footsteps of the (federal) Department of Defense which earlier had started to work with The Nature Conservancy on ecological diversity issues pertaining to military lands. In 1991, a memo of understanding was signed by the NNHP and the Vermont Military Department regarding the management of this forest. During the following two years, a steering committee was set up, including NNHP, the Vermont Military Department, The University of Vermont, and the town of Colchester, providing advice on project goals and strategies.

The goal of the project is to restore and maintain a mosaic of old-growth presettlement white pine-oak forest and pitch pine-oak-heath woodland. Strategies designed to meet these goals include 1) conducting rare plant and invertebrate inventories of all intact sandplain forest; 2) delineating management units for mature forest and early successional woodland; 3) preparing a detailed site base map for the restoration area; 4) developing fire management plans for early successional units; 5) establishing monitoring plots in all study area management units to collect baseline and long-term measurements of vegetation, soils, and other pertinent biotic and physical habitat parameters; monitoring will provide useful information concerning changes in this community in different successional stages over time; and 6) collecting pitch pine seeds from the site and other nearby sites in order to determine genetic variation both within and among the populations, and to attempt germination.

STATUS AND OUTLOOK

Project and restoration activities have been made possible through a five-year grant from the Legacy Funding Program of the Department of Defense. Activities that have taken place so far include rare plant, butterfly, and beetle inventories, delineation of management units, a prescribed burn, establishment of monitoring plots, and collection of pitch pine seeds from three different sites.

Factors Facilitating Progress

A realization among the project participants that this is the best remaining example in Vermont of what was once a much more common vegetative community, and a commitment to attempt to restore and preserve the ecosystem, has helped this project proceed. Cooperation and communication between the Vermont Military Department and other cooperators have been very advantageous. The latter has also provided additional financial support. The assistance of the Vermont Department of Forests, Parks and Recreation with the prescribed burn, and the collaboration of the University of Vermont on the invertebrate inventory and pitch pine seed studies, have also been helpful to the project. In addition, the support of the Colchester Department of Parks and Recreation and the permission of the Colchester Fire Department for the burn, have facilitated the project's progress.

Obstacles to Progress

It has been challenging to convince all the involved parties, especially the town of Colchester, that despite being highly degraded and not comparable with other sandplain communities outside the state, the sandplain community at Camp Johnson is the best remaining example in the state and is therefore highly significant. The town of Colchester would like to develop a portion of the base and has plans to construct a road through the base and the sandplain community.

Contact information:

Mr. Robert Popp
Nongame and Natural Heritage Program
103 S. Main Street
Waterbury, VT 05671-0501
(802) 241-3718

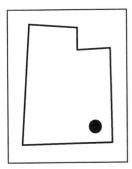

CANYON COUNTRY PARTNERSHIP

Location:
Southeastern Utah

Project size:
15 million acres

Initiators:
Bureau of Land Management,
U.S.D.A. Forest Service,
National Park Service,
Utah Division of Wildlife,
Utah Division of State Parks and Recreation,
State Trust Lands,
Counties of Carbon, Grand, San Juan, and Emery

PROJECT AREA DESCRIPTION

The regional geologic features of the Interior Colorado Plateau create a unique landscape of sandstone canyons, mesas, buttes, spires, cliffs, arches, and shale badlands. Within this region are three mountain ranges. At lower elevations, large open tracts of desert shrub and grassland vegetation can be found, including bunchgrasses, blackbrush, and salt desert shrub. One of the most important features at these elevations is cryptobiotic crust, a living ground layer which protects the highly erodible soil surface and acts as a natural fertilizer through nitrogen fixation. At higher elevations, dominant vegetation includes sagebrush and pinyon-juniper communities, Gambel oak, Douglas-fir, blue spruce, aspen and ponderosa pine forests, and subalpine forests of Engelmann spruce and fir. The highest mountain peaks support alpine tundra vegetation.

There is a high rate of endemism within the plateau, with some very rare plants growing in narrowly restricted habitats. There are several federally listed threatened and endangered plant and animal species in the region, including Jones cycladenia, bald eagle, and peregrine falcon. The Colorado River contains several federally listed threatened or endangered fishes, such as the bony-tailed chub, Colorado squawfish, and razor-back sucker.

ECOSYSTEM STRESSES

Disturbances of the plateau's sensitive ground surfaces by livestock and recreational use have led to increased soil erosion, disrupted nutrient cycling, and increased salinity of the Colorado River. Dramatically-increased recreational activity has been particularly problematic in remote areas that were once difficult to access. Areas untouched by livestock for 200 years are now susceptible to serious damage within three months by uncontrolled recreation. Furthermore, overgrazing, exotic plants (such as cheat-grass and tamarisk) and fire suppression have altered the region's natural vegetative community composition. Historic mining activity and present-day oil and gas exploration and drilling have also left a mark on the land. Finally, increasing residential development has led to habitat disruption and fragmentation.

PROJECT DESCRIPTION

In 1993, under the Bureau of Land Management's (BLM) new direction toward ecosystem management, the agency embarked on a large-scale resource management plan for the Colorado Plateau region. To better understand the principles of ecosystem management and to identify those who should be involved in its planning, the BLM held an ecosystem management conference in May 1993. During this conference, local natural resource managers and a variety of stakeholders realized that the cumulative impacts of increased recreation and other activities on this fragile landscape could have serious ecological as well as economic ramifications.

As a result of this conference, the Canyon Country Partnership was formed in 1994. The partnership consists of representatives from the BLM, National Park Service, U.S.D.A. Forest Service, Utah Division of Wildlife, Utah Division of State Parks and Recreation, State Trust Lands, and local government representatives from Grand, Carbon, San

Juan, and Emery counties. Sub-committees in the partnership deal separately with scientific and data issues including geographic information systems (GIS), data sharing, and data standardization between land managers. The public is involved through issue committees to work on strategies with the partnership on issues such as recreation.

The partnership is an interagency, grassroots initiative which grew out of the vision and efforts of local agency managers. The goals of the Partnership are two-fold: 1) collaboration among members of the partnership, including the sharing of information and resources so that management decisions can be made with an understanding of potential impacts to adjacent lands; and 2) sustainable land manage-ment, that is, preserving ecosystem functions while providing products that society needs and desires.

STATUS AND OUTLOOK

Since 1994, the partnership has mostly worked on fostering co-operative relationships between members and seeking funding for scientific research.

Factors Facilitating Progress

Collaboration continues to grow between members of the partnership, despite concerns listed below.

Obstacles to Progress

The project has proceeded slowly thus far for several reasons. First, the lack of institutional funding and support for the collection of scientific data (both ecological and socio-economic) has been a barrier for sustainable land management planning efforts. Second, skepticism on the part of local residents and officials concerning the federal government's role in the partnership has been considerable. Finally, the Federal Advisory Committee Act (FACA) has been a stumbling block for collaborative efforts. As a result, the partnership has amended its charter so that, rather than reaching consensus, it now only functions as a discussion group.

Contact information:

Mr. Joel Tuhy
Colorado Plateau Public Lands
  Director
The Nature Conservancy
PO Box 1329
Moab, UT 84532
(801) 259-4629; (801) 259-2551
Fax: (801) 259-2677

CHATTOOGA RIVER PROJECT

Location:
*Northeasten
Georgia, southern
North Carolina,
western South
Carolina*

Project size:
120,000 acres

Initiator:
*U.S.D.A. Forest
Service*

PROJECT AREA DESCRIPTION

This river basin is primarily in public ownership, with the actual project area comprised of three National Forests: the Sumter, Chattahoochee, and Nantahala. The ecosystem, located at the interface of the piedmont and southern Appalachian divisions, is primarily mountainous, with elevations ranging from under 1000 to over 4800 feet. The forests are predominantly mixed hardwood with some pine, the latter mostly planted in this century. The region is considered unique from an ecological perspective; for example, the habitat boundaries of many northern and southern species are located there.

The Chattooga basin is home to approximately a dozen federally listed threatened and endangered species, including the small-whorled pogonia, persistent trillium, peregrine falcon, several bats, and over 100 proposed or sensitive species. In addition, the Chattooga is prime black bear habitat. The Chattooga River is a federally designated Wild and Scenic River (WSR) and is considered the "crown jewel" of the eastern WSR system. Current uses of the land are resource extraction (timber, grazing, historic mining), recreation (rafting, hunting, sightseeing), urban and rural settlement, and wilderness or other reserves.

ECOSYSTEM STRESSES

Timber management (on both public and private lands), disturbance of the fire regime, and land conversion to urban use are the most significant stresses. Water and air pollution, potential increases in recreation due to the 1996 Olympic Games in Atlanta (only two and one-half hours away),

localized water development, exotic species invasion (kudzu), and fragmentation due to continued development pressures are other existing or potential stresses.

PROJECT DESCRIPTION

The effort had its roots in the U.S.D.A. Forest Service's (USFS) New Perspectives program. In the Fall of 1991, the USFS was requested by environmental interests to consider the Chattooga as a program site due to its unique ecological qualities. (One group even submitted an ecosystem management plan for the forest.) The following summer, ecosystem management was adopted as a new Forest Service direction, with the USFS Southern Region selecting the Chattooga as a pilot area.

The three-year-long project was primarily information and data oriented. Its stated goal was to develop an integrated and ecological approach to managing the Chattooga watershed, which involved coordinating across state boundaries, three national forests, and multiple ranger districts. Meeting public demands for forest uses within the context of sustainability and collaborating with Forest Service researchers were additional goals. Strategies to attain these goals included 1) identifying and developing strategies for water quality problems; 2) inventorying and classifying the ecosystem and its biodiversity, using geographic information systems (GIS) technology; 3) addressing land-ownership issues (inholdings mostly); 4) increasing public involvement; and 5) developing plans to move into implementation (e.g., modifying forest plans).

Operating solely as an internal USFS

effort, the project was guided by a board of directors (three forest supervisors, three district rangers, one researcher) who received recommendations from six working groups focussing on the following issues: water quality, ecosystem classification, GIS, biodiversity, land acquisition, and developing desired conditions and public involvement. These entities were to disband at the end of the project, although a USFS research advisory committee was expected to review all management proposals within the watershed for another year.

While these groups are staffed solely by USFS personnel, other institutions or individuals were involved, including nongovernmental organizations, private consultants, a local rafting outfitter, the city of Clayton, Georgia, and the U.S.D.A. Natural Resources Conservation Service. The project incorporated outside researchers through contract and cost-share agreements with five universities and several other researchers. For example, more than 130 interviews were conducted with the public to assess the public's view of desired future conditions; the findings were used to develop alternatives to elements of the forest plan.

STATUS AND OUTLOOK

The effort ended in late 1995. Several products were developed, including an ecological classification guidebook, a GIS database of the area, and several reports on specific natural features of the area. An adaptive management element of this effort is being implemented through district-level projects, such as small restoration efforts and projects addressing water quality and sedimentation problems. Clear-cutting has been vastly reduced in the region and replaced with systems that reflect natural disturbance patterns. The traditional favoritism toward pine is being reduced in favor of mixed native hardwoods species.

Cooperation between the three states has improved. For instance, prescribed burns now cross state boundaries, an impossibility prior to the effort's inception. Efforts are under way to integrate the project's findings into the forest plans for all three national forests, which need to be amended in order to reflect project recommendations, develop consistency, and achieve ecosystem management goals.

Finally, the USFS has launched partnerships to improve water quality using community-driven approaches. For instance, a local outfitter collects twenty-five cents from each rafter, which are matched by U.S. Environmental Protection Agency funds, to assist landowners with water quality improvement projects.

Factors Facilitating Progress

Because of its status as a pilot project, funding was initially relatively secure, and aided in this effort's progress, although the funding level was not as high as desired. The top-level directive and de-facto permission to proceed with such an effort, in effect, a change in traditional USFS practices, were also instrumental. Contracting with outside specialists brought expertise and novel approaches that are not always available within the agency.

Emphasizing GIS and ecological classification from the inception of this effort is considered by the USFS to have been on target; the availability of this data is helping districts implement projects using an ecosystem approach.

Obstacles to Progress

Primarily because this effort reflects a "new way of doing business," coordination among the three forests has been problematic, especially in personnel, procurement, and budgeting, both internally and with other federal agencies. Internal resistance was significant initially, with a significant amount of internal education required to bring personnel on board. The project also suffered from significant funding cuts in USFS research in 1995.

Achieving consistency in the amended forest plans is another concern as the plans will be modified by separate processes, without a guarantee of uniformity or even of incorporating project recommendations. However, even though follow-through is not guaranteed, steps are under way to transfer and implement the project's tools and information to forest districts. Some observers have stated that Congressional pressure to meet timber goals could make it difficult for natural resource managers to effectively undertake ecosystem management, despite genuine efforts to do so.

Contact information:

Mr. David Meriwether
USDA Forest Service
1720 Peachtree Road NW
Atlanta, GA 30367
(404) 347-4663

CHEQUAMEGON NATIONAL FOREST LANDSCAPE ANALYSIS & DESIGN

Location:
Northern Wisconsin

Project size:
850,000 acres

Initiator:
U.S.D.A. Forest Service

PROJECT AREA DESCRIPTION

The Chequamegon National Forest displays a mosaic of glacial features such as clay, sand and loess plains, outwash, and moraines. In presettlement times, northern hardwood forests could be found on richer soils, whereas jack pine barrens and red and white pine forests were found on poorer soils. Peatland complexes were also common. Around the turn of the century, heavy logging and associated fires took a heavy toll on the area's forests. As a result, many mesic sites are currently occupied by sugar maple forests, and the drier sites now support aspen and pine plantations.

The forest is home to six federally listed threatened and endangered species, including Fassett's loco weed, gray wolf, and American burrowing beetle. The forest also supports some eighty-five state listed species. The forest is managed by the U.S.D.A. Forest Service (USFS) for multiple uses and benefits including timber and wildlife management, mining of metal ores, recreation, and wilderness.

ECOSYSTEM STRESSES

Pine barrens (listed as globally imperiled) have become overgrown because of intense pine planting programs in the 1930s and fire suppression programs since that time. Fire suppression has also reduced the natural regeneration of red and white pine, and may have changed peatland ecosystems.

Timber management and logging also affect the area, resulting in some of the forest being managed for early successional species such as aspen, paper birch, balsam fir, and jack pine. These species are important natural resources for the pulp and paper-based regional economy.

Roads, trails, and the development of second homes on privately-owned in-holdings can result in fragmentation of the forest and increased human presence in more areas of the forest. Species sensitive to human presence, such as the loon, bald eagle, wolf, and bobcat, can be affected.

PROJECT DESCRIPTION

The 1986 forest plan for the Chequamegon National Forest included an extensive economic and recreational analysis, but lacked a forest-wide ecological assessment. Old growth identified on 3 to 5 percent of the forest was not representative of the forest as a whole. In response, an ecosystem group within the Chequamegon National Forest initiated a landscape analysis and design project in 1992. The ultimate goal of this project is to develop a network that includes the full range of ecosystem diversity in the forest, and that protects remnants as well as the best examples of those ecosystems. Reserves will consist of core and buffer areas. Modified management will be proposed for the areas surrounding the reserves. The network design will be based on a field inventory and evaluation of natural areas in the forest, combined with a spatial analysis of these areas. The Nature Conservancy (TNC) and the Wisconsin Department of Natural Resources' Natural Heritage Section have provided scientific expertise. TNC has also provided financial support.

The resulting reserve network design and several alternatives will be factored into the proposals for the revised forest plan, which is expected to be completed in 1998 after an extensive public comment period. Reserve management strategies, such as prescribed burns and cessation of timber harvests, will be suggested.

STATUS AND OUTLOOK

Over the last several years, an ecologist has conducted a field inventory using multiple tools such as a timber database, high-altitude photos, and an ecological classification system. Maps have been developed for the natural areas in the forest. A geographic information systems (GIS) database is under development and will provide the tool for spatial analysis. The USFS has embarked on a partnership with TNC and the University of Wisconsin-Madison to conduct spatial analysis, applying a variety of reserve design models to their inventory information with the goal of designing alternative reserve scenarios. Forest plan alternatives are expected to be drafted in late summer 1996.

Through the forest plan revision process, approximately 100,000 acres have been identified as high-quality examples of the forest's ecosystems. These areas have been temporarily deferred from timber management or other intensive activity.

Factors Facilitating Progress

The endorsement of the project by Jerry Franklin, a well-known and well-respected ecologist, has resulted in much support within the Chequamegon National Forest. In addition, the involvement of TNC has provided the project with scientific credibility to the environmental community. The development of public relations and marketing of the project have proven invaluable in gaining support as well. The project was awarded the U.S. Department of Agriculture's Honor Award for Environmental Protection.

Obstacles to Progress

Not all Chequamegon National Forest personnel agree on the concept of a reserve system; value systems among personnel are sometimes conflicting.

Inadequate funding recently became a problem. This project was already functioning on a very low budget, and budget cuts have thrown the project's future into doubt. Specifically, the federal FY96 Continuing Resolution, a temporary congressional funding measure which was expected to be the operating budget for the remainder of the fiscal year, contained severe cuts for this project from ecosystem management sources, while the timber program line item received a nearly-equivalent increase.

The revised forest plan must consider social, economic, and ecological factors. Therefore, the pure scientific basis of any reserve system may be altered to address non-ecological issues and considerations.

Contact information:

Ms. Linda Parker
Ecologist
USDA Forest Service
Chequamegon National Forest
1170 4th Avenue
Park Falls, WI 54552
(715) 762-5169

CHESAPEAKE BAY PROGRAM

Location:
Maryland, Virginia, Delaware, Washington, D.C., Pennsylvania, West Virginia, New York

Project size:
41 million acres

Initiators:
U.S. Environmental Protection Agency, state agencies

PROJECT AREA DESCRIPTION

Considered the largest and most productive estuary in the United States, the Chesapeake Bay is 195 miles long, has 1,750 of navigable shoreline, and has more than forty significant rivers and countless streams feeding into it. Created when the Susquehanna River Valley was flooded following the last ice age, the bay contains a range of aquatic environments, from freshwater to nearly full-strength saltwater. Today, the bay supports 300 species of fish, 45 shellfish, and 2,700 plants, in addition to many eastern land animals. Its vast watershed stretches from central New York, through central Pennsylvania, virtually all of Maryland, and Washington, D.C., to eastern West Virginia, northern Virginia, and western Delaware. Its topography and vegetation range from the mixed hardwood Appalachian Mountains to 1.2 million acres of wetlands and beaches. Numerous federal and state listed threatened and endangered species are present.

Human uses of the watershed are many, predominantly farming, significant urban and rural development, industry, and recreation. Major cities are Washington, D.C., Baltimore and Annapolis (Maryland), Richmond (Virginia), and Harrisburg (Pennsylvania).

ECOSYSTEM STRESSES

Much of the land surrounding the bay has been converted to agriculture and urban development. Nutrient runoff into the bay, from agricultural and urban non-point sources, is the stress of greatest concern. Nutrient over-enrichment inhibits the growth of submerged aquatic vegetation which provides critical habitat in shallow areas. In deeper areas, excess nutrients rob the water of oxygen, killing fish, shellfish, and other aquatic life. Other significant stresses include pesticide runoff; continuing land conversion for rural settlement and other urban uses, causing habitat destruction, erosion, and associated non-point source pollution; point source pollution from sewage, industries, and shipping traffic; overfishing of the bay's resources; and hydrologic alteration due to loss of forest and plant cover, and river and stream impoundments.

PROJECT DESCRIPTION

Three years after Hurricane Agnes devastated the mid-Atlantic states in 1972 and raised consciousness about the fragility and importance of the bay, Congress directed the U.S. Environmental Protection Agency (EPA) to conduct a five-year, $25-million study of the Chesapeake Bay, its resources, uses, and stresses. Five years later, the Chesapeake Bay Commission (CBC), a tristate body with representatives from Maryland, Virginia, and Pennsylvania, was formed to coordinate approaches to state legislation regarding the bay. In 1983, EPA issued its report, identifying ten areas of environmental concern. Three areas, nutrient enrichment, toxic substances, and declines in submerged aquatic vegetation, were given priority because little research had been devoted to these. Other concerns range from wetland alteration to shellfish closures.

As a result of the report and pressure from citizen environmental groups, the first Chesapeake Bay Agreement was signed between EPA, the three states and Washington, D.C., and the CBC, officially creating the Chesapeake Bay Program. The program's mission was to develop and implement cooperative plans to improve and protect water quality and living resources in the bay. An executive council was created to endorse policy initiatives and promote implementation efforts.

Over the ensuing nine years, multiple committees were set up to address various tasks and issues facing the bay, the program, and stakeholders in general. A second bay Agreement was signed in 1987 and amended in 1992 to reaffirm its goals and set forth specific action plans. Each of six issue areas laid out in the 1987 agreement (living resources, water quality, population growth, public education and participation, public access, and governance) has a four-tiered hierarchy of goals, objectives, commitments, and targets. The states, and to a lesser degree the federal government, are responsible for implementation.

An extensive monitoring program is included in the program, focusing initially on submerged aquatic vegetation as a primary biological measure of progress in restoring the bay, but now including the status of all living resources as measures in achieving program goals.

STATUS AND OUTLOOK

The Chesapeake Bay Program is now in its second decade and

well into implementation. An administrative structure has developed over the last twelve years, resulting in over fifty committees, subcommittees, advisory committees, work groups, and task forces. Research has been a primary focus of the program's efforts, although on-the-ground restoration has commenced and state legislation has been enacted on several fronts to help protect and restore the bay. More broadly, there has been widespread acceptance by the public and political entities that the bay watershed functions as an integrated system.

More recently, the focus of clean-up efforts has shifted to the bay's tributaries. Maryland, Pennsylvania, and Virginia have developed tributary strategies, with Maryland having appointed implementation teams for each of its ten subbasins. For example, Maryland and Virginia developed their strategies through a multi-year, "bottom up," public process, involving a series of meetings to first inform the public and stakeholders about the issues, and then to develop specific strategies.

Factors Facilitating Progress

The voluntary nature of the program has been credited with encouraging participation and commitment at the political and administrative levels, especially from high-level agency decision-makers. A solid administrative structure with wide representation has also aided the process.

Funding from state and federal sources has been quite good, and research efforts have been very thorough; early emphasis on monitoring and research have been key. Participation from many sectors—government to citizens groups—has been instrumental in moving the program ahead and keeping it true to its intentions.

Obstacles to Progress

Differences in administrative structures between the states has hindered program acceptance and progress. While commendable in their detail, goals and objectives were initially considered hard to quantify, thus making measures of success hard to illustrate. Growing population pressures and development will continue to stress the bay watershed: for example, certain commercially-sought fish and crustaceans continue to decline. Funding of clean-up efforts, at all levels and sectors, continues to be a concern, as funding is not guaranteed beyond year-to-year appropriations and special grants. Finally, it is unclear how new, more conservative gubernatorial administrations in Pennsylvania and Virginia will affect the program's future.

Contact information:

Chesapeake Bay Foundation
162 Prince George Street
Annapolis, MD 21401
(410) 268-8816

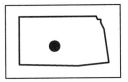

CHEYENNE BOTTOMS WILDLIFE AREA

Location:
Western Kansas

Project size:
41,000 acres

Initiator:
Kansas Department of Wildlife and Parks

PROJECT AREA DESCRIPTION

The Cheyenne Bottoms basin has long been recognized as an area of great diversity, especially of birds, with over 320 species counted in recent years. It is one of the last major wetland systems in Kansas, and is a stopping point for many migratory shorebirds and water-fowl, including all five species of sand-pipers. This flat elliptical basin is a palustrine emergent marsh charac-terized by cattail and bulrush. The sur-rounding uplands are dominated by marsh saltgrass, western wheatgrass, prairie cordgrass, and spikerush com-munities. Six federally listed threatened and endangered species can be found there, all birds: whooping crane, bald eagle, piping plover, snowy plover, peregrine falcon, and least tern. State listed species include the white-tailed ibis and eastern spotted skunk. The area is officially listed as one of only eleven Western Hemisphere Shorebird Reserves, and has been designated as a "Wetland of International Importance" under the Treaty on Wetlands of International Importance.

Twenty-thousand acres of the basin are owned by the State Department of Wildlife and Parks (DWP), an addi-tional 7,000 acres are owned by The Nature Conservancy (TNC), and the remaining 14,000 acres are in private ownership. The state and TNC lands are managed for wildlife, primarily migratory water birds. Private lands are in agriculture, primarily grazing, alfalfa, wheat, and sorghum.

ECOSYSTEM STRESSES

Lack of water has long been a stress to the Bottoms, especially in the last fifty years with increases in agriculture, and even more so since the 1960s with the advent of center pivot irrigation in western Kansas. Irrigation has lowered water tables in the Wet Walnut Creek and Arkansas River drainages to the point of ending base flows (85-90 percent reduction in flows). Increasing agricultural and municipal water pressures, and existing and proposed flood control structures in the Wet Walnut drainage are future threats.

Fire suppression has resulted from farming and the presence of homes in the basin; needed upland burning programs cannot take place as a result.

PROJECT DESCRIPTION

In the late 1940s, and following twenty years of efforts by conservationists and the state to create a National Wildlife Refuge, the state purchased the bulk of the 20,000 acres it now owns in Cheyenne Bottoms. During the 1950s, dikes, dams, and an inlet system were built to conserve water and provide more waterfowl hunting opportunities. By the late 1980s, agricultural ground-water pumping was preventing water from entering the Bottoms, and thus interfering with the state's legal right as a "senior" water rights holder. In 1990, the State Division of Water Resources declared an Intensive Groundwater Use Control Area (IGUCA) in Wet Walnut Creek, effectively reducing the amount of water that "junior" water right hol-ders could pump (up to 67 percent). The purpose was to reduce water removal so that the aquifer would rise, allowing base flows to return. The IGUCA went into effect in January 1992.

At the same time, the state has been undertaking an $18 million renovation effort since 1989 to improve water management capability in the face of still-declining water availability.

In 1991, TNC purchased major tracts adjacent to the wildlife area. Since then, the state and TNC have coordinated management efforts, although each has developed separate management plans for their lands. DWP's management plan was developed in the early 1990s with input from two public meetings and using results of work performed by engineering consultants on the renovation effort. An advisory panel consisting of eight citizens was created in 1994 to make recommendations to the area manager, who ultimately decides on all land management decisions.

The plan's goals are to provide migratory and nesting habitat for waterfowl and shorebirds, as well as public recreation opportunities which do not interfere with the habitat goals. Specific on-the-ground strategies include storing water when available, mechanical vegetation control or burning, and dike and dam manipulation, among others. Monitoring of vegetation control results and levels of bird and invertebrate habitat use are included in the plan.

STATUS AND OUTLOOK

The IGUCA is still in effect on Wet Walnut Creek and its effect will be evaluated at the end of the five-year review period in 1997. The renovation effort is expected to be completed in 1997. The advisory panel is being expanded to nine members and will be filled by representatives of interest groups (agriculture, conservation, recreation). Finally, the state's operating and maintenance budget for Cheyenne Bottoms has increased in response to the installation of several pump stations.

Factors Facilitating Progress

Recent public attention and awareness of wetlands, grants from the Wetlands Conservation Council (under the North American Waterfowl Management Plan, a federal program), state commitment through funding (from pesticide and irrigation equipment taxes), and TNC's land acquisitions have been the most important factors benefiting the project. The cooperative efforts of professionals, from many disciplines as well as public and private agencies, has been another benefit.

Obstacles to Progress

Lack of personnel and equipment for vegetation control has hindered this effort's progress. The shortage of water will be a continuing problem unless base flows are reestablished in the nearby waterways. Conflict between the agricultural community and the state has always been significant, beginning with the initial purchase of land in the 1940s when farmers' land was condemned. This opposition was solidified with the imposition of the IGUCA, which restricted farmers' water use.

Contact information:

Mr. Karl Grover
Area Manager
Kansas Department of Wildlife
*   and Parks*
Cheyenne Bottoms Wildlife Area
Route 3
Great Bend, KS 67530
(316) 793-3066

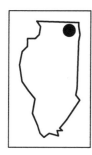

CHICAGO WILDERNESS

Location:
*Northeastern
Illinois*

Project size:
200,000 acres

Initiators:
*U.S. Fish and
Wildlife Service,
The Nature
Conservancy,
Field Museum of
Natural History,
Cooke County
Forest Preserve
District*

PROJECT AREA DESCRIPTION
The Chicago metropolitan area encompasses roughly six counties in northeastern Illinois. Although highly developed or industrialized, not all of this region is relegated to urban sprawl. Several hundred natural areas, comprising some 200,000 acres, can be found in the area. Some of the best remnants of tallgrass prairie and oak savanna landscapes east of the Mississippi can be found here. These remnants support a rich diversity of life, including five globally imperiled ecosystems and almost 200 of Illinois' endangered and threatened species.

ECOSYSTEM STRESSES
Exotic species are threatening to drastically change species composition of the prairie remnants. Fire is a critical ecosystem process in prairies, and is currently lacking in most remnants due to suppression. Another major stress is increasing habitat fragmentation and habitat loss due to increasing development encroaching on natural areas.

PROJECT DESCRIPTION
Starting well before the initiation of this project, The Nature Conservancy (TNC) in the Chicago area developed an extensive volunteer network consisting of some 5,000 volunteers. These volunteers (or "land stewards") work in preserves owned by a variety of agencies, such as county forest preserve districts and soil and water conservation districts. TNC provides general administration, coordination, and training. Many land stewards have moved on to become the land managers for the owners of the preserves.

Once these land managers came together, they realized that management of natural areas could be improved with more communication and coordination between landowning agencies. In order to do this, agency executives needed to be involved. As a result, executives of thirty local agencies and organizations met in September 1994. Examples of participants include TNC, the Field Museum of Natural History, the Northeastern Illinois Planning Commission, the National Park Service, the U.S. Environmental Protection Agency (EPA) Region 5, the EPA Great Lakes National Program Office, the U.S. Fish and Wildlife Service, and all six forest preserve and conservation districts.

During this meeting, all agencies and organizations that were present formed an informal partnership. An executive committee was elected, representing all levels of government and organizations. In addition, five working teams were created, each with its own focus (land management, science, policy/strategy, education/outreach, and marketing). Within each team, members consider how the team's particular focus area can add to the understanding and management of the ecosystems. For instance, the science team is defining conservation goals, determining which studies need to be done, and how to measure success. Overall, project planning takes place in the executive committee and in the working teams. Decisions are made by consensus.

STATUS AND OUTLOOK
As described earlier, the management plan for the Chicago Wilderness is currently under development. The partnership was formalized in April 1996, when the partnership and a number of projects advancing the group's mission were announced at a

press conference. Funding is being lined up for these projects. Many management efforts are carried out by the individual partners or small groups of partners.

Factors Facilitating Progress

Although this project is still young, it is already clear that the involvement of many committed people is very beneficial to the effort. In addition, an understanding and appreciation by all involved of the critical need for natural areas in an urban and suburban environment has been a major advantage.

Obstacles to Progress

It will be a challenge to maintain the involvement of this many partners, working through a complex matrix of projects.

Contact information:

Ms. Laurel M. Ross
Bioreserve Program Director
The Nature Conservancy
Illinois Field Office
8 S. Michigan Ave., Suite 900
Chicago, IL 60603
(312) 346-8166 ext. 14
Fax: (312) 346-5606
E-mail: lross@mcs.com

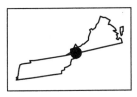

CLINCH VALLEY BIORESERVE

Location:
Southwestern Virginia, northeastern Tennessee

Project size:
1.4 million acres

Initiator:
The Nature Conservancy

PROJECT AREA DESCRIPTION

The project area includes the watersheds of the Clinch and Powell rivers. A mountainous area with many caves at lower elevations, the region falls within the Ridge and Valley, and Cumberland Plateau provinces. Mostly rural, the area is generally forested, predominantly in deciduous oak-hickory forests with some red spruce. There are approximately twenty-five federally listed threatened and endangered species present, including freshwater mussels and three types of bats. The Nature Conservancy (TNC) has ranked 136 globally-rare species in the area.

ECOSYSTEM STRESSES

Abandoned mines are a major concern due to the seepage of toxic metals such as copper and zinc into the rivers. A second concern is the loss of riparian buffer zones along streams due to farming and livestock activity. Finally, nonpoint source pollution (especially sewage) from poor management practices threatens water quality in the watersheds.

PROJECT DESCRIPTION

The project was initiated in 1984 when TNC acquired their first preserve on the Clinch River. The organization was attracted to the area because Pendleton Island has the highest concentration of mussel diversity in the world. In 1989, TNC opened an office on the Clinch preserve and subsequently purchased five other reserves. In 1990, they initiated cave education programs for landowners, as well as incentive programs for farmers to reimburse them for fencing materials protecting riparian areas.

Also in 1990, TNC began gathering various partners together to hold roundtable discussions in order to begin cooperative management efforts in the watersheds. These partners, now totaling seventy, include agencies, universities, landowners, and nonprofit organizations. A major milestone came when the U.S. Fish and Wildlife Service opened a Clinch River Office. Other agencies also took action in the area: the U.S. Environmental Protection Agency selected the region as one of its five pilot sites for ecological risk assessments; the U.S. Geologic Survey included this area in its national water quality study; and the Tennessee Valley Authority designated the Clinch River as one of its special river action team projects.

The overall goal of the TNC project is to restore populations of rare species to levels adequate for long-term viability. The strategies designed to meet this goal fall into several categories: protection, research, and community development. Specific actions to achieve the objectives include 1) identification of the sources of stress on populations; 2) restoration of native vegetation along river banks; 3) reclamation of abandoned mine lands near some of the most significant sites; and 4) acquisition of core areas.

STATUS AND OUTLOOK

Since 1995, TNC has engaged in an aggressive effort to foster and encourage sustainable development in the region. Examples of this effort are the use of draft horses for timbering as an alternative to machinery, and agricultural diversification into organic crop production. The results of these projects will be seen in the long term, while results from stream-bank restoration projects and improvements in water quality have already been observed.

Factors Facilitating Progress

Through the work of research partners, life cycles of the rare species now are better understood so that protection efforts can be better targeted. People in the area have realized that the coal industry is decreasing and are therefore open and supportive of the sustainable development approach.

Obstacles to Progress

It has been difficult to gain acceptance for the concept that ecological results cannot be observed in the short term. Coordination among many partners has proven logistically challenging. Ensuring funding for the project has been an obstacle as well, especially since the Chesapeake Bay falls within the same region and is a competitor for limited funding.

Contact information:

Mr. Bill Kittrell
The Nature Conservancy
102 South Court Street
Abingdon, VA 24210
(540) 676-2209
Fax: (540) 676-3819

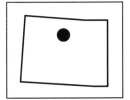

COLORADO STATE FOREST ECOSYSTEM PLANNING PROJECT

Location:
North-central Colorado

Project size:
70,768 acres

Initiator:
Colorado State Board of Land Commissioners

PROJECT AREA DESCRIPTION

The Colorado State Forest is located in north-central Colorado along the eastern boundary of Jackson County and North Park. The state forest is surrounded on three sides by public land: Routt National Forest to the north and south; Roosevelt National Forest (including the Rawah and Neota Wilderness areas) to the south and east; and Rocky Mountain National Park to the southeast. The western boundary abuts private land and Bureau of Land Management (BLM) lands.

At elevations ranging from 8,100 feet to the 12,900-foot high Clark Peak, much of the project area is high mountain valley. Big sage and western wheat grass are two common cover types in the lower-elevation valleys of the forest. The dominant forest cover is comprised of Engelmann spruce and subalpine fir. The second most common type of cover is lodgepole pine. No federally listed threatened or endangered species are known to exist in the project area. However, the boreal toad, wood frog, wolverine, river otter, and greater sandhill crane are listed by the state of Colorado.

ECOSYSTEM STRESSES

Grazing and range management are among the stresses to the ecosystem, possibly contributing to an increase in undesirable forbs and grasses and a decrease in willow. Also, grazing may be responsible for a reduction of willow species. Roads and timber harvesting have fragmented habitat. Suppression of the natural fire regime has led to high density stands of lodgepole pine with large amounts of mistletoe infection. Finally, the ecosystem of the

Colorado State Forest is and will continue to be stressed by the number of competing uses on a fixed acreage, including increases in recreational use and proposals for commercial development.

PROJECT DESCRIPTION

As the trustee of state trust lands, the Colorado State Board of Land Commissioners (Land Board) initiated the project. The Land Board has been interested in capturing additional revenues from the state forest and further desired that the various agencies responsible for managing different facets of the forest coordinate their management efforts. Finally, the Land Board was in need of more and better information on which to base renewal of several leases for grazing and other activities.

A sustainable future for Jackson County is of concern to many local citizens. Preserving local values and lifestyles is a high priority. Because most county residents depend on extractive industries for their economic security, a task force of environmental, industry, and local government representatives has been formed to facilitate public input into the planning process. Besides the open houses that Land Board convened at critical junctures of the project, community residents are free to have input into the process at any time through contact with task force members who represent their interests. The task force uses a consensus approach to decision making. Those issues for which consensus cannot be attained are tabled until more information becomes available or the decision-making context changes. The process for involving the public in agency activities was based in part on the process used by the Elliott State Forest effort in Oregon (see page 135).

A noteworthy component of the project is its proposed monitoring activities. Monitoring will focus predominantly in three areas: 1) ecosystem components, including selected fauna that are indicator species (northern goshawk, boreal owl, boreal frog, and aquatic macroinvertebrates); 2) ecosystem structure, including species composition in upland ranges, riparian areas, wetlands, and forest ecosystems; and 3) ecosystem functions, such as water and air quality and landscape-level changes.

Results from the monitoring will then be compared against explicit goal statements developed for ecosystem components, structure, and function. The information collected will be integrated with the existing information infrastructure and put into a geographic information systems (GIS) database developed by the College of Natural Resources at Colorado State University.

STATUS AND OUTLOOK

The current phase of planning and data collection is slated to end in mid-1996. By that time, the monitoring system will be in place. Attention will soon turn to a search for financing of ongoing monitoring activities. A draft strategic management plan was due in August 1995. The plan will pass through a public comment period before being finalized.

Factors Facilitating Progress

The task force has been instrumental in helping the Colorado State Forest Service navigate social and economic issues in Jackson County. The level of support by state and local land management agencies has been and will continue to be crucial to the project's success. In addition, the willingness of the State Land Board to experiment with a task force and undertake and provide ongoing support to the project were also identified as factors facilitating progress.

Obstacles to Progress

Development pressures on the state forest, such as a recent proposal for a ski area near Cameron Pass stirred much debate, controversy, and stakeholder polarization. Other challenges involve the constitutional mandate for the Land Board to generate revenues from public land. Most of the ecological, social, and economic factors in the state have dramatically changed since the revenue-focused mandate was enacted. Agency and stakeholder "turf" battles have also slowed the project, as certain stakeholders perceive a loss of power over resource decisions in the project area.

Contact information:

Mr. Jeff Jones
Special Program Coordinator
Colorado State F ₁ st Service
203 Forestry Building
Colorado State University
Ft. Collins, CO 80523
(303) 491-7287
Fax: (303) 491-7736

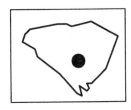

CONGAREE RIVER CORRIDOR WATERQUALITY PLANNING ASSESSMENT

Location:
Central South Carolina

Project size:
22,200 acres

Initiators:
*National Park Service,
U.S. Geologic Survey,
National Biological Service*

PROJECT AREA DESCRIPTION

Located approximately twenty miles southeast of Columbia, South Carolina, the project area is the Congaree Swamp National Monument, the last remaining old-growth bottomland hardwood forest in the southeast United States. The site is adjacent to the Congaree River, a dendritic and braided free-flowing river, and has minimal topographical relief. Bobcat, turkey, alligator, otter, and more than 130 species of birds can be found there, including two federally listed threatened species, the red-cockaded woodpecker and the American woodstork.

Preservation and recreation (hiking) are the principal uses of the National Monument, which was created in 1976 and is administered by the National Park Service (NPS). It was designated as the South Atlantic Coastal Plain Biosphere Reserve by the United Nations in 1983. Agriculture and timber are the dominant land uses in the surrounding area, along with limited rural housing.

ECOSYSTEM STRESSES

Stresses in the National Monument appear to have been minimal, but increasing upstream residential development and associated point and non-point source pollution are growing threats. The most significant current stress is from feral hogs uprooting significant tracts of the forest floor, the cumulative effects of which are unknown.

PROJECT DESCRIPTION

Several federal agencies had been conducting or planning to begin studies involving certain water quality aspects of the Congaree River. NPS has been assessing potential contaminant sources to the fragile floodplain since 1993, as part of a cooperative agreement with East Carolina University to prepare a water resources management plan. Other collaborative efforts concerning the Congaree River watershed include the U.S. Geological Survey (USGS) National Water Quality Assessment Program-Santee River Basin Study which began in 1994; and an ecological profile assessment of all major waterways in Congaree Swamp National Monument beginning in 1995 by the National Biological Service-Southern Science Center (NBS).

All of the federal agencies expressed interest in working in a collaborative effort once they became aware of each other's mutual interests. Current "reinventing government" and other efforts to foster interagency cooperation also played a factor in initiating this effort.

An unofficial interagency panel with representatives from NPS, USGS, NBS, and East Carolina University came together in early 1995 to discuss the effort. In June of that year, they agreed to develop a strategy to complete a comprehensive water quality assessment on the Congaree River watershed and develop strategies to address the overall health and vitality of the Congaree River corridor. The assessment phase is envisioned to last two to three years, with participants contributing different elements of research to provide a relatively complete snapshot of the floodplain's health. At that point, long-term protective strategies will be developed.

The panel consists of ten to fifteen federal agency and university representatives. Initially, conference calls and electronic mail were the mode of communication, with informal negotiation and exchange of letters documenting commitments.

STATUS AND OUTLOOK

The panel met in August 1995 and arrived at mutually-agreeable strategies based on current and contemplated project work. Field work was expected to begin in the Fall.

Factors Facilitating Progress

The willingness of agency representatives to look beyond normal agency boundaries and explore the possibilities of running cooperative efforts has been described as a benefit to the process. Participating organizations are able to use each other's research in a much more cooperative manner than had occurred previously. The net result has been eliminating duplicative efforts and leveraging contributions. Participants have been characterized as enthusiastic about working toward a common goal, in spite of heavy workloads in the face of what will be a very large study. Finally, the effort would not have progressed as rapidly nor as far in this initial phase without the use of up-to-date communication technology (conference calling, electronic mail).

Obstacles to Progress

Problems have been limited to the administrative realm. Primarily, coordinating schedules between twelve or more participants from five different agencies and offices, and in many different locations, has proven challenging albeit rewarding once accomplished.

Contact information:

Mr. Richard A. Clark
Resource Management Specialist
National Park Service
Congaree Swamp National
  Monument
200 Caroline Sims Road
Hopkins, SC 29061
(803) 776-4396 ext. 307
Fax: (803) 783-4241

CORPUS CHRISTI BAY
NATIONAL ESTUARY PROGRAM

Location:
South coastal Texas

Project size:
*352,000 acres of water;
7,383,680 acres of adjacent land*

Initiators:
*Private citizens,
State of Texas,
U.S. Environmental
Protection Agency*

PROJECT AREA DESCRIPTION
Centered around Corpus Christi, Texas, the twelve-county project area includes three estuaries: the Nueces-Corpus Christi, Baffin-Laguna Madre, and Copano-Aransas systems. Most of the terrain is gently sloping coastal plain with a barrier island (Padre/Mustang) that delimits the estuaries and reduces tidal exchange with the Gulf of Mexico. Inland native plant cover is mesquite brush with limited areas of prairie, much of that converted to agriculture.

The bays are generally shallow with sand, silt, and shell bottom. Sea grasses form extensive beds in the shallowest and undisturbed regions of the bays. The area is home to several federally listed threatened or endangered species, including several turtles (Kemp's Ridley, loggerhead, green, leatherback, hawksbill), whooping crane, piping plover, and brown pelican. In addition, twelve state listed species are present, including the opossum pipe fish, sheep frog, Texas tortoise, and two dolphins.

Landownership is mostly private, in large tracts owned by a few individuals or large corporations. One-quarter of the land is in row crop farming (sorghum and cotton primarily) and 50 percent is cattle rangeland; oil and gas production overlaps both of these. The remaining 25 percent is in urban and industrial use, with some wetlands and other undeveloped lands. Federal lands include Padre Island National Seashore and Aransas National Wildlife Refuge; state lands include submerged coastal areas and other scattered parcels.

ECOSYSTEM STRESSES
Although stresses have yet to be fully characterized, agricultural practices, urban development, hydrologic alteration, and non-point source pollution are potential significant stresses to the project area. About half of the native inlands have been converted to agricultural, urban, and industrial uses. Agricultural practices may contribute to nutrient and pesticide loading, and alter some freshwater input to the estuaries and bays. Urban-related stresses likely include dredging and filling of coastal wetlands for residential use, non-point source nutrient and pesticide runoff, oil/grease pollutants, trash dumping, and two public water supply reservoirs. Other, less significant stresses include cattle grazing, commercial and recreational overfishing, industrial point source pollution, exotic species (brown mussel, nutria), and other recreation.

PROJECT DESCRIPTION
In April 1992, local concerned citizens with the Coastal Bend Bays Foundation, Gulf Coast Conservation Association, and academic community, with the support of the Texas Natural Resource Conservation Commission and the governor of Texas, initiated the nomination process for Corpus Christi Bay (CCB) to become a National Estuary Program (NEP). The nationwide program is authorized by the Clean Water Act and administered by the U.S. Environmental Protection Agency (EPA). Following its acceptance by EPA in 1992 and a start-up year beginning in December 1993, the CCBNEP officially began in September 1994. Five committees were established in Spring 1994 to form a management conference.

The goal of the program is to develop a Comprehensive Conservation Management Plan (CCMP), which will be carried out under state and local auspices. As with all NEPs, the CCBNEP seeks to involve all stakeholders in a consensus-building approach, through their participation on the five committees (policy, management, citizen, local government, scientific/technical advisory), whose members are representatives of state agencies, local government, local industry and agriculture, recreational user groups, environmental organizations, commercial fishing, and local academic institutions. A sixth committee, financial planning, was formed in FY96 to aid in the implementation phase.

The CCBNEP's first priority is to use existing data to characterize the status and trends of important components of the study area. Management actions will then be developed, reviewed by stakeholders, and subject to extensive public review before being revised and then submitted to EPA for approval in September 1998. A monitoring program will be designed to measure progress in meeting program goals, and will be coordinated with existing monitoring efforts.

During the four-year duration of the program, EPA provides 75 percent of the funding, with the state providing the remainder. Thereafter, the state is responsible for funding and implementation, anticipated to last twenty to forty years.

STATUS AND OUTLOOK

The most significant reported outcomes have been setting up

committees and the overall program, increased communication among stakeholders, and a shift to a multi-species management approach. The first "All Conference Workshop" was held in February 1995. It served to provide insight to over 200 committee participants on the CCBNEP's habitat and degradation, and to create a common vision statement and operating principles. A second workshop was scheduled for March 1996.

The program is currently receiving the results of twelve studies on the area's historic and current condition. Reports have been received for the characterization of living resources, non-point source pollution, and status and trends of red and brown tides in the CCBNEP study area.

It is still unclear whether the necessary level of consensus on major issues has been reached. However, positive feedback was received at a recent meeting of the thirteen action plan task forces, each of which address a specific issue, such as agricultural non-point source pollution. At this first meeting, each task force began considering the severity of each problem and the universe of possible management actions to address that problem.

Attention has been brought to some of the environmental issues, but increased public awareness of these issues is considered necessary. A public outreach component of the program includes making presentations at club meetings, using the media, producing a

program newsletter, and other activities to increase public awareness of bay issues.

Factors Facilitating Progress
The willingness of the management conference members to attend meetings and volunteer for project development and review subcommittees have been the most important factors reported. The CCBNEP's administrative structure, clarity of its goals, and project leadership have also been cited as positive factors.

Obstacles to Progress
While the committees and their members have worked well together, there have been jurisdictional concerns among state agencies. Maintaining interest in the citizen's advisory committee has been challenging, because there are few substantive issues for their consideration at the moment, even though this committee will become more important later. Finally, it is unclear how a relatively new state administration will affect the project's continued progress. Likely reductions of EPA funding could reduce the scope of the project over the next two years and subsequently into its implementation phase.

Contact information:

Dr. Hudson DeYoe
Corpus Christi Bay National
*   Estuary Program*
Texas AMU-Corpus Christi
Campus Box 290
6300 Ocean Boulevard
Corpus Christi, TX 78412
(512) 985-6767 ext. 6301
E-mail: Deyoe@tamucc.edu

Location:
Southern California

Project size:
*Entire basin—
30,000 acres; to be
protected in Area of
Critical Environ-
mental Concern—
14,000 acres*

Initiators:
*Bureau of Land
Management,
The Nature
Conservancy*

DOS PALMAS OASIS

PROJECT AREA DESCRIPTION

Dos Palmas Oasis is a desert wetland system located within the lower Colorado Desert portion of the Sonoran Desert, near Salton Sea State Park. This desert wetland system consists of the lower portion of a watershed, including the surrounding mountains. It is a relatively flat basin bisected by the San Andreas earthquake fault zone. The fault zone causes water to reach the surface via numerous artesian springs. Palm oases are generally located at the headwaters of these springs, whereas marsh occurs where the waters have spread out into a broad area. Surrounding the marsh is mesquite bosque habitat, which consists of a dense zone of leguminous trees. Two federal and state listed endangered species occur in this area: the desert pupfish and the Yuma clapper rail. The California black rail and the flat-tailed horned lizard, both candidate species, can also be found at Dos Palmas.

Approximately 50 percent of the area is owned by the Bureau of Land Management (BLM); the other 50 percent is privately owned with fewer than ten residences. Most of the area is undeveloped open space.

ECOSYSTEM STRESSES

In the recent past, fish farming pulled water from native habitats and has introduced exotic fish into the wetlands, creating competitors and predators for the endangered pupfish. In addition, non-native salt cedar has invaded and altered vegetative composition and structure in the wetlands. Salt cedar competes with native vegetation for water, resulting in drought stress and subsequent replacement of native plants.

PROJECT DESCRIPTION

Approximately fifteen years ago, BLM went through a planning procedure for all its desert lands in California. In response to the presence of endangered species, BLM designated its lands in the Dos Palmas area as an Area of Critical Environmental Concern (ACEC). In 1982, a site-specific management plan was prepared and partially implemented. After purchasing a 1,200-acre ranch in Dos Palmas in the Fall of 1989, The Nature Conservancy (TNC) signed a cooperative management agreement with BLM. Subsequently, TNC funded a consultant to prepare an ecosystem management plan that could meet BLM standards and gain BLM approval. The plan is currently going through BLM's internal review process. That plan, which covers the ACEC lands, will be the basis for all actions by either BLM or TNC. Currently, BLM and TNC meet four to six times per year to review progress toward goals and to assign new tasks. Decisions are made by consensus.

The overarching goal of the project is to maintain and enhance wetland habitat for the pupfish and rail species. In order to reach that goal, salt cedar will be controlled or eliminated, wetlands that were altered by previous fish farming will be restored or enhanced, and important wetland habitats will be acquired. Salt cedar removal as well as restoration of previous fish farming areas will be attempted using existing and experimental methodologies. Pupfish and rails are monitored annually.

STATUS AND OUTLOOK

This project has already accomplished several tasks. BLM has acquired an additional 1,500 acres through exchanges with TNC. Additional lands have been purchased with Land and Water Conservation Funds. As a result of cooperative management between the BLM and TNC, salt cedar has been eliminated from one wetland area. Also, a pupfish refugium has been created.

Restoration of a former fish farm is due to begin in 1996 and will include recontouring and reestablishing a natural-appearing wetland between fifty and 100 acres, depending on the cost. Finally, BLM is taking an increasingly active role in the management of the Dos Palmas Preserve.

Factors Facilitating Progress

TNC's land acquisition has been helpful to the project's progress. It has also been beneficial that TNC was willing to pay a consultant to write the ecosystem management plan for the Dos Palmas Preserve in accordance with BLM agency standards. The time commitment of a few individuals and their willingness to take responsibility has also been beneficial to this project. In addition, joint efforts to control salt cedar have facilitated project progress.

Obstacles to Progress

Changes in staff and lack of resources have been problematic. The atmosphere in the 104th Congress does not indicate any future positive changes in BLM staffing or funding. TNC and BLM currently have grant funding for their work on the Dos Palmas site. However, TNC may not be able to continue to allocate resources to this project indefinitely. Professional and agency values (i.e., a single-species focus of federal personnel as opposed to a community ecology or ecosystem focus) is a continuing concern.

Contact information:

Mr. Cameron Barrows
Southern California Area
  Manager
The Nature Conservancy
PO Box 188
Thousand Palms, CA 92276
(619) 343-1234
Fax: (619) 343-0393

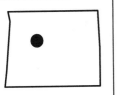

EAST FORK MANAGEMENT PLAN

Location:
Northwestern Wyoming

Project size:
54,000 acres

Initiator:
Wyoming Game and Fish Department

PROJECT AREA DESCRIPTION

The drainage area of the Wind River ecosystem covers 350,000 acres of grasslands, mountains, and badlands. The 54,000-acre East Fork site is defined largely by two former ranches purchased by the Wyoming Game and Fish Department (WGFD) in the last twenty years. The East Fork Management Area extends from Horse Creek drainage north to the Continental Divide and south to Crow Creek on the Wind River Indian Reservation. Elevation ranges from 6,200 to over 12,000 feet.

In lower elevations, western wheat, prairie june grass, Idaho fescue, and green needle grass mix with limber pine. The vegetative cover in higher elevations includes blue grasses, aspen stands, and some firs and spruces. The grizzly bear and peregrine falcon are the only federally listed threatened and endangered species known to reside in the area. The presence of wolverine and lynx cannot be ruled out in the less accessible areas.

ECOSYSTEM STRESSES

Hunting, hiking, camping, and motorized recreation all take place on WGFD's property and nearby U.S.D.A. Forest Service (USFS) and Bureau of Land Management (BLM) lands. Yet strictly enforced regulations and permitting procedures prevent major stress on the ecosystem. Grazing and timber harvesting are not permitted on WGFD lands and the site is closed to all human presence every year from December 1 to May 1. On some parts of the former ranches, overgrazing has resulted in periodic flash floods.

The most significant stresses are due to activities outside WGFD lands. Timber harvesting in Shoshone National Forest to the north could disturb migrating elk. Regionally, land conversion to residential development may alter elk migration corridors in the north and west.

PROJECT DESCRIPTION

WGFD began managing for elk in the 1940s after the herd's winter foraging began damaging private lands. WGFD erected elk fences while simultaneously purchasing land in the area. Large-scale acquisition began with the purchase of the 17,000-acre Inberg-Roy property in the mid-1970s.

The project's primary goal is to perpetuate the region's wildlife by preserving sufficient habitat. In the mid-1980s, WGFD began to emphasize ecosystem-based approaches to land management. Preservation of water resource quality and restoration of riparian areas damaged by grazing are also goals. In 1991, the Spence-Moriarty property (37,000 acres adjacent to the Inberg-Roy) was purchased. Now administering enough winter habitat, WGFD removed the thirteen miles of elk fence. WGFD sets herd-unit objectives based on carrying-capacity and hunter demand.

A technical committee, consisting of WGFD representatives, USFS and BLM officials, industry groups, conservation organizations (Rocky Mountain Elk Foundation, Trout Unlimited, Sierra Club), and local officials, is drafting a management plan for the entire 350,000-acre Wind River ecosystem.

STATUS AND OUTLOOK

Three years of range data collected on the new property were still being analyzed in early 1996. This data, used in reissuing grazing permits, will also be used to generate recommendations in a final report detailing the completed enhancement work. The report is to be published in 1996. A road plan has also been developed.

With the absence of grazing, willows and cottonwoods are returning to riparian areas. WGFD also provided input on mineral exploration activities on adjacent federal properties that could impact the project site.

Finally, a cooperative aspen-regeneration project on the forest has resulted from this effort, with funds contributed by WGFD and the Rocky Mountain Elk Foundation. Management of the regeneration project area will be turned over to the USFS.

Factors Facilitating Progress

Progress on the project is attributed to stakeholder support. The stakeholders' participation, professionalism, and mutual respect make it difficult for splinter groups to derail planning and management activities. State and federal agencies such as USFS and BLM also support the East Fork project. Their support, as well as that of local politicians and county commissioners, is necessary for continued progress.

Also contributing to this support were public meetings held by WGFD in 1995 to describe activities on the property and to discuss funding concerns.

Obstacles to Progress

While agencies have been cooperating well, their different missions and mandates result in bureaucratic issues that hinder the effort's progress. Paperwork stemming from compliance with the National Environmental Policy Act (NEPA), other environmental regulations, and environmental impact studies drains WGFD resources away from project activities.

Contact information:

Mr. Chuck Clarke
Habitat Management Coordinator
Wyoming Game and Fish
*   Department*
260 Buena Vista
Lander, WY 82520
(307) 332-2688

EASTERN UPPER PENINSULA ECOSYSTEM MANAGEMENT CONSORTIUM

Location:
Northern Michigan

Project size:
3,880,000 acres

Initiator:
Michigan Department of Natural Resources

PROJECT AREA DESCRIPTION

The project area included in this effort encompasses several distinct ecosystems in the eastern one-third of Michigan's Upper Peninsula (UP), from the west boundary of the Hiawatha National Forest eastward, including Drummond island. The land is mostly flat, with some areas of rolling sand hills, although the elevation change never exceeds 300 feet. The region is a mix of glaciated landforms over limestone bedrock, including outwash plains, moraines, and old lake basins; bedrock outcrops, dunes, and various wetlands are also found there. The area is characterized primarily by various mixes of northern hardwood and coniferous forests mixed with marsh, bog, open plains, and agricultural lands. The climate of the entire region is highly influenced by the Great Lakes.

Federally listed threatened and endangered species are numerous, including the piping plover, bald eagle, common tern, osprey, gray wolf, and pitcher's thistle. State listed species include the Lake Huron tansy, dwarf lake iris, pine marten, and wood turtle.

Other than timber harvesting, this is one of the most undeveloped regions of the eastern United States. Current uses of the project area include timber management, recreation (camping, hiking, hunting, other), wilderness, and scattered rural development. Approximately two-thirds of the land base is owned or managed by participants in this ecosystem management consortium. Fifty percent of the region is in public ownership, including the Hiawatha National Forest and Lake

Superior State Forest (each at one million acres), Tahquamenon Falls State Park, Seney National Wildlife Refuge, and Pictured Rocks National Lakeshore. The remainder is privately held, including large industrial forests.

ECOSYSTEM STRESSES

Timber harvesting is the most significant stress to the project area, as most of the area is forested. Timber practices have been present on a large scale for over 100 years, and virtually none of the present forest has been left uncut from earlier times. Fire suppression has been a common practice, which has led to changes in plant communities, and results in a build-up of fuel (leading to fires more damaging than normal). Other stresses, though less significant, include exotic species such as purple loosestrife; alteration of hydrology due to development of wetlands and coastal areas, and roads; recreation and tourism pressures; and road construction.

PROJECT DESCRIPTION

The project began in July 1992 to address questions that kept arising on issues that cross ownership boundaries (biodiversity, recreation, wildlife management). Initiated by the Forest Management Division in the Michigan Department of Natural Resources (MDNR), the consortium was designed to facilitate communication among the various stakeholders in the region, including federal agencies (U.S.D.A. Forest Service (USFS), Fish and Wildlife Service, National Park Service), three industrial timber landowners (Mead Corporation, Champion International Corporation, Shelter Bay Forests), and The Nature Conservancy, and MDNR.

Regular consortium meetings are held every six to eight weeks: progress toward goals and specific tasks is reviewed, information exchanged between members, and future direction and projects are set. As special projects arise, various subgroups are set up to address those tasks.

The effort's mission statement is, "To facilitate complementary management of public and private lands for all appropriate land uses through large-sale, landscape-ecological approach to maintaining and enhancing sustainable representative ecosystems in the eastern Upper Peninsula." No specific management plan is being developed, nor is the consortium considered a planning committee. Rather, specific activities which serve to further the effort's mission are undertaken on a continual basis. The general direction of the effort is to promote the continued gathering and exchange of information, particularly geographic information systems (GIS) information, to allow stakeholders to make more informed decisions.

STATUS AND OUTLOOK
To date, the consortium's activities have focused on classifying the land base into ecosystems

and gathering information to describe these lands, their current uses, and the natural processes occurring there. For example, a map of ecosystem units compatible with USFS hierarchical ecological classification was developed in 1994.

More broadly, the development of trust among the diverse landowners has been cited as a specific outcome. Increased communication and cooperation among stakeholders have been described as positive outcomes, along with reduction of conflict among stakeholders. Restoration of degraded areas has begun, previously identified stresses are being reduced, and management is shifting away from single species or resources to management of the ecosystem or ecological landscapes.

Factors Facilitating Progress
A need to work together, the development of trust among stakeholders, a common set of goals, the informal nature of the coalition, and the exchange of useful information have been described as factors that have benefited the process. Support by stakeholders, including federal and state agencies, corporations, and other stakeholders, has been especially beneficial.

Obstacles to Progress
Because the consortium is unofficial, with no dedicated staff or funding, lack of time, personnel, and resources have been cited as barriers to the effort's progress. Additional studies of the area's resources would help the consortium and its members make more informed land management decisions. Finally, incomplete agreement or understanding of new scientific concepts—biodiversity, ecosystem management—by consortium members and the scientific community at large has caused difficulties in interpretation and decision making.

Contact information:

Mr. Les Homan
District Forest Planner
Michigan Department of Natural Resources
PO Box 77
Newberry, MI 49868
(906) 293-5131
Fax: (906) 293-8728

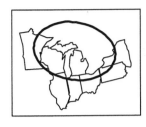

ECOSYSTEM CHARTER FOR THE GREAT LAKES–ST. LAWRENCE BASIN

Location:
Minnesota,
Wisconsin, Illinois,
Indiana, Michigan,
Ohio, Pennsylvania,
New York; Ontario
and Quebec, Canada

Project size:
189 million acres

Initiator:
Great Lakes
Commission

PROJECT AREA DESCRIPTION

The Great Lakes ecosystem includes the drainages of five major lakes—Superior, Michigan, Huron, Erie, and Ontario—whose waters flow into the St. Lawrence River to the east. The lakes cover a surface area of about 94,000 square miles, with the entire watershed having very diverse climate, soils, topography, and vegetation as a result of its size and location. Northern areas, once covered by vast coniferous forests, have a colder climate and relatively thin, acidic soils underlain by granite. Southern areas have been glaciated, are warmer, and were once characterized by deciduous forests, prairies, and swamps.

The lakes themselves affect the region's climate, first by acting as a giant heat sink which results in milder climate of surrounding land compared to other areas of the same latitude. Second, the lakes act as a giant humidifier, increasing the moisture content of the air and precipitation throughout the area, especially to the south and east.

Although part of the same ecosystem, each of the lakes has distinct features. For example, each lake drains different lands with their different characteristics; water resides in each lake for varying amounts of time (on average, 2.6 years for Erie, 191 years for Superior); and each lake is of a different size, depth, and elevation, with different characteristics as a result. The surrounding land is characterized by extensive agricultural and urban development, with a mixture of public and private lands on both sides of the border. Several dozen threatened and endangered plants and animals are

found in the system, including species found across the United States and others endemic to the region.

ECOSYSTEM STRESSES

Clearing of the Great Lakes forests during settlement in the 1700s and 1800s for timber and agriculture was the first, large-scale stress to the system. Large-scale logging eliminated habitat, eroded stream banks, and disrupted stream bottoms, all resulting in erosion and smothering of spawning grounds. Conversion to agriculture resulted in wetland losses, plowing of native prairies, and burning of other forests. Wetlands are still being lost to agricultural, industrial, and residential development.

The connection of waterways through many canals, starting in 1825, has not only disrupted the hydrology, but led to increased shipping and the introduction of extremely harmful exotic species, such as the alewife, sea lamprey, Eurasian River ruffe, and zebra mussel.

Increased industrialization, farming, shipping, urbanization, sewage, and various forms of point and non-point source pollution led to tremendous water quality problems, peaking in the 1960s and 1970s. Commercial fishing declined precipitously due to pollution as well as overfishing and exotics. Other severe effects were felt by wildlife, such as bird egg-shell thinning caused by DDT and other pesticides.

PROJECT DESCRIPTION

When the ecosystem charter effort was initiated by the Great Lakes Commission in 1992, the ecosystem management concept had already been adopted by many government and citizen-based institutions, regional agreements, and policies in the Great Lakes Basin. The commission began the effort in response to the need for a single, clearly-defined document defining goals for ecosystem management in the basin, to move beyond the largely conceptual level into implementation, and to prevent the multitude of efforts from being compromised.

A thirty-five member ecosystem advisory committee was established in late 1992, with representatives from several different state and federal natural resource agencies, nonprofit organizations, universities, industries, and regional commissions. The committee developed the ecosystem charter, drawing on more than sixty existing treaties, laws, policies, and agreements from the region and beyond. It was then refined through two public comment periods, state roundtables, workshops, and presentations. A twelve-member advisory committee, mainly from academic and policy institutions, provided guidance on the project.

The ecosystem charter has three priorities: 1) promoting and assessing efforts to implement ecosystem management in the basin; 2) communicating an ecosystem management vision for the basin; and 3) advocating the interests of the basin ecosystem and its inhabitants. As a voluntary, nonbinding document, the charter is intended to guide the activities of each organization that signs it. The charter contains a twelve-point vision statement, followed by seventeen principles organized under several subcategories.

STATUS AND OUTLOOK

The charter was finalized in late 1994, with more than 160 signatories from nearly all basin states endorsing it as of February 1996. In Fall 1995, the one-year anniversary of the document's public release, the Great Lakes Commission queried the charter signatories and other groups outside the basin engaged in similar efforts, asking them to describe how they have been using and/or benefiting from the charter. Some of these were profiled by the commission in its newsletter in late 1995. The commission also has an on-line World Wide Web homepage for this effort.

Factors Facilitating Progress

Broad outreach by the Great Lakes Commission to organizations and agencies in the basin from the outset of this effort has been described as instrumental to the effort's progress, especially in drafting and revising the ecosystem charter. The document appears to have been readily accepted by signatories because it was based on existing laws, agreements, and policies. Gaining community "ownership" has been both a deliberate strategy and a result of the effort. Finally, institutional cooperation has been significant and beneficial. Several of the more recent signatories joined the efforts because of the charter's benefits to other signatories, now evident after more than a year of existence.

Obstacles to Progress

Obtaining the endorsement of groups at both ends of the environmental spectrum (for example, selected environmental organizations on one end, certain industries on the other) has proven challenging. Because the ecosystem charter and the basin itself are so broad-based, the charter's goals are not easily measurable. Thus, demonstrating certain kinds of progress as a direct result of this effort will be difficult. In general, the charter does not directly resolve stresses to the basin, partly because of its voluntary nature, and again because of its very large scale.

Other Efforts in the Region

A number of other ecosystem-based efforts have been initiated in the Great Lakes region, including the Joint Strategic Plan for Great Lakes Fisheries (by the Great Lakes Fishery Commission); Great Lakes Charter (focusing on water quality and quantity—distinct from this effort); Great Lakes Ecoregion Team (U.S. Fish and Wildlife Service); International Joint Commission; Lake Superior Basin Biosphere Reserve Feasibility Study (National Park Service); and various projects by nonprofit organizations. These efforts focus on specific resources (e.g., fish, water quality) or landownerships. The effort featured here incorporates language or concepts from many of these efforts, and was designed to bridge the gap between them.

Contact information:

Ms. Victoria Pebbles
Great Lakes Commission
Argus Building, 400 Fourth
Ann Arbor, MI 48103-4816
(313) 665-9135
Fax: (313) 665-4370
E-mail: pebbles@glc.org

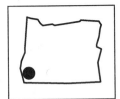

ELLIOTT STATE FOREST MANAGEMENT PLAN

Location:
Southwestern Oregon

Project size:
93,000 acres

Initiators:
*Oregon State Land Board,
Oregon Department of Forestry*

PROJECT AREA DESCRIPTION

The Elliott State Forest covers an area in southwest Oregon from the slopes of the Coast Range to within ten miles of the Pacific Coast. The forest is currently dominated by Douglas-fir, western hemlock, and western redcedar. Riparian areas add to the overall vegetative diversity by supporting bigleaf maple and red alder. Willow, Oregon myrtle, and Pacific yew are present but occur with less frequency. The northern spotted owl, marbled murrelet, and peregrine falcon are among the forest's endangered or threatened species. Hardwood forests, cliffs, talus, wet and dry meadows, wooded swamps, and bogs are all important to the forest's overall diversity. Of the 162 non-game species of fauna, 140 are found in riparian zone habitats.

ECOSYSTEM STRESSES

Because 85 percent of the forest has been managed in the past for timber, timber harvesting, road-building, and fire suppression represent the major human-induced disturbances. From 1978 to 1993, stands eighty years or older have decreased by 34 percent because of harvesting, significantly reducing habitat for the northern spotted owl. Periodic flooding of riparian forests, wind-generated blowdowns, insect and disease outbreaks, and especially fire are among the natural stresses that have played a major role in defining the forest's biodiversity.

The Douglas-fir stands' early stage of succession is due to the effects of wildfires and timber harvesting that have occurred in the forest. The varying conditions of the forest's thirty-eight miles of rivers and streams are the result of natural erosion, road-building, stream cleaning, and harvesting activities.

PROJECT DESCRIPTION

The listing of the northern spotted owl was the impetus for the Oregon State Land Board to direct the Oregon Department of Forestry (ODF) to use ecosystem-based approaches on the land the ODF manages. In 1992, the ODF linked up with other state agencies, universities, and federal agencies, including the U.S. Fish and Wildlife Service (FWS), to agree on management practices that were within National Environmental Policy Act (NEPA) guidelines. The resulting plan for the Elliott State Forest is a sharp departure from the previous, timber-focused management plans. The "integrated forest management plan" considers all the forest's resources and examines its resources and health in conjunction with the forests managed by neighboring federal and private landowners. The plan is currently in draft form; a final plan was expected in March 1995.

STATUS AND OUTLOOK

The planning process is now complete and a management plan has been developed. A Habitat Conservation Plan (HCP), the first combined HCP for both the northern spotted owl and the marbled murrelet, was approved by FWS in 1995. As part of the process, FWS approved an incidental take permit for potential impacts to both species as a result of harvesting practices. The plan is also the first prepared by a state for lands the state manages.

The plan has already allowed the Department of Forestry to achieve certain goals: compliance with all legal requirements of the Endangered Species Act, the formation of partnerships with other natural resource agencies, and movement toward approaches that protect biodiversity through ecosystem-based initiatives.

Because the ODF has a better understanding of the Forest's endangered species and their locations, timber harvesting is occurring in endangered species habitats only at levels permitted in the HCP. Harvest levels have fallen from 50 million board feet in 1990 to a single salvage sale in 1993. Although the final plan will allow for some timber harvesting, the timber industry, trade groups, and counties would like to see higher levels of harvest than what is permitted under the HCP.

Factors Facilitating Progress
Rather than developing a plan and then asking for public input, stakeholders were proactively included throughout the development process. Honesty about the impacts of the new management approaches and articulating the constraints under which the Oregon Department of Forestry manages state lands have helped stakeholders accept the change in management practices.

Obstacles to Progress
Previously, the ODF managed the forest in isolation; it was challenging to establish working relationships with several other agencies. Also, ODF staff had little experience with ecosystem-based approaches or developing a management plan with heavy public participation. Furthermore, the planning process does not occur on a "stable platform": the political and legal environments continuously shift, scientific knowledge is incomplete, and new understanding of the ecosystem constantly emerges.

Contact information:

Jill Bowling
State Lands Program Director
Oregon Department of Forestry
2600 State Street
Salem, OR 97310
(503) 945-7348

ESCANABA RIVER STATE FOREST

Location:
Northern Michigan

Project size:
420,400 acres

Initiator:
Michigan Department of Natural Resources

PROJECT AREA DESCRIPTION

The Escanaba River State Forest (ERSF) includes a variety of landscapes. Limestone, as well as granitic and sandstone bedrock are found there. Glacial features such as outwash plains, drumlins, kames, and moraines also occur. The dominant forest type is aspen. Also significant are stands of northern hardwoods and swamp conifers. Fourteen percent of the ERSF supports brush, grass, bogs, marshes, sand dunes, and water.

A total of 286 species of birds, mammals, and reptiles live in the ERSF. These include federally listed threatened and endangered species such as the bald eagle and the gray wolf. The common loon, osprey, fisher, beaver, black bear, and bobcat are examples of non-listed species that occur in the area.

The ERSF is used primarily for the production of forest products and for recreational activities, such as hunting, cross-country skiing, snowmobiling, and sight-seeing. Mining of iron ore and iron oxide pigments takes place as well. A few relatively small areas (ranging from 40 to 1,500 acres) have been set aside as natural areas in which human activity is minimal.

ECOSYSTEM STRESSES

The suppression of fire in certain forest types may impede the regeneration of tree and herbaceous species. The large population of white-tailed deer in some areas is also problematic. Deer browsing may lower successful tree regeneration after timber sales. The deer overpopulation is due to active timber markets which have created excellent deer habitat, a long string of

relatively mild winters, and public demand for good viewing and hunting. All-terrain vehicles pose another threat. They may disrupt recreational experiences of others, create new roads, and degrade existing forest roads and trails. Roads can contribute to overuse, since they allow more hunters and other recreationists to enter deeper into the forest. Finally, roads may also be responsible for erosion and subsequent siltation of streams.

PROJECT DESCRIPTION

The Escanaba River State Forest Comprehensive Resource Management Plan was developed in response to the 1983 Statewide Forest Resources Plan. The former was initiated by the formation of a forest interdisciplinary team (FIDT) within the Michigan Department of Natural Resources (MDNR). The FIDT represents the Divisions of Forest Management, Wildlife, and Fisheries, and other divisions to a lesser extent. It is charged with the development and implementation of the plan. The plan's goal is to identify management opportunities and to provide for that combination of products, services, and amenities that will be of greatest public benefit.

The planning process encompassed the identification of management issues, data collection and analysis, and the proposal of management strategies. Although some of the proposed strategies apply to the entire ERSF, many strategies have been specifically written for ecological management units. The boundaries of these units are based on natural landscape patterns, integrating climate, landform, soil, and vegetation. Management objectives vary, but include the designation of at least 5 percent of the forest as old growth, the identification and refining of ecosystem management techniques, and the production of a regulated forest age structure. Management strategies include the sale of timber on 6,500 acres annually, the development and application of geographic information systems (GIS), use of best management practices for erosion control, and a prescribed burn program. Monitoring focuses on timber harvests and wildlife populations. The entire plan will be reviewed once every ten years.

STATUS AND OUTLOOK

The plan has forced resource managers to consider biodiversity, multiple species management, old growth, and ecosystem management concepts. After implementation started in 1990, about 6,500 acres and 105,000 cords of timber have been offered for sale annually. This is somewhat less than

historical averages and less than anticipated due to lack of time and personnel.

Wildlife managers have been attempting to implement the population and habitat strategies identified in the plan with mixed results. In addition, 29,000 acres of the forest have been set aside to date as potential old growth, with more to follow. GIS equipment has been acquired and a pilot project begun to train employees in GIS and computer mapping of compartments and timbers sales.

Factors Facilitating Progress

Having a full-time planner assigned to the project has been very helpful. In addition, public interest has been a plus. Monitoring of the timber harvest using a database has also been advantageous to the project.

Obstacles to Progress

Funding and personnel may not be adequate to address many of

the ambitious recommendations stated in the plan. Many concepts and ideas were outlined in the plan without commitment to them or follow-through in the form of funding, resources, and training. Some disciplines were not interested in the planning process and participated reluctantly or to a much lesser extent than others. Finally, there was a unwillingness to commit specialized assistance to the project.

Contact information:

Lee Evison
Forest Management Division
Escanaba River State Forest
Michigan Department of Natural
  Resources
6833 Highways 2, 41, and M35
Gladstone, MI 49837
(906) 786-2351
Fax: (906) 786-1300
E-mail: evisonl@dnrserver1
  .dnr.state.mi.us

FISH CREEK WATERSHED PROJECT

Location:
Northeastern Indiana, northwestern Ohio

Project size:
70,400 acres

Initiators:
Indiana Department of Natural Resources, Ohio Department of Natural Resources, U.S. Fish and Wildlife Service, The Nature Conservancy

PROJECT AREA DESCRIPTION

Fish Creek is a thirty mile long pristine creek in the St. Joseph River basin in northeastern Indiana and northwestern Ohio. This aquatic ecosystem hosts 43 different species of fish and 31 species of mussels, 3 of which are federally listed as endangered: white cat's paw pearly mussel, club shell mussel, northern riffle shell mussel. In addition, the purple lilliput mussel, rayed bean mussel, eastern sand darter, and copper belly watersnake have been proposed for federal listing. Approximately seventy-five state listed species are present in the watershed, including mussels, fish, amphibians, reptiles, birds, and plants. In presettlement times, the watershed was covered by beech-maple and oak-hickory forests. Currently, only 15 percent of these forests are left. The remaining area is used primarily for row crop agriculture.

ECOSYSTEM STRESSES

Agriculture has posed a large stress on the ecosystem. Disc plowing, together with an alteration of the hydrology of the watershed, has led to increased amounts of runoff and subsequent siltation of the creek. Associated with the sediments are nutrients and pesticides.

PROJECT DESCRIPTION

In 1988, a study focusing on fresh water mussels in the St. Joseph River basin showed that Fish Creek contained the last remnant populations of freshwater mussels that once occurred throughout the entire basin. Within the basin, only Fish Creek had a high enough water quality to support these populations. After this study, the Indiana and Ohio Departments of Natural Resources, the

U.S. Fish and Wildlife Service, and The Nature Conservancy (TNC) formed the Fish Creek Partnership. The goal of the partnership is to secure the habitat of populations of freshwater mussels in Fish Creek by maintaining or improving water quality. The partnership hopes that Fish Creek populations may serve to repopulate the entire St. Joseph River basin.

Once the partnership was formed, a local presence was needed in the watershed to facilitate contact with local landowners. This occurred in 1992 when TNC received a Section 319 grant from the U.S. Environmental Protection Agency to hire a project manager and open a project office. The partnership has since expanded to include additional Ohio and Indiana state agencies, federal agencies such as the U.S.D.A. Natural Resources Conservation Service (NRCS) and the Agricultural Stabilization and Conservation Service, county surveyors, local Soil and Water Conservation Districts (DeKalb, Steuben, and Williams Counties), the Maumee River Basin Commission, and Purdue University.

In order to control soil erosion in upland areas, the partnership promotes conservation tillage practices through one-on-one contact with farmers, and through assistance in the purchase of conservation tillage equipment. Conservation tillage can reduce soil erosion up to 90 percent. The partnership also works with the NRCS and the local Soil and Water Conservation Districts to implement Best Management Practices to control soil erosion. In the riparian corridor, the partnership works with the landowners to identify gaps in the corridor forest. After these have been identified, the areas are reforested. From 1993 through 1995, 200 acres have been reforested out of the 350-acre target area.

In addition to the formal partnership, an advisory group has been formed, consisting of local landowners and a few project partners. This group meets two to three times a year to discuss the progress of the project, and to obtain local input on how better to address soil erosion.

STATUS AND OUTLOOK

It is still too early to determine if any of the ecological goals of the project have been realized. However, progress is being measured through monitoring of soils (erosion), land use changes, and aquatic health. Monitoring results are also used to adjust current management strategies.

Factors Facilitating Progress

So far, the project has been successful according to project leaders. Factors that have facilitated this success include the presence of a project manager on site, the involvement of and communication with landowners, and a non-confrontational approach.

Obstacles to Progress

Somewhat problematic is the absence of adequate scientific knowledge concerning fresh water mussels. For instance, tolerance levels to sedimentation are unknown, and therefore it is difficult to determine what an acceptable level of sedimentation should be.

Contact information:

Mr. Larry Clemens
Project Manager
The Nature Conservancy
Fish Creek Watershed Project
  Office
Peachtree Plaza, Suite B2
1220 North 200 West
Angola, IN 46703
(219) 665-9141
Fax: (219) 665-9141

FLORIDA BAY

Location:
Southern Florida

Project size:
640,000 acres

Initiator:
Florida Department of Environmental Protection

PROJECT AREA DESCRIPTION

Located between the southern mainland and the Florida Keys, Florida Bay is functionally the southern part of the Everglades ecosystem. The bay depends on water flowing out from the Everglades for its ecological integrity. Therefore, while the bay is a large, shallow estuary, its description by necessity includes two watersheds to the north: Taylor Slough, now the only source of water for the bay, and the "C111" canal, notable in that it drains water *away* from Florida Bay. Turtle sea grass is the dominant vegetation in the bay, with sawgrass, tree islands, and mangrove trees dominating the coastal and remnant wetlands of the Everglades region. The Everglades and Florida Bay are home to several dozen federally listed threatened and endangered species, including the American crocodile, Florida panther, manatee, and several sea turtles.

Eighty-five percent of Florida Bay is included in Everglades National Park. Recreational fishing occurs within park boundaries, with commercial fishing in the remaining 15 percent only. Most of the non-protected mainland is in row crop agriculture and urban uses.

ECOSYSTEM STRESSES

While Florida Bay itself has not been directly modified, most of the land up-river from the bay has been highly modified. Because of land conversion to agricultural and urban uses, the most significant stress to both the former wetlands and the bay is hydrologic disruption, due to a vast drainage and flood control effort known as the Central and South Florida Project.

The bay suffers primarily from hydro-period disruption (i.e., it does not receive the proper amounts of water at the right times). Also affecting hydro-period has been the construction of roads on the mainland, such as the east-west Tamiami Trail (now an interstate highway) and U.S. Route 1. As a result of inadequate water flow into the bay, over 100,000 acres of the bay have been lost to hypersaline conditions, which has resulted in an algae ooze covering the area and smothering the sea grass. Aquatic life—diversity and quantity—has declined precipitously, even causing a collapse of commercial shrimp fishing as far away as the Dry Tortugas Islands, 100 miles to the west.

PROJECT DESCRIPTION

In 1993, the Florida state legislature consolidated two state natural resource agencies into the Department of Environmental Protection (DEP). The legislation carried a mandate for the DEP to "manage and protect systems in their entirety." As a result, DEP created six water basin management areas in February 1994. Florida Bay was selected as one of the management areas because it is a high visibility region, with its decline very apparent to researchers and the public.

A conceptual management plan was written in a relatively short time, completed in March 1994. Developed entirely by DEP personnel, the plan identified three goals: 1) coordinate and integrate research on Florida Bay; 2) improve intergovernmental coordination; and 3) develop a land acquisition program in order to restore hydroperiod and water quality. Involvement from other stakeholders during plan development was limited, although the list of affected parties is long, including many federal agencies (Departments of Interior and Commerce, Corps of Engineers, U.S. Environmental Protection Agency), various user groups (e.g., fishers), environmental organizations, and several private landowners.

STATUS AND OUTLOOK
It is unclear what the future of this effort will be, as the document does not include implementation strategies, and most of the project area is under federal jurisdiction. The overall program's future depends in large part on the state legislature. The original 1993 mandate was very broad and lacked a specific directive on how to achieve ecosystem management. Further definition from the legislature or governor will be necessary.

Factors Facilitating Progress
A committed DEP staff is credited with making this effort move forward despite significant obstacles. Despite lack of direction and the project's limited potential, a program was developed from the ground up with few resources and significant time limitations.

Obstacles to Progress
Florida DEP has no regulatory authority over most of this ecosystem due to federal jurisdiction over most of the bay; for example, the National Park Service (NPS) is likely to be receptive to DEP recommendations only in as much as they agree with the NPS's own ecosystem management effort in the region. DEP's Florida Bay effort is designed to address interagency communication and public concerns, and appears to have had some success in this regard. The plan, however, is not expected to resolve these problems, especially between scientists and decision-makers for instance, who have widely divergent opinions about land management in south Florida.

The success of any restoration effort in Florida Bay depends on the success of similar efforts in the upstream Everglades. Some parties believe that it may be impossible to fully restore the ecosystem; that perhaps only remaining functions and remnants of the ecosystem can be preserved; and preserving remnant habitat may itself be impossible without adequate restoration of modified habitat. They argue that the region has been irreversibly altered and that growing population demands will place pressures on the ecosystem that are incompatible with its survival.

Finally, DEP's effort will require additional resources—funding, personnel—to succeed in achieving its goals.

Contact information:

Florida Department of
*   Environmental Protection*
3900 Commonwealth Boulevard
Tallahassee, FL 32399
(904) 488-4892

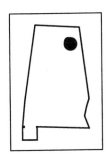

GEORGIA MOUNTAIN ECOSYSTEM MANAGEMENT PROJECT

Location:
Northeastern Alabama

Project size:
1,250 acres

Initiator:
Tennessee Valley Authority

PROJECT AREA DESCRIPTION

The project area, located in Marshall County and owned by the Tennessee Valley Authority (TVA), is located within the southern Cumberland Plateau physiographic region. Typical of the region, plateau and upper slope soils are derived from Pennsylvanian age sandstone. The topography is hilly and the project area is mostly forested. While the forest has been harvested at least once, the general appearance suggests mature or late successional conditions. Forests range from hardwoods on upper elevation slopes to mixed pine/hardwood on plateau soils, and occasional bottomland hardwoods in valleys and along streams. The project area is the only publicly-owned land in the area, much of it bordering artificial lakes.

Human use of the greater ecosystem was once primarily agriculture, but is now mostly in rural settlement, with subsistence farms and commuter 'ranchettes' predominating. A few caves are also present in the area, some inhabited by federally listed endangered candidate species.

ECOSYSTEM STRESSES

The ecosystem landscape has been largely converted to agriculture and, subsequently, rural settlements; the remaining landscape consists of heavily-modified forests. While the area has undergone significant hydrologic development (dams and reservoirs), the major stresses to the project area have been poor forest management and agricultural practices on private uplands, causing non-point source impacts on streams and floodplain communities, including wetlands. Also, road shoulders and drainage ditches have

not been maintained or managed correctly, and livestock have unrestricted access to plateau streams. On TVA lands, the greatest impacts have resulted from off-road vehicles (ORV) abuses, poorly constructed or maintained forest roads, and fires, the latter resulting mostly from informal campsites. Exotic species, primarily Japanese honeysuckle and Eurasian privet, are rampant in disturbed areas.

PROJECT DESCRIPTION

This effort, initiated in April 1994, is a localized but highly-focused result of the efforts of a single TVA employee. The effort came about primarily because of public pressure and concern about how the land was being managed by TVA: a few well-organized individuals had circulated petitions widely and applied other pressure that resulted in TVA changing its practices on this tract of land.

The effort is recognized by TVA as not being a true ecosystem plan, since the entire ecosystem is in multiple ownership; it may never be possible to manage it as a whole. Rather, this effort is characterized as an ecosystem-based approach to managing natural resources. Much of the defining characteristics of the plan—ecosystem boundaries, goals—have been defined by TVA and modified with public involvement. Some of the project's ecosystem boundaries were set by socioeconomic factors, with part of the system cut off by a major highway.

Formal outside participation has been limited to public involvement, consisting generally of public meetings, comment periods, informal presentations, and meetings with local citizens or sportsmen's groups. While an official plan has not yet been produced, the effort's goals are currently being defined as 1) planning for management of natural resources in the context of the landscape; 2) maintaining and enhancing biodiversity at different scales (species, community, landscape); 3) involving the public in plan design and implementation; 4) protecting rare or unique species and populations, water quality, and public recreation among others; and 5) developing a methodology for ecosystem management for other TVA lands.

STATUS AND OUTLOOK

A baseline inventory has been completed using geographic information systems (GIS) technology, aerial photography, and field work. Several rounds of public meetings have helped to determine future desired conditions. A draft management plan was completed in 1995; a draft environmental assessment will be completed in 1996 before additional implementation is carried out.

Implementation began in Fall 1995, with community-based

restoration efforts as the first steps. The plan's anticipated temporal scope is twenty-five years.

Because of a pine-beetle outbreak on one of the managed (planted) pine stands, the original proposed treatment—to gradually thin and convert the stand to native bottomland hardwoods—has effectively been eliminated. This intermediate step was foregone in order for a salvage operation of the dead trees to take place.

Factors Facilitating Progress

The TVA employee's initiative, effort, dedication, and vision have been instrumental to this effort. He has access to excellent resources, from TVA-based GIS and aerial photography assistance to extensive library and electronic resources. The support and flexibility of higher management has also been a benefit. Finally, public involvement has been instrumental: the public was described by the project coordinator as intelligent and willing to work with TVA toward a common vision. Working with the public "on their terms" has been especially helpful.

Obstacles to Progress

Continuing uncertainties surrounding federal budgets and downsizing efforts have made this project's future unclear. The scale of the project area is too small for the complete ecosystem to be managed in its entirety. Rather, the effort is considered to be an ecosystem-based approach to managing the lands, still an important distinction from past management practices. Expanding these efforts to private lands is expected to be difficult, especially in terms of public education of the many landowners, relatively small lot sizes, and the fact that most landowners focus on production potential of their land.

Contact information:

Mr. J. Ralph Jordan
Senior Natural Resource
  Management Specialist
Tennessee Valley Authority
Ridgeway Road
Norris, TN 37828
(423) 632-1604
Fax: (423) 632-1534

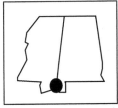

GRAND BAY SAVANNA

Location:
Southwestern Alabama, southeastern Mississippi

Project size:
60,000 acres

Initiators:
The Nature Conservancy, U.S. Fish and Wildlife Service

PROJECT AREA DESCRIPTION

The Grand Bay Savanna is a highly specialized coastal wetland located between Mobile, Alabama, and Pascagoula, Mississippi. This largely pristine ecosystem is bounded by the Gulf of Mexico and the Mississippi Gulf Coast to the south and east, and stretches inland to the Pascagoula River (north of Interstate 10). Considered one of the most diverse habitats in North America, the ecosystem forms a gradient from open long-leaf and slash pine savanna and "flatwoods," through pond cypress savanna, brackish and saltwater marshes, out to the Mississippi Sound and protective barrier islands beyond. It is rich in waterfowl, wading birds, shore birds, and fish, with over twenty plants considered rare by both states. It contains one federally listed endangered species, the Mississippi sandhill crane, and two candidates for the federal list. This alluvial system provides water recharge for the regional aquifer and estuary, the latter supporting commercial and sport fishing.

Little human development has occurred other than minimal agriculture and timber, both of which were more extensive historically, but failed due to the poor soil and drainage conditions. Landownership is primarily private, with public lands being acquired by the U.S. Fish and Wildlife Service (FWS).

ECOSYSTEM STRESSES

As a fire-dependent habitat, fire suppression has been the most significant stress to the ecosystem. Suppression occurred where conversion of land to other uses took place, as well as from general policies reflected by public awareness efforts such as the "Smokey The Bear" campaign. Although sparsely populated now, urban development and associated non-point source pollution are the greatest potential threats, due to pressures from nearby Mobile and Pascagoula. Less serious but still present are point source pollution stresses from industrial (chemical) complexes and hydrologic alteration of the wetland.

PROJECT DESCRIPTION

Mississippi has already recognized the savanna as a significant wildlife resource area. Also, it is included in the Gulf Coast Joint Venture, as authorized by the North American Waterfowl Management Plan. First efforts to protect the area began in 1989 when 2,649 acres of private land were transferred from private ownership to FWS for protection. In 1990, The Nature Conservancy (TNC) designated the savanna as one of its "Last Great Places." In addition to FWS and TNC, participants in the effort include the Mississippi Department of Wildlife, Fisheries, and Parks; the Alabama Department of Conservation and Natural Resources; and local county officials from Jackson, Mississippi, and Mobile, Alabama, and others.

FWS is currently acquiring land for the new 13,000-acre Grand Bay Savanna National Wildlife Refuge along the Mississippi-Alabama border, at the core of the ecosystem. Similarly, the Mississippi Department of Marine Resources is acquiring land to be protected in the Bangs Lake area.

Region-wide project goals have not yet been finalized, although FWS has developed specific objectives associated with the creation of the refuge, including establishing a nesting site for the endangered Mississippi sandhill crane.

The biggest change in management in the area will be restoring the natural fire regime. TNC's effort is expected to last approximately twenty years. Coordination between the institutions has been through informal meetings; no formal coordination body has been formed. While a monitoring plan has not yet been finalized, it is expected to include monitoring factors such as burning, point and non-point source pollution, visitation by humans, and the amount of land under conservation easements.

STATUS AND OUTLOOK

FWS acquisition for the new refuge is approximately 50 percent complete and state acquisition efforts are also under way.

The area has been designated by TNC as one of its official bioreserves and has been included in its "Last Great Places" program, thus elevating the area to a higher priority level within the organization.

Factors Facilitating Progress

Interest by FWS in creating a new refuge and acquiring land has been especially important to this effort. Congressional support has played a key role in securing federal land acquisition funds through the Land and Water Conservation Fund. Interest and dedication of resources by TNC, including geographic information systems (GIS) resources from its southeast regional office, clearly allowed this effort to begin and proceed.

Obstacles to Progress

Lack of resources—funding, personnel, and expertise—presents the greatest obstacle to this effort's progress. Opposition by a small but highly-vocal group of private landowners to federal acquisition efforts has also presented obstacles. More fundamentally, there is a distrust by the public of any federal agency. TNC is attempting to alleviate the situation by acting as a go-between and by engaging in public education efforts to inform residents of the socioeconomic or cultural importance of preserving this area, such as its water quality, fisheries, and recreational potential.

Contact information:

Mr. Pat Patterson
The Nature Conservancy
P.O. Box 1028
Jackson, MS 39215-1028
(601) 355-5357

GREATER YELLOWSTONE ECOSYSTEM

Location:
Northeastern Wyoming, southern Montana, eastern Idaho

Project size:
19 million acres

Initiators:
National Park Service, U.S.D.A. Forest Service

PROJECT AREA DESCRIPTION

Described as one of the last relatively intact temperate zone ecosystems in the world, the Greater Yellowstone Ecosystem (GYE) consists of forested mountains surrounded by undeveloped prairies and basins. The GYE is dominated by a high plateau at its center, essentially Yellowstone National Park (YNP).

The ecosystem was first delineated in 1979 based on the ranges of animals such as the grizzly bear, elk, trumpeter swan, and mountain lion, as well as geologic, hydrologic, and other abiotic features. Located in the park at 8,000 feet altitude, Yellowstone Lake is the largest high elevation lake on the continent. The GYE is the largest thermal basin in the world, with over 200 geysers and 10,000 geothermal features. The headwaters of three major rivers, the Yellowstone-Missouri, Green-Colorado, and Snake-Columbia, are found there.

Eighty percent of the ecosystem is forested, mostly in lodgepole pine, along with limber and whitebark pine, Douglas-fir, subalpine fir, and Engelmann spruce. Over seventy mammals are present, including large and wide-ranging animals such as moose, bison, antelope, grizzly bear, mountain sheep, and deer. Over 300 bird, 24 amphibian and reptile, 10 fish, and numerous invertebrate species inhabit the GYE. Several federally listed threatened and endangered species can be found there, including the Kendall Warm Springs dace (a fish), the migratory whooping crane, grizzly bear, peregrine falcon, and the recently reintroduced gray wolf.

The GYE is in public ownership primarily, with more than twenty-eight federal, state, and local agencies managing its resources; the U.S.D.A. Forest Service (USFS) and National Park Service (NPS) control over 80 percent of the GYE. Private landowners include industrial forest owners, ranches (working and residential), and limited rural development. Primary uses of the area are wilderness, recreation, timber, grazing/ranching, mining (hard rock, oil/gas, geothermal), and hydrologic.

ECOSYSTEM STRESSES

Resource extraction has been the predominant source of stress for most of this century on GYE lands outside of Yellowstone National Park. Logging, oil and gas development, hardrock mining, and grazing top the list. Several water development projects, for impoundments, hydroelectric facilities, channelization, and irrigation, alter the hydrology of many streams and rivers.

All of these activities destroy, alter, and fragment habitat; produce toxic wastes; cause erosion; reduce water quality through point source and non-point source pollution; and result in the introduction or spread of non-native species, all with negative consequences on the GYE's flora and fauna. While these activities have declined since the late 1970s, they are still occurring on a significant scale. Motorized and non-motorized recreation and subdivision development pose additional threats.

PROJECT DESCRIPTION

High levels of resource extraction in the GYE throughout this century, along with the imperilment of high visibility species and public and scientific concern for the ecosystem's health, eventually led, in 1986, to Congressional pressure on NPS and USFS to coordinate their activities in accordance with ecosystem management principles.

As a result, the Greater Yellowstone Coordinating Committee (GYCC), created in 1964 but generally inactive since then, was revived in 1986 and a memorandum of understanding (MOU) was signed between NPS and USFS. Later, the Bureau of Land Management and U.S. Fish and Wildlife Service were included on the GYCC. The MOU laid out a two-step plan, with the creation of a vision document to describe desired future conditions of the GYE and how they can be achieved as the first step. Subsequent to public review, the second step of the plan was to amend USFS and NPS plans, guides, and statements to achieve those goals.

The first-step report was released in August 1990, and contained three goals, several subgoals, and ninety-five coordinating criteria. The goals were to 1) "conserve a sense of naturalness and to maintain ecosystem integrity;" 2) "conserve opportunities that are biologically and economically sustainable;" and 3) "improve coordination."

Following enormous outcry from special interest organizations,

particularly resource extraction interests, and political opposition, the report was drastically revised and shortened, with the original emphasis on coordination giving way to an emphasis on the separate and distinct missions of the NPS and USFS.

The initial GYCC plan was developed internally by USFS and NPS, with limited public involvement (public meetings, comment periods).

STATUS AND OUTLOOK

Since the final GYCC report was released in 1991, little has happened in terms of coordinated management among public agencies. No overriding management plan has been developed for the GYE. Many of the existing plans of each agency for their lands or jurisdiction remain in conflict with one another and do not reflect attempts to incorporation ecosystem management principles. Over the last two years, there have been limited attempts to revive the ecosystem management focus, but these have been restricted to data-oriented efforts such as developing a region-wide geographic information systems (GIS) database.

Factors Facilitating Progress

Public attention and the uniqueness of the area were the most important factors facilitating the initiation and early stage devel-

opment of the GYCC effort. In particular, studies of high visibility megafauna, the wildfires of 1988 throughout the ecosystem, and in general, the high level of attention that the park had been receiving, were especially important.

Obstacles to Progress

Politicization of the GYCC process appears to have crippled the effort, driven in large part by resource extraction interests. Continued lack of coordination between agencies, stakeholder conflict, disagreements over a common vision and appropriate uses of the GYE resources, conflicting agency mandates, and a lack of scientific information about those resources have been major factors preventing the development of a comprehensive coordination effort in the GYE. Finally, national attention of ecosystem management applications has shifted away from the GYE to other areas of the country such as the Everglades and North Woods.

Contact information:

Greater Yellowstone Coalition
P.O. Box 1874
Bozeman, MT 59715
(406)586-1593

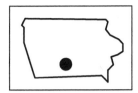

GREEN VALLEY STATE PARK ECOSYSTEM MANAGEMENT PLAN

Location:
Southwestern Iowa

Project size:
2,300 acres

Initiator:
Iowa Department of Natural Resources

PROJECT AREA DESCRIPTION

Green Valley State Park is located in the rolling hills of southwestern Iowa. In presettlement times, this area was covered by tallgrass prairie. However, by the time the Iowa Department of Natural Resources (IDNR) purchased this property in 1950 and 1951, only patches of prairie remained, surrounded by blue grass pasture. Currently, the vegetation of the park consists of brome grass, tallgrass prairie patches, and encroaching hardwoods and red cedars. The park also sports a 394-acre artificial lake. Although the original intent was to supply water to the town of Creston, it is now used primarily for recreation and as a cooling water supply for a nearby power plant. Included in the project area are approximately 1,300 acres of lands surrounding the 1,000-acre park. These lands are primarily covered by bluegrass pasture and row crops. The project area does not contain any federally or state listed threatened or endangered species.

ECOSYSTEM STRESSES

The greatest stress on the ecosystem is posed by livestock and crops on the lands surrounding the park. Through drainage and runoff, nutrients such as cow manure and crop fertilizers end up in the lake. In addition, pesticides affect the lake and fishery quality and harm the prairie vegetation on the outskirts of the park. Until the current introduction of fire as a management tool, fire exclusion was also a significant stress to the ecosystem, resulting in woody encroachment of the prairie vegetation.

PROJECT DESCRIPTION

In 1994, the IDNR adopted a new mission statement which implicitly required long-range ecosystem manage-ment planning for Iowa's state parks. In response, a departmental committee was established whose main task was to design a statewide framework for the preparation of plans for all individual state parks. Based on the committee's work, the individual ecosystem man-agement plans were to be developed.

Because of its highly disturbed nature, Green Valley State Park was chosen as the first park for which a plan was to be designed. A management planning team was formed consisting of an executive officer, park ranger, and IDNR staff representing the disciplines of forestry, wildlife management, and community ecology. The plan was written during the Summer of 1994 with the input of team members, the U.S.D.A. Natural Resources Conser-vation Service (NRCS), private individuals (through public meetings), and park visitors (through user questionnaires). Implementation of the plan started in the Fall of 1994.

The plan has divided the park into nine management units and makes separate management recommendations for each of these units. In order to maintain tall-grass prairie, woody encroachment will be removed by cutting and fire. In other areas, trees are to be planted. In order to protect the water quality of the lake, the IDNR is working with existing pro-grams, such as the Clean Lakes Project of the U.S. Environmental Protection Agency and NRCS programs. These programs can assist surrounding land-owners in the development and use of erosion control practices and methods to decrease the amount of nutrients from feedlots reaching the lake. Another strategy involves the acqui-sition of lands surrounding the park

park from willing sellers as funds become available.

The following factors will be monitored: lake water quality, siltation of the lake basin, eutrophication, plant species composition, and animal populations. Monitoring results will be taken into account when the ecosystem management plan is revised, which will occur once every five years.

STATUS AND OUTLOOK

Even though this ecosystem management plan began relatively recently, some ecological results have been observed. Controlled burns have occurred, resulting in a reduction of woody encroachment, especially of red cedar.

Several ecosystem management plans need to be developed for other state parks. As a result of the Green Valley pilot project, the planning process has been improved, with greater use of geographic information systems (GIS) graphics and data in subsequent plans.

Factors Facilitating Progress

Especially helpful during the planning process were the management planning team's multidisciplinary approach, administrative commitment, and staff interest.

Obstacles to Progress

The limited amount of knowledge concerning animals other than birds and mammals has been a problem. In addition, the perception of the public that a state park is supposed to be a forest represents a hurdle in the implementation of the plan. Lack of staff and funding have been and will continue to be a concern. For example, a natural areas staff person whose interests were particularly well suited to this effort recently left the IDNR for a position elsewhere.

Contact information:

Mr. Jim Scheffler
Wetland Project Coordinator
Iowa Department of Natural
  Resources
Wallace State Office Building
Des Moines, IA 50319
(515) 281-6157
Fax: (515) 281-6794

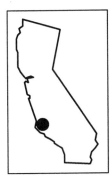

GUADALUPE-NIPOMO DUNES PRESERVE

Location:
California central coast

Project size:
3,897 acres (12-mile stretch of 18-mile dune complex)

Initiators:
California State Coastal Conservancy, The Nature Conservancy

PROJECT AREA DESCRIPTION

The Guadalupe-Nipomo dunes complex on California's central coast is a nearly eighteen-mile stretch of beaches, sand dunes, wetlands, and freshwater lakes. This dune ecosystem comprises three dune groups: the Callender, Guadalupe, and Mussel Rock dunes. Shrublands, arroyo willow woodlands, and an estuary at the mouth of the Santa Maria River add to the great diversity of natural communities of the Guadalupe-Nipomo Dunes. The dunes provide critical nesting habitat for shorebirds and are home to a number of federally and state listed threatened and endangered animal and plant species, including the California least tern, western snowy plover, Gamble's watercress, and marsh sandwort. Many rare species can be found there as well. Currently managed by The Nature Conservancy (TNC), the Guadalupe-Nipomo Dunes Preserve provides passive recreational opportunities such as hiking, surf fishing, horseback riding, and other non-motorized activities. Neighboring land uses include oil development, sand and gravel mining, agriculture, and livestock grazing.

ECOSYSTEM STRESSES

The dunes attract nearly 120,000 visitors each year. The trampling of rare plant species and the disturbance of rare shorebirds that nest on the open sand are serious concerns. In the midst of the preserve and adjacent to the Santa Maria River estuary sits a 2,300-acre oil field which, over time, has spilled an estimated 8.5 million gallons of oil. The effects of this spill on the ecosystem are presently unknown. On the edge of the preserve, increasing extraction from a sand mine may be exceeding the deposition of sand from the Santa Maria River, thus removing dune habitat from the preserve. Finally, urban development on neighboring lands has promoted the introduction of exotic plants, such as European beachgrass and veldt grass, which have reduced the abundance of native vegetation and corresponding habitat.

PROJECT DESCRIPTION

Of the four intact dune complexes remaining in the state of California, Guadalupe-Nipomo Dunes is the most diverse. Because of the absence of protective management for the dunes, the California State Coastal Conservancy and TNC began a land acquisition program in 1988 for the creation of the Guadalupe-Nipomo Dunes Preserve. The Coastal Conservancy purchased the Mobil Coastal Preserve from the Mobil Corporation, later transferring the preserve to TNC for long-term management. TNC also holds a twenty-five-year lease on Mussel Rock Dunes, owned by the Santa Barbara County Parks Department, and has an agreement with the Off-Highway Vehicle Division of California State Parks giving TNC prescribed rights to manage its Oso Flaco Lake property as a natural area. Consulting with each of these partners, TNC's goal is to balance human use with protection and restoration of the natural communities and native species of the dunes.

In 1991, TNC developed a management plan which instituted new regulations on activities as well as an entrance fee for the preserve. Developed and implemented with little public input or notice, this plan met with strong public protest, and public officials rescinded the fee in response. This was a major setback for TNC: the fee was to have provided income that would make this project self-sustaining. TNC realized from this experience that to protect natural systems, management plans must be both land and culturally based.

STATUS AND OUTLOOK

As of late 1995, TNC has been engaged in a user-analysis study which was to result in a revised management plan. Outreach efforts are being expanded, for instance, to include cooperative activities with school groups. Community outreach is a vital element of this study in order to learn about the needs and desires of the people who use and live near the preserve. Working with a local steering committee, the Coastal Conservancy and TNC are also developing a visitors center in the small town of Guadalupe near the preserve, which was expected to be open by Spring 1996. It is hoped this center will provide a stimulus for the local economy and foster a sense of stewardship toward the dunes. The exotic grass locations on the dunes have been mapped, and TNC is seeking funding for an exotic plant species eradication program.

Factors Facilitating Progress

The construction of boardwalk trails and signs have been successful in protecting rare plant communities and least tern colonies. Furthermore, education and outreach programs have increased support from local communities.

Obstacles to Progress

Animosity still remains toward TNC as a result of certain elements in the original management plan for the dunes; a great deal more local support is needed for further progress on this project. Furthermore, the project is in need of greater administrative and financial support. Finally, the project is described as being in need of an advisory committee which should include all stakeholders, in order to improve communication and coordinated management efforts. Such a committee may emerge from the revised management plan.

Contact information:

Ms. Nancy Warner
Field Representative
The Nature Conservancy
PO Box 15810
San Luis Obispo, CA 93406
(805) 545-9925
Fax: (805) 545-8510

GULF OF MAINE RIVERS ECOSYSTEM PLAN

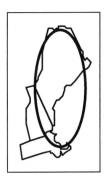

Location:
*Maine,
northwestern
Massachusetts,
eastern New
Hampshire*

Project size:
44,233,600 acres

Initiator:
*U.S. Fish and
Wildlife Service*

PROJECT AREA DESCRIPTION

The Gulf of Maine Rivers ecosystem consists of a network of rivers, forests, lakes, wetlands, and estuaries. This watershed includes the entire state of Maine, portions of the states of New Hampshire and Massachusetts, and the estuarine and marine waters of the Gulf of Maine. This area provides vital natural habitat for the U.S. Fish and Wildlife Service (FWS) trust resources such as anadromous fish, endangered and threatened species, and migratory birds. The watershed also sustains major forest-related fishing, tourist, and recreation industries. Portions of the Canadian provinces that are in the Gulf of Maine watershed are not addressed in this ecosystem management plan.

ECOSYSTEM STRESSES

Threats to the integrity of the Gulf of Maine Rivers ecosystem include wetland loss, blockages of rivers by dams, habitat fragmentation, pollution, and other cumulative effects of development. In the future, the threats to specific geographical focus areas identified in the plan will be further researched and defined.

PROJECT DESCRIPTION

This project is one of the many watershed-based initiatives that are part of the FWS transition to ecosystem-based land management. The project boundaries for this ecosystem, as well as the boundaries of the other fifty-one projects, were set by the national and regional offices of FWS in early 1994. An ecosystem team was set up, with a rotating team leader, and consisting of regional office liaisons and field station project leaders who are responsible for day-to-day operations in the watershed.

The team met initially in May 1994. Through consensus and with the help of trained facilitators, the team developed a draft ecosystem plan for the Gulf of Maine Rivers project in August 1994. The plan includes a mission statement and six goals reflecting the region's trust responsibilities. Examples of the goals include, among others, 1) a healthy aquatic community; 2) a healthy, diverse, and functioning wetlands and associated habitats; and 3) an informed public which values fish and wildlife. Seven resource priorities are further identified, defining action strategies to address specific geographical focus areas. Potential partners have been named to aid in the implementation of each of the resource priorities. These partners include nonprofit organizations, local governments, and community groups.

STATUS AND OUTLOOK

The ecosystem plan was revised in Spring 1995 and the project has moved into an implementation phase. The team continues to meet quarterly. Agencywide, FWS has reorganized its structure along geographic lines away from a programmatic structure and line of authority. Although the plan is a long-term initiative which will be realized over many years, one of the immediate successes is increased communication between the FWS offices within the project area. There is an increased understanding of the different responsibilities of the offices and less duplication of effort.

Finally, a regional-level ecosystem coordinator has been designated to assist the team leaders in their administrative responsibilities.

Factors Facilitating Progress

Facilitated team meetings are helpful and enabled the project team leader to concentrate on developing the plan. Facilitation also helped in clarifying and resolving some of the initial concerns of team members regarding infringement on prior responsibilities.

Obstacles to Progress

A major obstacle in reaching the overall FWS goal of an ecosystem-based approach is the current program-by-program funding structure of Congress. It is difficult to receive funding on a program basis while managing on an ecosystem level.

The ambitious initial deadlines set by the national office for completion of the plan were also problematic, since these coincided with an already active summer field season.

A lack of communication on the intentions of the ecosystem management plans caused some confusion at first. The field believed that the first draft of the plan was only to address activities that would received new funding; ongoing FWS priorities and activities were omitted. On the other hand, budget planners in the national office understood that the plans addressed all activities, and that omitted activities were done so deliberately and thus could be subject to cuts in funding. This problem has been corrected with the Spring 1996 plan revision, which is expected to include all FWS activities and priorities.

In some of the affected states, state fish and wildlife agencies with whom FWS cooperates initially felt excluded from the priority-setting process. But the states eventually recognized that this ecosystem planning process was meant to address federal trust responsibilities for FWS, and was not intended to exclude states in all areas of activity.

Finally, the 1995-1996 federal government shutdowns have slowed progress on some activities.

Contact information:

Mr. Vic Segarich
Ecosystem Team Leader
U.S. Fish & Wildlife Service
Nashua National Fish Hatchery
151 Broad Street
Nashua, NH 03063
(207) 827-5938
(603) 886-7719

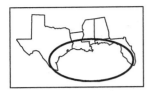

GULF OF MEXICO PROGRAM

PROJECT AREA DESCRIPTION

The U.S. Gulf of Mexico coast line is more than 1,600 miles long. About two-thirds of the continental U.S. drains into the gulf, with the Mississippi River watershed accounting for most of that drainage. Gulf coastal areas include sand beaches, freshwater and saltwater marshes, barrier islands, mangrove swamps, and seagrass beds. Many federally listed threatened and endangered species occur in the area, such as the Mississippi sandhill crane, Alabama beach mouse, and several sea turtles. In addition, the gulf provides essential habitat for a large portion of North American migratory waterfowl.

Shrimping, recreation, and oil and gas offshore production occur in the gulf. Agriculture, silviculture, aquaculture, industrial uses, and urban/rural development occur in the gulf's watershed.

ECOSYSTEM STRESSES

Non-point source pollution generated throughout the Mississippi River watershed ends up in the gulf, including pesticides, toxic substances, and nutrients. In addition, over one million pounds of trash and debris are picked up on gulf beaches annually. Coastal and shoreline erosion is another serious problem. Parts of Louisiana retreat sixty-five feet per year or more, while shoreline erosion rates of fifteen feet per year can be found in many other areas of the gulf. Freshwater diversion projects initiated for flood control, navigation, recreation, and water supply have led to reduced levels of freshwater entering coastal areas, with subsequent saltwater intrusion.

Location:
Gulf of Mexico coast in Florida, Alabama, Mississippi, Louisiana, and Texas

Project size:
410 million acres

Initiator:
U.S. Environmental Protection Agency

PROJECT DESCRIPTION

The Gulf of Mexico Program was established in August 1988 by the U.S. Environmental Protection Agency (EPA) in response to signs of serious long-term environmental damage appearing throughout the marine ecosystem of the gulf. The program is an interagency effort which aims to 1) protect, restore, and enhance the coastal and marine waters of the Gulf of Mexico and its coastal natural habitats; 2) sustain living resources; 3) protect human health and the food supply; and 4) ensure the recreational use of Gulf shores, beaches, and waters occur in ways consistent with the economic well-being of the region. The major strategies of the program are to: identify threats to the gulf; suggest solutions (in action plans); and identify and, if needed, fund the appropriate agencies and institutions that could implement these solutions as individual projects.

Examples of projects promoted by the program are education and outreach, garbage pick-up programs, and shellfish bank restoration (through reduction of sediment reaching the shellfish banks). Monitoring varies from project to project. In some, wetland vegetation is monitored. In others, water quality, toxicity, and pathogens are measured.

STATUS AND OUTLOOK

Many partnerships have been developed since the program's inception. For instance, in 1992, a "Partnership for Action" document was signed by governors of all five gulf states, representatives from eleven federal agencies, and the chair of the Citizens Advisory Committee to the program. In February 1995, the federal partners, now numbering eighteen, signed a more detailed partnership document outlining individual tasks and contributions. The Gulf of Mexico Program is drawing up a similar partnership document with the states. Action plans have been developed addressing many Gulf of Mexico issues, including marine debris, habitat loss, freshwater inflow, and nutrient enrichment.

The program has funded approximately 200 projects so far, including projects that demonstrate the use of wetlands for filtration of domestic, agricultural, and urban waste water to reduce impacts on shellfish-growing waters. In addition, the program

has facilitated restoration of 600 acres of coastal habitat in cooperation with the Tampa Bay Estuary Program and the state of Florida.

Another result of the Gulf of Mexico Program is the development of new data and information management tools, such as the Gulf of Mexico Program electronic bulletin board system and user-friendly geographical information systems (GIS).

Factors Facilitating Progress

Multi-agency, multi-state partnerships are the strength of the program. They have resulted in an exchange of information as well as in shared funding. Working through consensus building has also been beneficial to the program.

Obstacles to Progress

The Gulf of Mexico ecosystem is very large. To address all threats in this ecosystem would require significant amounts of funding. Therefore, priorities must be set and addressed. In addition, because of the large geographical area involved, the project must deal with a wide array of political, geographical, cultural, and economic interests.

Contact information:

Dr. Douglas A. Lipka
Director
Gulf of Mexico Program
Building 1103, Room 202
Stennis Space Center, MS 39529
(601) 688-1172
Fax: (601) 688-2709

HUDSON RIVER/NEW YORK BIGHT ECOSYSTEM

Location:
New York, New Jersey; small portions of Connecticut and Massachusetts

Project size:
11.5 million acres

Initiator:
U.S. Fish and Wildlife Service

PROJECT AREA DESCRIPTION

The project area lies between Montauk Point, Long Island, and Cape May, New Jersey. These watersheds drain the southern half of Long Island, the Atlantic coast drainage of New Jersey, the entire drainage of New York Harbor and the Hudson River drainage area.

These watersheds are complex in physiography and include marine zones, barrier beaches, coastal plains, mountainous highlands, and the Hudson River valley. The headwaters of the Hudson are found in the Adirondack region of New York State. Vegetation ranges from three of the most significant pitch pine barrens in the world, to remnant habitats of grassland, oak-dominated deciduous forests, and significant tidal and freshwater wetlands. The region contains twenty-nine federally listed threatened and endangered species and approximately 200 state listed species.

ECOSYSTEM STRESSES

The greatest threat to the area is land conversion to urban uses, especially outward from the New York City metropolitan area, resulting in fragmentation of habitat and loss of open space. A second major stress is non-point source pollution from urban and agricultural areas. This includes runoff from roads, sewers, farms, and storm water. Runoff is exacerbated by degraded riparian areas.

PROJECT DESCRIPTION

This project is one of the many watershed based initiatives that are part of the U.S. Fish and Wildlife Service's (FWS) transition to ecosystem-based land management initiated in March 1994. In May of that year, a Hudson River/New York Bight ecosystem team began meeting to define priorities and action steps and to write an ecosystem plan. The ecosystem team consists of one representative from all FWS offices within the ecosystem. The New York Department of Environmental Conservation and the New Jersey Department of Environmental Protection became involved in project planning in September 1994. They have helped to set priorities and identify partners within the states for the various action steps.

Within the ecosystem plan, six priority goals are identified, which are defined by community or habitat types that are considered priority areas. These include barrier beaches, grasslands and forests, coastal lagoons and rivers, tidal and freshwater habitats, and pine barren communities. The plan contains a total of twenty-one action steps (three to four for each priority goal). Monitoring is based on these action steps. Ecological indicators of habitat and species return will be used to measure progress in restoration activities. For other action steps, more quantitative measures are used, such as number of acres transferred from lawn to natural habitat.

STATUS AND OUTLOOK

Several action steps specified in the plan were existing FWS activities. As a result, the team predicts that this effort will produce results in the near future. More recently developed action steps, currently in the data compilation and mapping phase, will require a few years before management plans are in place and on-the-ground actions underway.

The ecosystem plan was revised in September 1995, with a net increase in action strategies (others were dropped). An accomplishment report was also produced as part of the revision. Many acres of wetlands and grasslands have been restored; mapping is still ongoing. Agencywide, FWS has reorganized its structure along geographic lines away from a programmatic structure and line of authority.

Factors Facilitating Progress

In FWS Region 5, some FWS offices traditionally have had a coastal focus, thus facilitating the progression to the current watershed approach. Facilitated team meetings also helped to move the process along efficiently. The team was extremely interested and willing to see this project move forward.

Obstacles to Progress

Initially, the team functioned under tight deadlines and could not gain all the staff level input and perspectives that would have fostered a fuller complement of information in the plan. A second potential obstacle is streamlining the process so that reporting requirements do not become overly burdensome. With federal downsizing efforts, it may be difficult for offices with many mandated activities to find time and funds to work on priorities identified in the plan.

Finally, the 1995-96 federal government shutdowns, combined with severe funding limitations and recent congressional budget battles in Washington, have hindered this effort's progress significantly. As of early 1996, it was still unclear how much funding, if any, this project would receive. The upheaval caused by these factors, as well as challenges of reorganizing the agency's structure, could become demoralizing to personnel.

Contact information:

Ms. Elizabeth Herland
Ecosystem Team Coordinator
U.S. Fish & Wildlife Service
Walkill River National Wildlife
  Refuge
P.O. Box 383
Sussex, NJ 07461
(201) 702-7266
Fax: (201) 702-7286

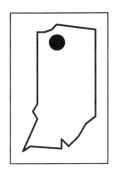

INDIANA GRAND KANKAKEE MARSH RESTORATION PROJECT

Location:
Northwestern Indiana

Project size:
26,500 acres

Initiators:
U.S. Fish and Wildlife Service, Indiana Department of Natural Resources

PROJECT AREA DESCRIPTION

In presettlement times, the Kankakee River watershed was dominated by the Grand Kankakee Marsh, a flat, 500,000-acre expanse vegetated by sedges, cattails and wild rice. Scattered throughout the marsh, sand ridges could be found, supporting black oaks, blueberries, and prairie-fringed orchids. Upland areas adjacent to the marsh were characterized by tallgrass prairie, beech-maple forests, and oak-hickory forests.

The project area contains twelve federally listed threatened, endangered, or candidate species and more than 200 state listed species. Examples of the former include the Karner blue butterfly, prairie-fringed orchid, Mitchell's satyr butterfly, Indiana bat, bald eagle, and peregrine falcon. During spring and fall migrations, the area is frequented by migratory waterfowl, including nearly the entire population of sandhill cranes.

ECOSYSTEM STRESSES

The most important stress on the ecosystem is land conversion to agriculture and concomitant loss of natural habitats such as marsh, prairie, and forest. In 1918, the Kankakee River, once a 240-mile long, meandering stream, was straightened and is now only 85 miles long. This was followed by drainage of virtually the entire marsh and the conversion of land to row crop agriculture. Except for some remnants, all the prairie land has been converted to agriculture as well. Approximately 50 percent of the forests have undergone a similar fate. The remaining forests are used for timber management. Part of the area is used for residential, municipal, and industrial purposes.

PROJECT DESCRIPTION

As early as the 1930s, efforts were made to restore the Grand Kankakee Marsh. Although most of these efforts have been unsuccessful, local residents have remained interested in restoration.

In January 1993, the U.S. Fish and Wildlife Service (FWS) and the Indiana Department of Natural Resources arranged an organizational meeting aimed at initiating a project under the North American Waterfowl Management Plan. Many interested organizations, such as Ducks Unlimited and Waterfowl USA, were invited. At subsequent meetings, more partners were added to the project, which now numbers fourteen different organizations, businesses, and agencies. The partners all contributed funding, land donations, or in-kind services to the project, at a total value of $2.3 million. In April 1994, they submitted a proposal requesting $1.5 million under the North American Wetlands Conservation Act to be added to the partner's share. The grant was funded in September 1994, and, in conjunction with the partners' contributions, will be used to fund the first two years of the project.

The ten-year project goal is the restoration and protection of 26,500 acres of wetlands and associated uplands in the watershed. The strategies that are employed to reach this goal include acquisition and easements, restoration of wetlands and upland prairies, ongoing fundraising, and development of public awareness of the project. Land acquired in the context of this project will be owned and managed by the various project partners. For instance, a parcel may be owned and managed by the

Lake County Parks Department in accordance with project goals.

STATUS AND OUTLOOK

Even before the funds were appropriated, thirty properties had been identified for potential acquisition and restoration. Properties are prioritized based on restoration potential and cost of acquisition. Six properties totalling about 1800 acres have been acquired or are in the final stages of purchase. Wetland restoration has begun on these properties, to be completed by Summer 1996.

The initial two-year project goal for land protection has been surpassed. As a result, a second proposal requesting additional funds under the North American Wetlands Conservation Act was to have been submitted in April 1996. Finally, an additional $976,000 has been contributed to the project by partners, including actual funds for land acquisition and restoration and in-kind land and construction donations.

Factors Facilitating Progress

Because the entire ecosystem is addressed rather than a single issue, many diverse organizations have become interested and are now participating in the partnership. The cooperation of a large number of partners and a good source of local funding have been very helpful and are considered a success in themselves. In addition, the media has assisted the partners by keeping the public informed about the merits of the project.

Another facilitating factor is the land base itself. Ever since the Indiana Grand Marsh was drained, the drainage system has been deteriorating. It has become more and more expensive to maintain the land as cropland. Thus, many properties are available for acquisition. In addition, the potential of the area to react positively to ecosystem management is helpful.

Obstacles to Progress

It is uncertain how the current Congress will influence the funding of the North American Wetlands Conservation Act. Even if funding remains at the same level, increasing requests for grants throughout the country may cause the amount of funding for each individual project to decrease. Since land values are higher than was originally estimated by the partnership, securing continued non-federal funding is of the utmost importance.

Contact information:

Mr. Jim Ruwaldt
Assistant Field Supervisor
U.S. Fish & Wildlife Service
620 S. Walker
Bloomington, IN 47403-2121
(812) 334-4261 ext. 213
Fax: (812) 334-4273
E-mail:
   James_ruwaldt@mail.fws.gov

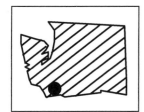

INTEGRATED LANDSCAPE MANAGEMENT FOR FISH AND WILDLIFE

Location:
Southwestern Washington initially; statewide eventually

Project size:
839,010 acres

Initiator:
Washington Department of Fish and Wildlife

PROJECT AREA DESCRIPTION

The headwaters of the Lewis and Kalama rivers originate in the volcanic drains of Mt. St. Helens and Mt. Adams. The Lewis-Kalama watershed boundary stretches from the Kalama River and Mt. St. Helens to the north to Lookout Mountain in the south. Mt. Adams and Bachelor Island define the watersheds' eastern and western boundaries respectively. The project site also includes the Indian Heaven Wilderness Area and part of Mt. St. Helens National Monument.

Douglas-fir, hemlock, and western red cedar grow at lower elevations. Red alder, willow, big-leaf maple, and cottonwood are also found throughout the watersheds. The northern spotted owl and bald eagle are among the federally listed threatened and endangered species. The Larch Mountain salamander is listed as a sensitive species by the state.

ECOSYSTEM STRESSES

Eighty to 90 percent of the site has been used for commercial forest management. Timber harvesting constitutes the most significant stress to the watersheds by disrupting hydrologic patterns. Harvesting has also affected the integrity of wildlife habitat. Urban encroachment has occurred due to the growth of Vancouver, Washington, to the south and has reduced wildlife habitat. Three major dams have been the source of hydroelectric power generation in the region for fifty years.

PROJECT DESCRIPTION

The project began in 1989 as an internal effort led by the director of the Wash-

ington Department of Fish and Wildlife (WDFG). The director's decision to move away from species-by-species planning to landscape-level planning ("Integrated Landscape Management," or ILM) began as a statewide effort to inventory 129 priority species and map nineteen priority habitats (PHS). Geographic information systems (GIS) technology was used to create the maps for private and state forest lands, urban growth areas, and the Gifford Pinchot National Forest.

Preventing the listing of more species was a prime motivating factor. The goal of the Lewis-Kalama watershed pilot project is to demonstrate how the ILM process can be used to work with the public in managing fish and wildlife on a watershed basis. The project was started to institute a systematic planning process for taking PHS data and identifying the public's future desired conditions for managing fish and wildlife populations at the landscape level.

GIS/remote sensing data, public involvement, species plans, habitat plans, recreation plans, and the application of adaptive management are core elements of the department's approach. By compiling data on the status and trends of species as well as the current, potential, and future use of habitats, the department will develop options for public and private landowners to manage resources to their benefit while protecting species, habitat, and recreational opportunities.

STATUS AND OUTLOOK

Species and habitat data have been collected from the Lewis-Kalama watershed and a watershed plan has been produced. Since late 1995, WDFG has been working to develop site-specific action plans to resolve "hot spots" situations, where species-habitat compatibility is judged to be problematic. WDFG is meeting with public and private landowners to begin implementing the action plans.

In January 1996, an evaluation of the ILM planning process was completed and included recommendations for future changes in the process. This will lead to an operator's manual to be used by regional managers in applying the water shed planning process to Washington's sixty-one other watersheds. Finally, the most important requirement for continued project progress is agency commitment. The future of the ILM program was to be determined by the current budget process, during which the ILM program still must compete for appropriations with traditional department projects.

Factors Facilitating Progress

A highly-motivated public has been heavily involved in the planning process. The department regularly shares information and receives input from various groups including other state agencies, local environmental groups, hunting, fishing, and angling organizations, as well as industry groups such as Longview Fibre Company and Pacific Power and Light. An engaged public and key talented people in the agency are accredited for the project's progress.

Obstacles to Progress

It was difficult for the department staff to accept the shift from species-by-species to landscape-level management. Organizational difficulties such as the merger of two state agencies in March 1994 to form the current WDFG, allocating staff time, and overcoming resistance from employees have been significant obstacles in early stages of project planning. Past depart-

mental planning efforts have had mixed success and have caused some of the department's employees to be skeptical of the integrated landscape approach.

Finally, as the result of an initiative passed by voters in November 1995, from which the authority to appoint the department's director was returned from the governor to the state Fish and Wildlife Commission, a new director is expected by July 1996. The commission will also become more involved in the department's budget process, as well as goal and objective setting. It is unclear if and how these changes will affect the ILM program.

Contact information:

Mr. Rollie Geppert
Washington Department of Fish and Wildlife
600 Capitol Way North
Olympia, WA 98501-1091
(360) 902-2587

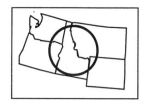

INTERIOR COLUMBIA BASIN ECOSYSTEM MANAGEMENT PROJECT

Location:
Eastern Oregon and Washington, Idaho, western Montana and Wyoming, northern Nevada and Utah

Project size:
144 million acres

Initiator:
President Bill Clinton

PROJECT AREA DESCRIPTION

The vast expanse from the crest of the Cascade Range to the Continental Divide defines the boundaries of the Interior Columbia Basin area. The Columbia River Basin, as well as parts of the Klamath Basin, the Great Basin, and Yellowstone National Park, are encompassed in this region. The varied topography includes high mountain alpine landforms, dissected plains, and the Columbia Plateau. Predictably, vegetation types vary widely. Ponderosa pine and mixed conifer forests in Washington and Oregon contrast to rangelands comprised of juniper, sagebrush, and bunch grasses. Wetter climates in Idaho and Montana yield a vegetation pattern of subalpine fir, white pine, and some lodgepole pine. Among the seventeen federally listed threatened and endangered species are the gray wolf, grizzly bear, Snake River salmon, and several plants. The area hosts as many as 200 candidate species to the list.

ECOSYSTEM STRESSES

The area has undergone tremendous alteration of its hydrologic system. A series of dams impound the Columbia River for power and recreation; a network of reservoirs and irrigation canals has been constructed to spur agricultural development and to control flooding. These stresses and overfishing have resulted in the decline of anadromous and other fish species.

The region's watersheds have been affected by excessive logging, road-building, and mining in concentrated spots. Stresses from a decade-long drought have exacerbated the risk of fire. Insect infestations and diseases have also plagued the forested lands recently. The proliferation of exotic species represents a substantial threat of further alteration of the ecosystem.

PROJECT DESCRIPTION

Changing social values concerning old growth and forest health, culminating with the Forest Summit in Portland, Oregon, in 1993, were a strong catalyst for the Interior Columbia Basin Project. Although President Clinton was unsuccessfully lobbied to include the Columbia Basin on the summit's agenda, he directed the Bureau of Land Management, the U.S.D.A. Forest Service, and other federal agencies to develop a scientifically-sound plan for the region's public lands. Agency staff on the project have conducted an inventory and assessment of what trends in resource use are occurring, how the trends and ecological conditions will change in the future, and what species, disturbance processes, and elements of the ecosystem are at risk.

The team created four long-term scenarios that highlight the social, economic, and ecological consequences and trade-offs if society chooses to 1) withdraw from public lands management and allow natural processes to occur without interference; 2) maximize the economic output of public lands; 3) focus on maintaining ecosystem processes and then distributing any excess benefits; or 4) continue with present management activities. Three additional scenarios are projected to be developed before the project's two-year charter expires in mid-1996. The report is targeted at decision makers in federal agencies.

STATUS AND OUTLOOK

The project's multi-disciplinary team has reduced the uncertainty in the region's natural resource decision making. The team is re-framing the question of resource management from "What would happen if . . . ?" to "Do we want this to happen?"

A draft environmental impact statement and several scientific documents are expected to be available for public review in June 1996. The project's future is unclear, as Congress was still debating in Spring 1996 whether or not to fund this initiative past June 1996 and if so, at what level.

Factors Facilitating Progress

A primary force for progress is the leadership of agency officials and their desire to avoid a repeat of the events that led up to the Forest Summit. Another has been the redefinition of "openness." Input from numerous federal and state agencies, interest groups, private landowners, and ordinary citizens is actively sought. The public is kept informed through presentations, open stakeholder meetings, the distribution of draft material, a computer bulletin board, and a toll-free telephone number.

Obstacles to Progress

Obstacles have included establishing working relationships with partners (largely due to constraints imposed by the Federal Advisory Committee Act), working within at least twenty federal land management jurisdictions, deciding the appropriate "turf" for science and management, and learning how to define and structure an open process. As reports are finalized, continued leadership is needed to buffer the political pressure to change the findings.

Finally, the Congressional budget debates in early 1996 has cast into doubt the future of this initiative.

Contact information:

Mr. Tom Quigley
U.S.D.A. Forest Service
Pacific Northwest Research Station
PO Box 3890
Portland, OR 97208-3890
(503) 326-5640

INTERIOR LOW PLATEAU

Location:
Central Kentucky, central Tennessee, northern Alabama

Project size:
30 million acres

Initiators:
State wildlife agencies, Tennessee Conservation League

PROJECT AREA DESCRIPTION

The project area is a physiographic province bounded by the Cumberland Plateau to the east, the Tennessee River to the south and west, and southern Ohio to the north. Local experts call it a "catch-all" area for various eastern landforms. It includes sub-units such as Bluegrass, Knobs Coal Fields, Penny-royal, and Highland Rim. It has moderate elevation and climate, as well as large rivers and tributaries, the latter generally having dictated landforms. A common, relatively consistent forest type has been oak-hickory (including savannas and bottomland hardwoods) and prairies.

ECOSYSTEM STRESSES

Stresses have not yet been officially identified as part of this effort. However, habitat destruction due to land alteration, land conversion to agricultural and urban uses, and a variety of pollution problems are believed to be the most significant stresses to the system.

PROJECT DESCRIPTION

This effort is an outgrowth of the Tennessee Biodiversity Program, which has focused on improving breeding bird populations. In early 1992, state personnel from this program met with their colleagues in similar programs from surrounding states, and realized that they had similar programs and thus an opportunity to coordinate their efforts. Officially, the effort began in May 1994, with an initially scheduled five-year duration. It is being coordinated by the nonprofit Tennessee Conservation League, the state affiliate of the National Wildlife Federation. While this effort is currently

acknowledged as a landscape-level plan for breeding bird populations only, it is viewed as a first step to an ecosystem management plan which could be extended to include other ecosystem attributes. In the short term however, the effort lacks the capability to look at all ecosystem factors.

The project's approach is on three spatial scales that combine realities of ecosystems with those of implementing management. First is the broad picture, that is, identification of important habitats and key locations to target to sustain and increase breeding bird populations across the physiographic province. The second scale is a watershed approach: identifying key areas in the watershed to maintain biological diversity. Third, site-specific conservation guidelines that identify important sites at the most local level for implementation by the local community are being produced in county planning manuals. Finally, for cooperators' lands, whether industrial or public, the project is developing customized management guidelines to match the landowners' traditional objectives and serve regional biodiversity goals.

In 1995, a habitat team was set up to identify stresses and develop goals and strategies to address these stresses. A steering committee is already in place, whose cooperators include representatives from state agencies (wildlife, forestry), federal agencies (U.S. Fish and Wildlife Service, Tennessee Valley Authority), industry (Westvaco Corporation, Willamette Industries, Champion International), and citizen groups (Tennessee and Alabama Ornithological Societies, Tennessee Conservation

League). These cooperators directly control one million acres within this ecosystem. The steering committee decided to involve the general public following an evaluation of the effort one year after its inception.

The first three years of the effort are expected to be spent developing information for education efforts, research priorities, management issues, and acquisition efforts. The following two years will be used to set up monitoring. By the end of the five-year period, active acquisition and management programs are expected to be in operation, with respective state wildlife agencies leading the effort thereafter.

STATUS AND OUTLOOK

Specific project goals have been identified, and include 1) using vegetative mapping through geographic information systems (GIS) technology to identify patch size and distribution in order to determine specific areas to manage and conserve; 2) ensuring ecosystem integrity by assuring biological sustainability for breeding bird populations; 3) cooperating with industrial private landowners in sustaining wildlife; 4) collecting baseline data; and 5) demonstrating the feasibility of private land involvement on multiple levels. More specific goals include a prioritization of habitats for management and restoration of glades, barrens, and bottomland hardwoods in the Ohio and

Tennessee valleys, and maintenance of existing tracts of upland hardwoods in the western Highland Rim of middle Tennessee.

GIS-based vegetative mapping and some analyses of breeding bird information have been completed for approximately 50 percent of the project area. Population objectives for breeding birds have been determined and important areas in which to target cooperative management objectives have been identified.

Progress has been made in communicating ecosystem management concepts to on-the-ground managers and getting them to think in broader terms about the impacts of their individual actions on the ecosystem.

Factors Facilitating Progress

The coordinating role that the Tennessee Conservation League has assumed has been an important asset, both for policy decision making and in facilitating technical research and monitoring between universities and the states. The league has been able to bring stakeholders to the table and has helped foster trust among participants. Typically-expected "turf wars" have been avoided. Finally, the league has conduits to government agencies (state and federal) and industry, key elements of this effort.

Obstacles to Progress

Administrative barriers were initially significant and contributed to delays in realizing

short-term goals (although most of these have been overcome). A small portion of southern Ohio is functionally part of the ecosystem, but Ohio's acreage was too small to justify the added administrative processes. Top administrators and technical personnel have different understandings of what ecosystem or landscape management entails. Similarly, no official definition of ecosystem management exists to guide planning efforts.

Limited resources and personnel have been problems as well. Technical difficulties with the GIS mapping have occurred. Finally, some pessimism exists as to the future of this effort, for several reasons. The primary reasons are the uncertainty of future incentive programs (e.g., the recent Farm Bill debate in Congress), convincing private landowners to consider non-game wildlife, and the limited impact that this effort will have on the ecosystem.

Contact information:

Mr. Bob Ford
Project Manager
Tennessee Conservation League
300 Orlando Avenue
Nashville, TN 37209-3200
(615) 353-1133
Fax: (615) 353-0083
E-mail:
  conserve.tcl@nashville.com

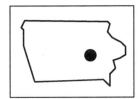

IOWA RIVER CORRIDOR PROJECT

Location:
Central Iowa

Project size:
50,000 acres

Initiator:
U.S.D.A. Natural Resources Conservation Service

PROJECT AREA DESCRIPTION

The Iowa River runs from north-central Iowa to southeastern Iowa where it joins the Mississippi River. The U.S. Fish and Wildlife Service (FWS) considers it part of the Upper Mississippi-Tallgrass Prairie ecosystem. The project area is defined by floodplain boundaries, and includes a part of the Iowa River that is not channelized but that winds back and forth across a three to four mile wide floodplain. Prior to the initiation of this project, over half of the project area was cropland. About a third of the area, in the riparian corridor, was woodland, dominated by oak and maple, walnut, willow, and cottonwood. The project area also includes the Otter Creek Wildlife Management Area, which measures approximately 3,000 acres.

The project area supports two active bald eagle nesting sites, and the state listed sandhill crane and river otter. The most important human uses of the project area are row crop agriculture, livestock production, timber management, and recreation.

ECOSYSTEM STRESSES

The project area has been impacted most by agricultural practices and land conversion to agriculture. Statewide, 98 percent of Iowa's wetlands and 99.9 percent of native prairie land have been converted to agriculture in the last 150 years. The project area floods frequently, since it is located in a floodplain and receives most of its water from the upper channelized parts of the river in a relatively short period of time. Although the 1993 flood was one of many floods, it was larger and lasted longer than previous ones. The

inundation of the area lasted six months and stressed both natural systems and the economy of the area. Crop losses in the project area exceeded $6.9 million.

PROJECT DESCRIPTION

After the 1993 record flood, many landowners recognized that agricultural use of lands along the Iowa River is not sustainable. Under the Emergency Wetland Reserve Program (EWRP) of 1993 and the Wetland Reserve Program (WRP) of 1994, the U.S.D.A. Natural Resources Conservation Service (NRCS) offered to buy easements on these properties to allow landowners to pursue more flood-tolerant land uses.

In addition, FWS is planning to offer buy-outs to landowners who enrolled in EWRP and WRP. The land acquisitions will become part of the National Wildlife Refuge system, but may be managed by the Iowa Department of Natural Resources. FWS and NRCS are also planning a cooperative effort with the remaining landowners focusing on sustainable land use practices, such as sustainable crop and timber management practices and improved grazing management.

A major goal of this project is to provide landowners with a broad menu of assistance options that represent sound floodplain management. Other goals include the management of public lands and easements to provide for the natural diversity and functions of the Iowa river system; to demonstrate the economic outcomes of alternative floodplain management and land uses; and to use private and public partnerships to accomplish these objectives. Nonprofit organizations are involved in this effort, including the Iowa Natural Heritage Foundation, Pheasants Forever, Ducks Unlimited, and The Nature Conservancy.

STATUS AND OUTLOOK

The land acquisition and easement components of the project are under way. One hundred ten out of 250 landowners responded positively to NRCS's offer. By the end of 1995, NRCS had easement documents and plans finalized for fifty to sixty properties. The FWS land-acquisition plan began in 1995 after approval of an environmental assessment.

As soon as 30 percent of the cropland comes out of production, immediate results are expected to be observed. There will be better opportunities for plants (e.g., hardwoods) and wildlife to establish and regenerate in the wetlands, and water quality is expected to improve. The development of the sustainable land use program is still in its infancy, and funds for it have yet to be made available.

Factors Facilitating Progress

So far, the project has proven very successful according to project leaders. The willingness of landowners to consider easement and buy-out options is not only economically sound, but has been invaluable in the effort to restore wetlands, forests, and prairie in the Iowa River corrior, and in the promotion of flood tolerant land uses as a whole. Furthermore, the cooperation between federal agencies, a state agency, nonprofit organizations, and private landowners has proven very fruitful.

Obstacles to Progress

Initially, landowner misconceptions needed to be overcome. Although agency cooperation facilitated progress, there were difficulties in coordinating between different agency processes. In addition, a very bureaucratic easement process went well beyond the period that some agency personnel felt it would take.

Contact information:

Mr. James R. Munson
Iowa Private Lands Coordinator
U.S. Fish & Wildlife Service
PO Box 399
Prairie City, IA 50228
(515) 994-2415
Fax: (515) 994-2104

KARNER BLUE BUTTERFLY HABITAT CONSERVATION PLAN

PROJECT AREA DESCRIPTION

Karner blue butterflies are found in the northern range of wild lupine habitat in portions of New Hampshire, New York, Michigan, Wisconsin, Indiana, and Minnesota. The pine barrens and oak savanna ecosystems are likely to be a mosaic of interspersed woody vegetation, such as pitch pine and scrub oak, and open grassy areas. The presence of wild blue lupine is a necessity, as it is the only plant that the Karner blue eats in its larval stage. Flowering plants serve as nectar sources for adult butterflies.

Nearly twenty plant and animal species that are endangered, threatened, or of special concern have a high association with the Karner blue. Among them are Blanding's turtle, eastern massasauga, prairie flame flower, western slender grass lizard, cobweb skipper, and frosted elfin butterfly. Land uses in central and northern Wisconsin include industrial forest production, state wildlife areas, county forest reserves, residential development, recreation, and agriculture.

ECOSYSTEM STRESSES

Historically, wildfire maintained the pine barren and oak savanna ecosystems by checking natural succession and creating a network of openings in the forest canopy. Habitat throughout the range of the Karner blue has been lost as a result of suppressed wildfire, silviculture, urbanization, and natural succession. The remaining habitat has been fragmented, preventing movement and dispersal of butterflies and resulting in small isolated populations. At one time, the butterfly was found in a nearly continuous narrow band across

ten states and one Canadian province, but it has been extirpated from at least four of these states.

PROJECT DESCRIPTION

The Karner Blue habitat conservation plan (HCP) focuses on central and northwestern Wisconsin. Concerns within the forest products industry over private land use prompted the Wisconsin Department of Natural Resources to undertake the HCP effort. The major thrust of the effort involves coordinating management practices across a mosaic of public and private landownerships. The HCP will serve as a basis for seeking an incidental take permit from the U.S. Fish and Wildlife Service so that land management activities on all properties, regardless of ownership type can continue.

The region's stakeholders are involved in the process at different levels of participation, depending on the extent of their interests and landownership patterns, for example. There are more than twenty partners at this stage.

STATUS AND OUTLOOK

Articles of partnership have been developed, and the partners have been meeting. Special provisions, such as protecting private property rights, have been included in the articles.

Factors Facilitating Progress

Stakeholder participation in the project planning has been the major factor facilitating progress.

Contact information:
Wisconsin Dept. of Natural Resources
Box 7921
Madison, WI 53707

Location:
Wisconsin

Project size:
9 million acres

Initiator:
Wisconsin Department of Natural Resources

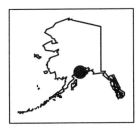

KENAI RIVER WATERSHED PROJECT

Location:
South-central Alaska

Project size:
1.4 million acres

Initiators:
Alaska Department of Fish and Game, U.S. Environmental Protection Agency, The Nature Conservancy

PROJECT AREA DESCRIPTION

Bordered on the west by Cook Inlet and on the east by Prince William Sound, the Kenai Peninsula extends out over 150 miles into the Gulf of Alaska. Diverse landforms are found throughout the Kenai River watershed. Mountainous regions provide the headwaters for the Kenai River which runs to the estuarine and wetland systems along the coast. The Kenai River flows uninhibited by dams and other structures; it has been called the home of the last great salmon run in the Northwest.

A combination of landform variety and maritime weather patterns account for the vegetative diversity in the region. White spruce can be found at higher elevations. Kenai paper birch, quaking aspen, stunted black spruce, and coastal western hemlock are all present in the watershed. Willow species dominate riparian areas. Commercial and sport fishing are the predominant economic and recreational activities. Additional land uses include timber harvesting, other forms of recreation, and federal and state parks. Approximately 44 percent of land along the Kenai River is privately owned.

ECOSYSTEM STRESSES

A spruce beetle infestation is stressing tree stands in the watershed. According to a study by The Nature Conservancy (TNC), timber harvesting could become a significant source of stress. The most severe impacts of timber harvesting include habitat fragmentation, sedimentation, alteration of biological assemblages, and alteration of surface water flow. However, the draining and filling of wetlands has had the greatest

negative impact on the area's natural systems. In addition to the impacts caused by timber harvesting, draining wetlands also causes eutrophication, altered water chemistry, and habitat degradation.

Sportfishing and the construction of private docks and structures have also contributed to habitat destruction along the banks of the Kenai River, negatively impacting salmon spawning and movement. Road construction, land development, and the removal of riparian vegetation are also significant stresses.

PROJECT DESCRIPTION

Residents within the watershed had been contacting TNC's regional office for information about managing lands in the watershed. In early 1993, the Alaska Department of Fish and Game (DFG) and the U.S. Environmental Protection Agency (EPA) asked TNC to become formally involved in conserving the Kenai River watershed. EPA asked TNC to identify conservation priorities for watershed lands and recommendations for effective public and private conservation actions. TNC has completed an in-depth stress assessment that determined and ranked the watershed's subsystems, stresses, and sources of stresses.

No other national environmental organization is active in the watershed, although there are several local river-based groups. A TNC representative, hired in 1995 with funding from EPA, is now located in the watershed, with an office opened in Soldtona. This representative is responsible for building coalitions among local stakeholders and landowners and providing technical assistance to their projects. The field representative will regularly meet with officials from the U.S.D.A. Forest Service, U.S. Fish and Wildlife Service (FWS), state agencies, timber industry, regional and village native corporations, and local landowners.

Specific projects will be further identified through contacts with community members and land managers. A land trust, the establishment of a resource center, and a watershed alliance were mentioned as possible activities for the field representative to develop. The TNC representative has also been charged through EPA funding to design and implement a community workshop. The workshop will bring together landowners, land managers, and users of the Kenai River to exchange thoughts, ideas, concerns, and information regarding the maintenance of the river system's health.

STATUS AND OUTLOOK

In general, there has been local acceptance of this effort. Currently, the TNC representative is focussing on three major thrusts: 1) continuing outreach efforts; 2) catalyzing creation of the land trust through efforts with the Kachemak Heritage Land Trust located in Homer; and 3) continuing preparations for the workshop, schedule for April 1996. A steering group of local stakeholders has been convened to work on workshop planning. Funding is being provided by EPA, Alaska DFG, and FWS; local sources of funding are being sought as well.

Factors Facilitating Progress

The livelihood of many residents depends on the river's resources; examples of salmon decline in other parts of the Northwest concern local residents. Public awareness and increasing involvement to protect the health of the river have created a receptive environment for TNC to under-take its work. A better understanding of the area's natural systems and the willingness of local residents to manage the river with a new understanding are required for future progress.

Obstacles to Progress

There have been no significant obstacles to date. While there is agreement by stakeholders around general principles and values regarding the Kenai River, different views on the causes of ecological problems, what management actions should be taken, and competing political agendas could cause difficulty as the project advances. Finally, funding is a continuing challenge.

Contact information:

Mr. Randall H. Hagenstein
Associate State Director
The Nature Conservancy of
Alaska
421 West 1st Avenue, Suite 200
Anchorage, AK 99501
(907) 276-3133
Fax: (907) 276-2584

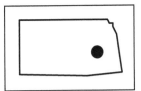

KONZA PRAIRIE RESEARCH NATURAL AREA

Location:
Northeastern Kansas

Project size:
8,616 acres

Initiator:
Kansas State University

PROJECT AREA DESCRIPTION

The Konza Prairie Research Natural Area consists primarily of a pristine native tallgrass prairie, interspersed with narrow strips of oak-hackberry forest in riparian areas. More than 450 plant species are present in this area, only a handful of which are non-native. The Konza Prairie contains the entire watershed of Kings Creek, and is therefore less susceptible to land management practices outside its borders. The prairie is situated within the Flint Hills, which represent the largest contiguous tract of unplowed native tallgrass prairie in the country. The Flint Hills run south in a fifty mile wide strip from the Kansas-Nebraska border to northeastern Oklahoma. Most of the Flint Hills are managed by private landowners for seasonal cattle grazing, including the lands bordering the Konza Prairie on the east, south, and west. Nevertheless, since these lands are very similar in species composition and ecosystem integrity, they function as effective buffers for the Konza Prairie.

ECOSYSTEM STRESSES

Although the Konza Prairie experiences some human induced stresses, none have a significant impact. The reasons for this are three-fold: 1) the Konza Prairie is used primarily for scientific research and public education; 2) it is nearly surrounded by a buffer area; and 3) the Prairie contains an entire watershed. Residential development of small tracts of land north of the Konza Prairie is of some concern, since it makes it more difficult to minimize human impact.

PROJECT DESCRIPTION

The Division of Biology at Kansas State University saw a need for a large-scale, long-term ecological research center in the tallgrass prairie region. In 1971, the current Konza Prairie Research Natural Area came up for sale. It was subsequently purchased by The Nature Conservancy (TNC) with the funds of an anonymous donor, to be managed by the University's Division of Biology.

The goals of the Konza Prairie management effort are the preservation of a large tract of tallgrass prairie, as well as biological research. The effect of fire and varying fire regimes on the prairie ecosystem has been a major research thrust. In 1979, the Konza Prairie was designated as a UNESCO (United Nations Educational, Scientific, and Cultural Organization) Man and the Biosphere Reserve. In 1980, the Konza Prairie became one of the first six sites in the Long-Term Ecological Research Network of the National Science Foundation (NSF). (This network currently consists of eighteen sites.) In 1987, the major research focus expanded with the reintroduction of a herd of American bison, a native grazer. Research now concentrates on the effect of grazing in combination with fire. The diversity of experimental grazing treatments will be broadened in 1996. Faculty from Kansas State University and many other regional universities conduct research at this extraordinary site.

Management strategies for the Konza Prairie Research Natural Area include the use of prescribed burning and grazing in a mosaic across the landscape to maintain natural processes and heterogeneity. Fire management is generally coordinated with the private owners of the lands surrounding Konza

Prairie. External human-induced impacts to the area are prevented or minimized, and activities that would introduce significant disturbance are prohibited. The removal of native plants and animals and the introduction of non-native plants or animals are also prohibited.

On a long-term basis, the following factors are monitored annually: species composition; diversity and populations of plants, birds, mammals, and insects; primary productivity; weather; water quality; health of the bison herd; invasion of woody vegetation; soil biota; and ecosystem nutrient cycling processes. Monitoring of wildlife populations on surrounding privately-owned lands is also occurring. Based on the results of monitoring, management strategies are refined or adjusted.

STATUS AND OUTLOOK

Since the inception of the Konza Prairie Research Natural Area, many of the research goals have been realized. In addition to numerous publications, a long-term database has been developed on tallgrass prairie ecosystem processes and patterns. The understanding of the role of frequent fires and native grazers in the ecosystem, biodiversity, ecosystem processes, and population dynamics, among others has increased significantly.

Factors Facilitating Progress

Several factors have contributed to the success of the research area. They include the pristine conditions of the ecosystem when the project started; the containment within the project area of a large watershed; the proximity of Konza Prairie to Kansas State University; the cooperative arrangement with TNC; and funding and commitments from NSF, Kansas State University, and the Kansas Agricultural Experimental Station. In addition, the commitment and cooperation of numerous scientific investigators has been helpful.

Obstacles to Progress

The confounding influence of natural grassland fires have posed an obstacle to research, as they affect influences fire frequency beyond the control of scientists. Other problems include adjacent landowner concerns and insufficient start-up funds.

Contact information:

Dr. David Hartnett
Kansas State University
Division of Biology
Ackert Hall
Manhattan, KS 66506
(913) 532-5925
Fax: (913) 532-6653
E-mail: dchart@ksuvm.ksu.edu

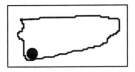

LAJAS VALLEY LAGOON SYSTEM

Location:
*Southwestern
Puerto Rico*

Project size:
48,000 acres

Initiator:
*U.S. Fish and
Wildlife Service*

PROJECT AREA DESCRIPTION

Located within the larger Caribbean watershed ecosystem (which includes Puerto Rico and the U.S. Virgin Islands), Lajas Valley runs roughly east-west on the island of Puerto Rico, connecting at both ends with the Atlantic ocean. The lagoon system is located in the flat valley floor and consists of four lagoons: Guanica, Anegado, Laguna Cartagena, and Boquerón. Boquerón is a Commonwealth of Puerto Rico refuge and Laguna Cartagena is protected as a National Wildlife Refuge; the other two lagoons are in private ownership or administered by commonwealth agencies. The rest of the valley is now predominantly in sugar cane and pasture (guinea grass) for cattle grazing. The mountains on either side of the valley, also part of the ecosystem, are partially forested in native hardwoods and exotic mesquite and acacia sometimes used for limited charcoal production.

Federally listed threatened and endangered species in and around Lajas Valley reflect the diversity of the landforms, from coastal to mountains, and include the yellow-shouldered blackbird, Puerto Rican nightjar, brown pelican, Puerto Rican crested toad, and several plants in the adjacent hills.

ECOSYSTEM STRESSES

The valley has endured most of the stresses: nearly all of it has been converted to agriculture. Agricultural practices and grazing have resulted in non-point source pollution, and to a lesser degree, point source pollution. Hydrological disruption of the lagoon system has been significant, another result of the land conversion. Exotic

species have largely replaced the native vegetation in the valley. By themselves, exotics are not responsible for the ecosystem's decline since the habitat has already been altered (e.g., by agricultural development). Finally, some of the land has been converted to urban uses, with corresponding point source pollution and road construction. The mountains have been far less impacted, but still suffer from clearing, development, and encroachment of exotic species.

PROJECT DESCRIPTION

The lead agency on this effort is the U.S. Fish and Wildlife Service (FWS), which initiated this effort in 1994 as part of the agency's new ecosystem management initiative. In particular, Lajas Valley was included in FWS's Caribbean Ecosystem Management Plan because of unique resource values and significant public interest in the valley, including local citizen conservation groups working to protect specific lagoons. This effort is expected to last five years, resulting in cooperative agreements with commonwealth agencies and nonprofit institutions.

A draft management plan for the Caribbean ecosystem includes five major goals, with Lajas Valley as the focus of one of the goals. Specifically, the plan calls for preventing further decline of the unique lagoon ecosystem and its restoration to pre-1970s condition by the year 2000. Specific objectives are drawn up for each goal, and include elements such as funding, administrative priorities, creating partnerships and/or committees with other institutions, and identifying specific projects. While FWS personnel put together the draft plan and did not officially seek stakeholder involvement, they are working to build partnerships with commonwealth agencies (Agriculture, Natural Resources), citizens, nonprofit conservation groups, municipalities (Lajas, Guanica), and the Puerto Rico Electric Power Authority (PREPA). FWS is expected to lead implementation of specific strategies in coordination with commonwealth agencies.

STATUS AND OUTLOOK

A draft Caribbean management plan has been prepared. Only limited implementation has occurred (removal of cattails on Laguna Cartagena National Wildlife Refuge); monitoring of ecosystem attributes has yet to be established. The future of the effort has been described as dependent on funding and political support from both commonwealth and federal sources (Congress, FWS).

Factors Facilitating Progress

Time constraints and deadlines have proven to be both problem- atic and beneficial to the process, the latter resulting in the plan being developed rapidly, proba- bly sooner than might have hap- pened otherwise. The plan helped to emphasize the need for projects that had been previously identified. Finally, the commit- ment and dedication of FWS personnel have been especially important to this process, as Lajas Valley is recognized by agency personnel as well as the general public as a particularly unique, threatened ecosystem.

Obstacles to Progress

As mentioned above, short deadlines imposed by the na- tional FWS office did not allow enough time to develop a more comprehensive plan. No addi- tional personnel and only limited resources have been provided for this effort thus far; the plan de- velopment has been an added responsibility for field personnel.

Political opposition is expected to be significant, especially if and when recommendations are made to remove agricultural areas from active production.

This resistance will be manifest from private landowners, lease- holders, and the commonwealth, which allows agricultural pro- duction on some of its land in the valley. Because the Lajas mayor did not win the primary, rela- tionships with a new mayor and administration will need to be reestablished in 1996; it is still unclear how a new administra- tion will affect this project.

Finally, lack of funding is a con- cern, from both FWS and the commonwealth; in particular, the commonwealth is likely to be asked to cost-share on the restoration efforts.

Contact information:

Mr. James Oland
Field Supervisor
U.S. Fish & Wildlife Service
Caribbean Field Office
PO Box 491
Boqueron, PR 00622
(809) 851-7297
Fax: (809) 851-7440
E-mail:
   R4FWE_MAPR@MAIL.FWS.
   GOV

LOWER RIO GRANDE ECOSYSTEM PLAN

Location:
Southern Texas

Project size:
38.4 million acres

Initiator:
U.S. Fish and Wildlife Service

PROJECT AREA DESCRIPTION

The Rio Grande River is separated into three distinct sections, the Upper river from its headwaters through Colorado, the Middle in New Mexico, and the Lower from El Paso to the river's mouth at the Gulf of Mexico near Brownsville, Texas. The lower river watershed is over 600 miles long from El Paso to the gulf, and approximately 50 miles wide on either side of the river, which forms the Texas-Mexico border. Based on habitat, the Lower Rio Grande ecosystem is further divided into three subsections: a lower subsection, referred to as "the Valley," and including coastal features; a middle, subtropical forest subsection, with scrub and slash forest; and the upper, scrubland Chihuahuan Desert stretching to El Paso, and featuring both low-land desert and highland mountains with coniferous and pine forests.

Over thirty federally listed threatened and endangered species occur in the Lower Rio Grande, including the ocelot, northern Aplomado falcon, brown pelican, and piping plover in the valley. Moving upriver, the star cactus, peregrine falcon, ashy dogweed, and Johnston's frankenia can be found. Also, more than fifty state listed species are present, including the Texas tortoise and black-spotted newt.

Row crop agriculture, particularly cotton, grain, and vegetables, occupies 95 percent of the valley, along with significant industrial and urban development. Moving upriver, grazing becomes more dominant. The middle subsection is far less urbanized, with a greater concentration of preserved lands, including wilderness areas.

Finally, the upper subsection is again dominated by urban and industrial development. Three national wildlife refuges, one national park, and several state preserves are located within the Lower Rio Grande region.

ECOSYSTEM STRESSES

Conversion to agricultural, urban, and industrial uses is the dominant stress in the valley as well as in the upper sub-section near El Paso. Associated stresses occur from agricultural practices, such as air pollution from burning sugar cane fields, and insecticide and herbicide pollution. Similarly, urban development has led to significant road construction and non-point and point source pollution from industry, especially near large border cities on both sides of the river.

As a result of increased trade resulting from the North American Free Trade Agreement (NAFTA), ten to fifteen new international bridges over the Rio Grande have been proposed in the valley, posing an additional threat. Hydrologic disruption has occurred from flood control structures built to protect against hurricanes, including two international dams on the Rio Grande. Also as a result of NAFTA, Mexico has proposed extending the Intracoastal Waterway into that country, which would cause significant habitat destruction and modification along the Gulf Coast.

In the middle subsection, overgrazing and brush clearing are the predominant stresses, along with lignite mining. Finally, illegal poaching of cacti for xeriscape landscaping, especially the star cactus, causes significant harm.

PROJECT DESCRIPTION

The U.S. Fish and Wildlife Service (FWS) initiated this effort in April 1994 as part of the agency's new ecosystem management initiative. An initial meeting of supervisors from FWS's Region 2 (TX, OK, NM, AZ) resulted in the formation of an ecosystem team of nine to ten members. The team developed a management plan for the region, including the Lower Rio Grande, which was given a high priority because of expected NAFTA impacts and the high number of endangered species.

While a Lower Rio Grande planning document was drawn up entirely by FWS personnel, the effort included partners from outside the agency, but only on an informal basis. This planning process included four open houses, where participants from the public and other institutions (state and federal) were asked to fill out a questionnaire and to comment on the draft management plan.

The Lower Rio Grande plan contains one extremely broad goal of protecting and enhancing biologic diversity. Several objectives were designated, along with specific action items to achieve these objectives. Even though FWS controls relatively little land in the area, the plan does not apply exclusively to wildlife refuges. In fact, some of the activities included in the plan have been characterized as being no different than those prior to the plan's inception, except that they are now better coordinated under an ecosystem rubric. Thus, the plan is more appropriately described as an internal FWS working document, subject to change.

A variety of stakeholders were involved informally in the document development or specific projects. These include several other federal agencies, Texas state agencies, agricultural interests, environmental and land conservation organizations, industry, private landowners, and the general public. According to FWS, involvement should have been solicited from the several major universities located in the region.

STATUS AND OUTLOOK

In addition to the working document, some on-the-ground projects have been initiated as part of this effort, several on non-FWS lands. Perhaps the greatest impacts, however, have been internal to FWS. The document is being used to guide FWS's "way of doing business" in the Lower Rio Grande, leading to improved communication among personnel and coordination among discrete projects; in the past, projects were independently developed and implemented.

Although some of the projects have yet to be completed, revegetation of up to 1,000 acres of land with native brush is occurring, as well as the completion of a base line fisheries inventory on the lowest segment of the Rio Grande.

Factors Facilitating Progress

The FWS's open mindedness and willingness to change have been viewed as instrumental in driving this process. Support from the national headquarters and regional office, as well as latitude in developing an ecosystem management program, have also been viewed as benefits. For example, the FWS Region 2 office created and filled a relatively senior-level position devoted specifically to implementing ecosystem management.

Obstacles to Progress

Funding has been the greatest obstacle to progress. Furthermore, the traditional line-item process has proven incompatible with the ecosystem management approach. A fundamental change in how funding is allocated; improved communication with Congress as to FWS's program direction; and different funding schemes all are needed. Finally, additional personnel will be required for this effort to move forward.

Contact information:

Mr. Art Coykendall
Wildlife Biologist
U.S. Fish & Wildlife Service
320 N. Main St., Rm. 225
McAllen, TX 78501
(210) 630-4636
Fax: (210) 630-1653

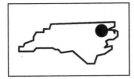

LOWER ROANOKE RIVER BIORESERVE

PROJECT AREA DESCRIPTION

In northern North Carolina, the lower Roanoke River begins at Roanoke Rapids (the fall line), where the river spills from the Piedmont. Moving east, the river empties into Albemarle Sound, the estuary that joins the Atlantic Ocean past the Outer Banks barrier islands. The primary boundaries of the project area are the floodplain walls on either side of the lower Roanoke, varying from one to five miles wide. Secondary project boundaries are the river's watershed. The larger ecosystem includes the estuary and islands, as well as the watersheds of the lower Roanoke and other rivers that empty into the sound. (These are not part of the project area.)

The lower Roanoke floodplain consists of swamp forest and hardwood bottomlands, including old-growth bald cypress and tupelo. Seven heronries, black bears, bobcats, and over 200 species of birds can be found there. A few federally listed threatened and endangered species are present, such as the bald eagle and several mussels.

Forestry and row crop agriculture are the primary land uses, the former for furniture-grade lumber. Hunting and fishing have a long and significant tradition. Several towns and paper mills are located in the region. Ownership is predominantly private, with significant holdings by large paper and timber companies. Other holdings are in state, federal, and private conservation.

ECOSYSTEM STRESSES

Hydrologic disruption of the river is the most significant stress. Three dams were built on the river forty years ago, giving rise to water quantity and qual-

ity problems. Irrigation for agriculture draws significant water from the river, with associated non-point source pollution. Fish populations have collapsed on the river, in part due to commercial overfishing, but possibly due to the hydrologic disruption as well. Forest clearing for agriculture, which increased soon after the dams allowed frequently-wet fields to dry out, has reduced neotropical bird habitat. Timber management, specifically forest roads which also disrupt hydrology and habitat, has been a significant stress. Other stresses include point source pollution from the paper mills and exotic species (Asiatic clam, cowbirds, and honeysuckle).

PROJECT DESCRIPTION

The Nature Conservancy (TNC) is the lead institution of this effort, having first become involved in the region in the early 1980s with several small land purchases. Also, preparations by Virginia Power to relicense its three dams on the river have led to greater awareness of water quantity and quality issues. Following efforts to involve other cooperators, TNC's effort was expanded with the inclusion of the lower Roanoke as part of their "Last Great Places" campaign. This was a result of a cooperative 21,067-acre land management agreement signed in November 1994 with Georgia-Pacific Corporation (GP), one of the major industrial landowners.

The GP-TNC Roanoke Ecosystem Partnership, or GREP, includes a cooperative management committee, made up of ten to fifteen participants with two permanent members from both GP and TNC, and several other invited

Location:
Northeastern North Carolina

Project size:
137 river miles, 1-5 miles wide

Initiator:
The Nature Conservancy

participants from academia, North Carolina State Wildlife Resources Commission, Natural Heritage Program, U.S. Fish and Wildlife Service (FWS), and several environmental organizations. Only the GP and TNC members have voting rights, so that each side effectively holds veto power. Unresolvable disagreements are to be settled by binding arbitration.

STATUS AND OUTLOOK

By mid-1996, an ecosystem management plan will be developed for Georgia-Pacific's co-managed lands. Goals are expected to include ensuring economic viability of these lands, maintaining or restoring the ecosystem, and mimicking pre-dam hydrologic conditions as much as possible without restricting current uses of the river.

The GREP agreement includes permanent conservation easements in addition to co-managing lands. The entire 21,067-acre area will be declared a black bear refuge, and Georgia-Pacific is using helicopters to harvest high-grade timber, reducing traditional forestry stresses. In all, nearly 50,000 acres have been protected in the lower Roanoke in some form, including the GREP lands, Roanoke River (State) Gamelands in Martin County, Roanoke River National Wildlife Refuge in Bertie County, and TNC's properties, including Devil's Cut Natural Area.

Research on hydrology, vegetation, animal populations, and alternative forestry practices is under way with several universities in the region, and will serve as a basis for developing the ecosystem management plan.

Factors Facilitating Progress

Agency support at all levels—federal, state, and local in the case of one county—has been instrumental. Ironically, the surfacing of environmental problems, including the declining fish populations and water quality, has raised awareness about the need for ecosystem restoration and conservation. Also, the dam relicensing process requires consideration of water quality and quantity effects of the dams. Long-term TNC presence and efforts carrying out public education, serving as a neutral third party mediator, and assisting the communities and companies in economic development and land management have allowed this project to progress. Finally, participation of wildlife agencies has been cited as important to the effort.

Obstacles to Progress

The most significant obstacle to this effort is ecological: restoring the river's hydrology will be very difficult due to the effects of the dams. In only forty years since the dams were built, the area has experienced significant alteration and stress. Public resistance to conservation efforts has been and will continue to be significant. Counties are concerned about a reduction in the land tax base due to the creation of conservation lands. Another source of resistance has been in the disruption of traditional hunting rights, which have been passed down through generations and whose exclusivity has been terminated on public lands.

Because of the region's poor economic condition, industrial development is being encouraged along the river. Conflict between this policy and preservation efforts is expected. Finally, as the federal, state, and TNC acquisition efforts are voluntary, securing additional lands for protection is not guaranteed.

Contact information:

Ms. A. Este Stifel
Director, Roanoke River Project
The Nature Conservancy
Suite 201
4011 University Drive
Durham, NC 27707
(919) 403-8558
Fax: (919) 403-0379
E-mail: estifel@tnc.org

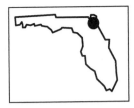

LOWER ST. JOHNS RIVER ECOSYSTEM MANAGEMENT AREA

Location:
Northeastern Florida

Project size:
1,777,280 acres

Initiator:
Florida Department of Environmental Protection

PROJECT AREA DESCRIPTION

The Lower St. Johns River Ecosystem Management Area encompasses the river's watershed from Lake George north to the river's mouth at Jacksonville, Florida. This segment of the river is essentially an elongated lagoon, having a low gradient and narrow floodplain, and containing numerous tributary streams and embayments. The entire lower St. Johns Basin is subject to tidal influences and short-term reverse flows.

The basin is relatively low and flat, with several ridge systems bordering the drainage area, and is characterized by many lakes, freshwater and saltwater marshes, and other wetlands. Bottomland hardwood forests commonly border the river and tributaries. The area is important for migratory and year-round waterfowl. The basin's condition ranges from small pockets of near-pristine areas to heavily urbanized regions, including highly industrialized sites along the river.

ECOSYSTEM STRESSES

The primary stress has been conversion to urban, suburban, industrial, and road uses. Development is considered uncontrolled in the region, with much of it abutting wetlands. Water pollution, mostly non-point source, is another major stress. As a result of land conversion and its proximity to sensitive areas, fire suppression is a significant secondary stress, affecting the composition and extent of the native vegetation which is fire-dependent. Other, less significant stresses include grazing, recreation, localized mining, and poor timber and forest management practices.

PROJECT DESCRIPTION

In 1993, the Florida state legislature consolidated two state natural resource agencies into the Department of Environmental Protection (DEP). In that legislation was a mandate for the DEP to "manage and protect systems in their entirety." As a result, DEP created six water basin management areas, including the lower St. Johns, in February 1994.

The first phase of this effort has been to draw up a conceptual recommendations document, using voluntary, interagency committees to reach consensus on what those recommendations would be. The committee members have included representatives from DEP, the federal government (U.S. Environmental Protection Agency, U.S. Fish and Wildlife Service, Army Corps of Engineers), the city of Jacksonville, the St. Johns River Water Management District, county and regional planners, large landowners, environmentalists, and citizens.

The agency's goals with regard to ecosystem management are to 1) better protect and management Florida's ecosystems; 2) base agency structure and culture on ecosystem management; and 3) develop a public ethic of shared responsibility for the environment. The lower St. Johns recommendations document will contain twelve to thirteen goals more specific to the basin. For example, to achieve the first broad goal described above, the initial focus of the basin document will be 1) to develop a comprehensive information base on the water basin, primarily an environmental inventory which will involve obtaining and standardizing new and existing data, and coordinating between agencies on information exchange; and 2) to control urban growth.

STATUS AND OUTLOOK

The conceptual planning document was published in June 1995. While the primary objectives of this phase of the project have been achieved according to DEP—in other words, greater and better interagency communication, having a clearer image of environmental inventory needs—this document does not include implementation strategies, and resulting impacts of this effort on private and public lands still unclear. The project's future depends in large part on the state legislature. While the original 1993 mandate was directive, it was very broad, and further definition from the legislature or governor will be necessary.

Factors Facilitating Progress

Leadership by the DEP secretary was instrumental in getting all six of the water basin efforts off the ground and progressing toward accomplishing their goals.

The secretary's vision and Governor Lawton Chiles's support were important to the initial progress of this effort as well. The legislative mandate, despite being broad, was instrumental in allowing the effort to progress.

Obstacles to Progress

Not having a clear directive from the legislature on how to achieve definition of ecosystem management has been a major obstacle to this effort. Further, the project area is not well understood, primarily because there is no environmental inventory of the watershed as yet.

Obtaining participation on the planning committees has been difficult because participation is purely voluntary. At first, the effort was not well known, and was viewed as an opportunity for participants to air grievances. In general, resistance to changing from traditional practices toward a new ecosystem management paradigm can be expected from various sectors (agency, general community). Challenges to the eventual plan are possible from

stakeholders, particularly industry and environmentalists. Legislative and implementation uncertainty present obstacles to visioning and continuity of DEP's effort. Finally, it is unclear who would ultimately implement management strategies and how effectively this project is being developed in conjunction with efforts of other institutions.

Additional needs for the project to progress include improving links with other agencies and combining expertise. Finally, increased citizen education on ecosystem management, and in particular this effort, are needed.

Contact information:

Ms. Jan Brewer
Environmental Specialist
Florida Department of
*   Environmental Protection*
Ste 200B, 7825 Baymeadows Way
Jacksonville, FL 32256-7577
(904) 448-4300
Fax: (904) 448-4366
E-mail: Brewer_J@JAXI.DEP.
*   STATE.FL.US*

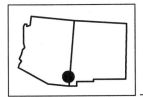

MALPAI BORDERLANDS INITIATIVE

PROJECT AREA DESCRIPTION

Located along the Arizona-New Mexico border, this extensive region consists of the Animas and Peloncillos mountain ranges, embedded in the lower elevation Chihuahuan, Sonoran, and Great Basin deserts. Within these "mountain islands and desert seas" exists a diverse array of vegetative communities including desert scrub and grasslands, pinyon, juniper, and evergreen oak woodlands, Apache and Chihuahuan pine, and remnants of southern Rocky Mountain vegetation, including ponderosa pine, Douglas-fir, and aspen. These communities support a host of state and federally listed threatened and endangered species, including the New Mexico ridgenose rattlesnake, Sanborn's long-nosed bat, lowland leopard frog, and gray wolf.

The borderlands region is divided almost equally between private and public ownership, including the Gray Ranch, a 500 square mile preserve and working ranch owned by the locally-based Animas Foundation with a conservation easement held by The Nature Conservancy (TNC). Current land uses on the largely unfragmented open spaces that characterize this region are almost entirely cattle ranching.

ECOSYSTEM STRESSES

A 100-year history of fire suppression in the region has been a major influence in the conversion and loss of native grasslands to woody shrublands. Also, pressures from human use and development threaten increased landscape fragmentation in the region. The loss of the unfragmented "open space" character of western rangelands and the loss of productive grasslands and regional

Location:
Southwestern New Mexico, south-eastern Arizona

Project size:
802,000 acres

Initiators:
Private landowners/ ranchers

ecological diversity have concerned environmentalists and ranchers alike in the borderlands region.

PROJECT DESCRIPTION

Concerned that their livelihood and "open space way of life" was threatened by the loss of productive native grasslands and potential development pressures that would lead to fragmentation of western rangelands, a grassroots coalition of private landowners and ranchers formed The Malpai Borderlands Group in 1990.

Working by consensus, the group's goal is "to restore and maintain the natural processes that create and protect a healthy, unfragmented landscape to support a diverse, flourishing community of human, plant and animal life in the Borderlands Region." The group works closely with local conservation districts, universities, and federal and state agencies. It has received significant support from TNC and the Animas Foundation. Recently, the foundation has proposed the development of monitoring protocols to guide ranchers and technicians in their monitoring efforts.

Restoring the area's natural fire regime has been the project's major thrust. In an attempt to encourage agencies responsible for fire suppression to let fires burn where possible, the group has developed strategies to help guide agencies in their response to fires throughout the project area. As a result, large-scale prescribed burns across public and private ownerships have been scheduled.

To prevent encroaching subdivision development, the group has developed a voluntary grass banking program. By donating a permanent conservation easement on their own property, ranchers can receive needed grass for their livestock from other ranching properties.

STATUS AND OUTLOOK

Although the project still remains in its planning stages, the accomplishments that have been realized by the group are unprecedented. Not only have landowners and state and federal agencies significantly increased their level of cooperation, communication and positive interaction, they now look beyond the borders of their own lands to focus on management of the entire landscape.

A 6,000-acre prescribed burn was successfully completed in the Baker Canyon watershed of the southern Peloncillo mountains, involving lands administered by three private ranches, the U.S.D.A. Forest Service, the Bureau of Land Management, and the Arizona and New Mexico state land offices. Planning is under way for a 15,000-acre prescribed burn (the

Maverick Burn) just north of Baker Canyon to be carried out in Summer 1996.

Four grass bank agreements have been completed with ranchers in the working area, in which they exchanged conservation easements prohibiting subdivision of their deeded land for grazing privileges on Gray Ranch, permitting each of them to rest their home ranches for up to three years while completing and implementing conservation plans.

Factors Facilitating Progress

Because the borderlands region has remained relatively unfragmented, the potential to use natural processes such as fire to restore and maintain the structure and composition of naturally occurring vegetation still exists. Furthermore, the people involved are committed to improving communication and cooperation, as well as maintaining the integrity of the landscape and the lifestyle it provides.

Obstacles to Progress

Due to the high number of sensitive species in the region,

responding to National Environmental Protection Act (NEPA) requirements has proven to be a difficult process for some landowners in the Malpai region. Furthermore, conventional notions of fire and the role of fire in these ecosystems have been in conflict with the project's "let it burn" policy.

Despite the great strides made in cooperation between stakeholders, the fragmented ownership pattern of the region still makes cooperative efforts difficult, particularly with the high turnover of agency personnel, which prevents landowners from developing working relationships with the agencies. Finally, the lack of adequate funding may restrict programs in the future.

Contact information:

Dr. Ben Brown
Program Director
Animas Foundation
HC 65, Box 179-B
Animas, NM 88020
(505) 548-2622
Fax: (505) 548-2267
E-mail: 6176022@mcimail.com

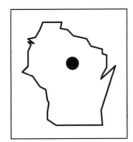

MARATHON COUNTY FORESTS

PROJECT AREA DESCRIPTION

The Marathon County Forests consist of seven tracts of forest in this central Wisconsin county. In presettlement times, the county's moraines, outwash, and drumlands were covered by northern hardwoods-hemlock forests and mixed pine-oak forests. However, the vegetation underwent a substantial change around the turn of the century when most of the area was logged and severe fires raged through the landscape. As a result, almost 75 percent of the Marathon County Forests now consist of an aspen-hardwood cover type. Other vegetation includes jack pine, white pine, red pine, northern hardwoods, and hemlock.

Wildlife species common to the Marathon County Forests include numerous species of birds, reptiles, amphibians, fish, and mammals. Some examples are migrant neotropical songbirds, migratory waterfowl, turkeys, deer, black bears, and fishers. In addition, the tawny crescent spot butterfly, proposed for federal listing, occurs in the area.

In addition to their ecosystem functions, the forests serve two main purposes: timber management and recreation. Timber sales are established on 500 to 600 acres per year. A variety of cutting techniques are employed, including clear-cuts, selection cuts, and shelterwood cuts. Recreational activities include bird watching, hunting, all-terrain vehicle use, mountain biking, cross-country skiing, snowmobiling, and horseback riding.

ECOSYSTEM STRESSES

Recreational trails and logging may lead to habitat fragmentation, which in turn may pose a threat to migrant neotropical songbirds. In addition, conversion of adjacent lands for housing, stresses the forests. An increasing number of people live in close proximity to the forests and "use the forests as their own backyards," resulting in an increase in garbage, damage to trees, and in illegal posting of county land. Furthermore, recreational uses impact the forests by disrupting habitat for species such as the black bear, causing trail erosion and reducing the overall wild character of some areas of the forests.

PROJECT DESCRIPTION

Management of Marathon County-owned forest land began in the mid-1940s. Since 1966, comprehensive forest plans have been written and revised once every ten years. Whereas management plans used to focus on the production of timber, deer, and grouse, the 1995 planning effort is attempting to take a broader, ecosystem approach to management.

In order to create a plan that takes the diversity of interests in the forests into account, an advisory committee has been formed to assist in gathering information, discuss issues, and recommend management options. This committee consists of landowners, representatives of recreational and environmental interests, loggers, timber industry, local government, and the Wisconsin Department of Natural Resources (DNR). The committee will advise a forestry recreation and zoning committee, which in turn makes recommendations to the Marathon County Board.

Although specific goals and strategies were still under development, an

Location:
Central Wisconsin

Project size:
27,000 acres

Initiator:
Marathon County, Wisconsin

example of strategies to be included in the plan is the regeneration and maintenance of pre-settlement forest types, such as hardwood-hemlock and mixed oak, where practical. Natural regeneration methods and plantings will take place.

Furthermore, in cooperation with the University of Wisconsin at Green Bay, a two-year migrant neotropical songbird census will be conducted in order to determine how current and past management practices affect these birds. Based on the results of these studies, timber sale designs, recreational development, land acquisition, and management agreements with adjoining landowners may be adjusted.

STATUS AND OUTLOOK
The first field season of the bird census and an experimental hemlock seedling establishment effort began in Spring 1995. The second field season was scheduled to begin in Spring 1996. A property master plan developed by the advisory committee is expected to be approved by mid-1996 after a review period in the first half of the year (public comment, approval by the country forestry committee, county board, and Wisconsin DNR). The DNR will prepare an environmental assessment as well.

Factors Facilitating Progress
The commitment of many people to the Marathon County Forests has been a very valuable component in the development of the forest plan. Their willingness to give time and thought to the planning effort has been very helpful. In addition, strong interest and encouragement for ecosystem management at both the state and the local levels have been very helpful.

Another important factor is the interest of state universities, and the willingness and availability of graduate students to help work on project proposals. The development of a geographic information systems (GIS) database has also been beneficial.

Obstacles to Progress
One of the difficulties in the management of these forests is the lack of easily analyzable data concerning the area. A second difficulty is continued development of housing on lands adjoining the forests, preventing the county forest administrator from acquiring these areas to lessen fragmentation. Compatibility and conflict between user groups is sometimes problematic. Finally, posing future restrictions on certain recreational activities may be controversial. The public is still focusing on forest outputs, rather than on maintaining ecosystem integrity.

Contact information:

Mr. Mark Heyde
Marathon County Forest
    Administrator
Marathon County Forestry
    Department
Courthouse
500 Forest Street
Wausau, WI 54403-5568
(715) 847-5267
Fax: (715) 848-9210

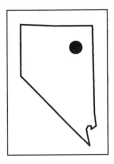

MARYS RIVER RIPARIAN/ AQUATIC RESTORATION PROJECT

Location:
Northeastern Nevada

Project Size:
332,800 acres

Initiators:
Nevada Division of Wildlife, Bureau of Land Management

PROJECT AREA DESCRIPTION

Located thirty miles east of Elko, Nevada, Marys River is one of the major tributaries to the Humboldt River, which is part of the Great Basin watershed, a desert environment characterized by big sage brush with perennial grasses such as bluebunch wheatgrass. Today, the area is home to more than 42 mammals, 87 birds, 6 fish, and 6 reptiles and amphibians. The area provides habitat for 7 candidate species for federal protection. Historically, the Lahontan cutthroat trout, a federally listed threatened species, occupied over 2,200 river miles of the Humboldt River watershed, including Marys River. Today, it exists only on 313 river miles, and Marys River is considered to have high potential for the trout's recovery.

The project area is primarily under control of the Bureau of Land Management (BLM), and all of the public lands along the river have been closed to livestock grazing, the primary land use in the area.

ECOSYSTEM STRESSES

Conversion of the land to grazing and historical grazing practices have degraded the region's habitat, especially along the river and its tributaries, which in turn has led to the reduction of the Lahontan cutthroat trout's habitat. The herbicidal and mechanical clearing of riparian willow thickets and the accompanying channelization of streams has altered stream hydrology, resulting in a scarcity of high-quality pools, suboptimal stream bank vegetative cover and stability, and a lack of desirable stream bottom materials due to excessive sedimentation and elimination of habitat.

PROJECT DESCRIPTION

The cutthroat trout was federally listed as endangered in 1970, then relisted as threatened in 1975 to facilitate management and restoration efforts. In accordance with the trout's listing requirements, the Nevada Division of Wildlife (NDOW) prepared a trout fishery management plan for the Humboldt River basin, providing management recommendations and guidelines for all public landowners.

As a result of the NDOW plan, which the U.S.D.A. Forest Service (USFS) and BLM both signed, thereby agreeing to cooperatively manage their lands, the current BLM Marys River Riparian/ Aquatic Habitat Management Plan (HMP) was completed in 1987. Its goals were to restore Lahontan cutthroat trout habitat with the objective of securing the delisting of the species and balancing use among various user groups within a multiple-use frame-work. While the plan did not specifically take an ecosystem management approach, BLM considers it to have similar intentions and it will serve as a basis for a comprehensive ecosystem-based management plan to be developed.

The 1987 plan only covered public lands, and because of the fragmented landscape within the ecosystem, watershed recovery was not considered possible without similar improvements on private lands. To address those issues, a major land exchange took place in 1991, in which a developer received 660 acres in Las Vegas in exchange for 47,000 acres, 65 miles of stream, and associated water rights in the Marys River watershed, boosting federal ownership to 80 percent of the land and 65 percent of the river. A master plan for the river was subsequently developed, with participation of a BLM multiple use advisory council already in place, representing a variety of interests (e.g., county commissioners, adjacent landowners).

On-the-ground activities include extensive monitoring (for example, plant and animal populations, water temperature, water quality, visitor days) and management actions (prescribed burning, riparian planting, range and grazing practice improvements).

STATUS AND OUTLOOK

Many areas have shown habitat improvements, some as a result of targeted restoration activities, but a significant amount occurring without any direct activities other than removing livestock.

Two more land exchanges are pending, which would result in additional protected habitat. The effort has been featured by the "Bring Back the Natives" program, a cooperative national effort of BLM, USFS, and the National Fish and Wildlife Foundation, to restore the health of aquatic ecosystems and their native species. Funding for the effort has come from federal and congressional sources, several gold mine operations or associations (as mitigation for mine damage occurring elsewhere), and several nongovernmental conservation organizations and associations.

Factors Facilitating Progress

Support for the effort, particularly the land exchanges, by BLM, local gold mines, other special interests, and to some degree, the Elko County government, is credited as the most important benefit to this project.

Obstacles to Progress

There has been local public skepticism as to the validity of this effort, primarily with regard to whether or not the habitat was degraded and in need of restoration. For example, there are claims that the trout's demise occurred only after the backcountry was opened up to off-road vehicle use (i.e., for fishing access), and not as a result of grazing. Cooperation of all federal and state agencies has not been readily forthcoming, hindering data exchange and the ability to effectively coordinate management activities across ownership boundaries. Continued funding for personnel, equipment, research, monitoring, and yearly water rights assessments are described as additional concerns for the future.

Contact information:

Mr. Bill Baker
Bureau of Land Management
Elko District
PO Box 831
3900 East Idaho Street
Elko, NV 89802
(702) 753-0200

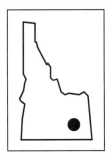

MCPHERSON ECOSYSTEM ENHANCEMENT PROJECT

Location:
Southeastern Idaho

Project size:
11,000 acres

Initiator:
U.S.D.A. Forest Service

PROJECT AREA DESCRIPTION

Mountainous terrain generally characterizes this 20,000-acre ecosystem on the Caribou National Forest in southeastern Idaho. Mountain maple is found within dense Douglas-fir forests. Choke cherry is the dominant shrub cover in the understory; pine grass grows on all the north-facing slopes. No threatened or endangered species are found there.

The ecosystem also serves as the municipal watershed for the town of Grace. Some sheep and cattle grazing occurs in the forest. Snowmobiling, motorbiking, and hunting are the primary recreational uses.

ECOSYSTEM STRESSES

Natural and anthropogenic influences have combined to severely stress the ecosystem. Fire history studies indicate the mean fire return interval to be around forty years with a mixed severity fire cycle. Most of the stands have missed two of those fire cycles due to suppression by the U.S.D.A. Forest Service (USFS) and by private landowners managing for grazing. As a result, tree densities are well above those considered to be sustainable.

A drought has afflicted the region for nearly eight years, further stressing the dense stands as they compete for nutrients and water. The weakened trees have been highly susceptible to bark beetle infestations, which have killed over 20,000 Douglas-fir trees within the national forest and adjoining Bureau of Land Management and state lands.

PROJECT DESCRIPTION

The drought, mortality from infestations, and thick stand densities have greatly increased the probability of a catastrophic wildfire in the forest. Concerned about the potential destruction of the community's watershed, USFS staff in the Montpelier Ranger District began the project in June 1993 in order to rehabilitate 11,000 acres through helicopter salvage harvest and the reintroduction of fire. The goals of the project are to reduce the probability of catastrophic wildfire, promote conditions that minimize the potential for a loss in resource values due to fires and human activities, and provide salvageable timber to local mills.

Although the Caribou National Forest has had a timber program since the 1960s, the salvage sale is the first timber harvesting activity in the project area during 100 years of USFS stewardship. The salvage sale is considered necessary to remove excess tinder and create a manageable situation into which fire can be reintroduced. Fire will be used to thin the denser stands. Other project activities include the rejuvenation of aspen stands and leaving snags in place for birds and wildlife habitat.

National Environmental Protection Act procedures were followed during plan development. The local Sierra Club, Idaho Conservation League, Idaho Department of Fish and Game, and timber industry were heavily involved. Staff from the U.S. Fish and Wildlife Service participated to ensure that a whooping crane rearing site on a local farm would not be disturbed by helicopters.

STATUS AND OUTLOOK

The salvage sale timber, totalling five million board feet, was removed from the site by early Fall 1995, slightly later than desired. As a result, the anticipated prescribed burn to remove slash materials and brush in order to enhance regeneration has been delayed until the Fall of 1996. At that time, weather conditions will have to provide a favorable burning window. The Forest Service considers the project a success because the timber removal has been satisfactory to all parties and implementation has occurred largely as planned.

Factors Facilitating Progress
The involvement of researchers, foresters, biologists, the timber industry, and environmentalists has allowed the project to proceed unobstructed.

Obstacles to Progress
There have been concerns from the Idaho Department of Fish and Game about the lack of big game hiding cover in the post-harvest stand densities after fire is reintroduced and impacts on

soil productivity.

Contact information:
Mr. Bruce Padian
USDA Forest Service
Caribou National Forest
Montpelier Ranger District
250 South 4th Avenue
Pocatello, ID 83201
(208) 236-7500
Fax: (208) 236-7503

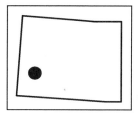

MESA CREEK COORDINATED RESOURCE MANAGEMENT PLAN

Location:
Western Colorado

Project size:
100,000 acres

Initiators:
Bureau of Land Management, Weimer Ranches (grazing permittee)

PROJECT AREA DESCRIPTION

The project area is a desert environment located at the confluence of the San Miguel and Dolores rivers. This rough, broken mesa country, characterized by sandstone uplifts, is dominated by sagebrush in the deeper soil areas, and piñon juniper in the rougher, steeper country. A few federally listed threatened and endangered species can be found there, including the peregrine falcon and bald eagle.

The project area is mostly federal land administered by the Bureau of Land Management (BLM), with private lands located in the deeper-soiled bottomlands where irrigation is possible. Ranching and livestock grazing are the predominant human uses.

ECOSYSTEM STRESSES

Livestock grazing and fire suppression, the latter due to livestock reducing the amount of fine fuels as well as aggressive firefighting, have been the major stresses to the ecosystem. Non-point source water pollution and hydrologic disruption have resulted from water diversions, livestock water pipelines and reservoirs, uranium mining, and the placement of roads adjacent to water courses. Exotic species such as tamarack, Russian napweed, and whitetop have invaded riparian areas in particular.

PROJECT DESCRIPTION

This effort was initiated in 1992 by the BLM and the project area's grazing permittee, Weimer Ranches, Inc., in order to develop an interdisciplinary approach to resource management, in contrast with the historic single resource approach (managing for livestock

grazing). A management plan was developed in May 1993, whose principle goal is to, "Emphasize the opportunity to look at the landscape mosaic and improve diversity in the ecosystem while protecting resource values with the participation of all interested parties." Strategies for achieving better management of the land include involving the public and stakeholders more in land management decision making, as well as specific grazing management and alternate fire suppression guidelines. Monitoring of ground cover and water flow and quality will be used to gauge the effectiveness of new management practices.

This effort is designed to involve Weimer Ranches in the decision-making process on grazing management taking place on public and co-mingled private lands on a much larger scale than previously allowed.

The management plan was co-developed by BLM and the residents of the area, with significant public participation primarily through open public meetings facilitated by a neutral BLM employee unfamiliar with grazing issues or practices, and through comment periods. Other activities in the plan development process have included field trips and open, semi-annual biological planning sessions.

In addition to BLM and Weimer Ranches, stakeholders involved in the process have included representatives from Colorado Division of Wildlife, local government, citizens, mining interests, other federal agencies (U.S.D.A. Forest Service, U.S. Fish and Wildlife Service, U.S.D.A. Natural

Resources Conservation Service), The Nature Conservancy, and local environmental organizations.

STATUS AND OUTLOOK

Major outcomes of this effort have included improved communication and cooperation among stakeholders, a better understanding by stakeholders of the effort's goals, and a common vision of what is needed for the land. There have been limited improvements in on-the-ground conditions, with glimpses of increases in perennial ground cover; in species of concern; improved water quality; and sustaining a traditional 'way of life' of the area.

Factors Facilitating Progress

The willingness and open-mindedness of BLM and Weimer Ranches to changes in land management, as well as their support and leadership, have been credited with helping this effort proceed. For example, the plan has required much effort and financial contribution on the part of the ranch, and the Weimers have participated in holistic management training to better work with BLM and gain new perspectives on resource management opportunities.

Obstacles to Progress

Greater participation by all stakeholders is desired, as attendance at meetings and field trips has dropped off from its original high levels, as those not directly affected by the decisions or land have lost interest.

Another challenge has been to vary management sufficiently to avoid repeating old undesirable practices. Finally, there is concern that the holistic resource management approach was adopted by BLM and Weimer Ranches without public input as to other possible rangeland management philosophies. On a larger scale, some question whether livestock grazing is an appropriate use of this semi-arid desert habitat.

Contact information:

Mr. Jim Sazama
Range Conservationist
Bureau of Land Management
Uncompahgre Basin Resource
  Area
2505 S. Townsend
Montrose, CO 81401
(303) 249-6047

MINNESOTA PEATLANDS

Location:
Northern Minnesota

Project size:
146,224 acres

Initiator:
Minnesota Department of Natural Resources

PROJECT AREA DESCRIPTION

Minnesota has approximately 7 million acres of peatlands scattered across the state. In these peatlands, a layer of peat from six inches to twenty feet deep covers an old glacial lake bed. The water table in the peatlands is very high, thus causing the saturation of soils and peat. The peatlands support large fens, bogs, and ovoid and teardrop islands. Vegetation includes sedges, bog birches, black spruce, tamarack, and white cedar. The eastern timber wolf, a federally listed threatened species, frequents the margins of the peatlands, but does not stray into the interior. Twenty-five state listed species occur in the peatlands, including the linear-leaved sundew, four-angled waterlily, northern bog lemming, English sundew, and yellow-eyed grass.

Eighteen peatlands are managed by the Minnesota Department of Natural Resources as Peatland Scientific and Natural Areas (PSNA). These PSNAs are core areas in which no disturbance is permitted. These areas are surrounded by additional peatlands that buffer the PSNA area. The largest PSNA is Red Lake Bog. It measures 83,000 acres and is ringed by 217,000 acres of buffer area. Red Lake Bog is crossed by only one road. The PSNAs are primarily used for long-term research, nature study, wildlife photography, and recreational hunting. Winter logging occurs in the surrounding areas.

ECOSYSTEM STRESSES

During the early 1900s, most of the peatland area was sold as agricultural land to settlers, who subsequently attempted to drain the area by digging a major ditch system. The land, however, was very unproductive and the settlers eventually abandoned the area. Their legacy—the ditch system—is now one of the main stresses to the ecosystem. Locally, the ditches have a profound influence on drainage, although the system pattern remains clear and only minimally affected. However, the ditches allow invasion by certain tree species, colonization by exotic species, and a higher number of beaver. In addition, timing and use of winter logging roads through the peatlands may pose a stress to the ecosystem.

PROJECT DESCRIPTION

After the settlers left, the state of Minnesota took title to all tax-forfeited property. The peatlands were left alone until an unsuccessful effort in the mid-1960s to protect these areas as state natural areas. In response to a large peat-gasification project proposed in the early 1980s, the Minnesota Department of Natural Resources (DNR) proposed legislation to protect all peatland core areas. The DNR organized informational meetings for local units of government, private individuals, and industry in order to obtain input on the identification of issues and solutions. Opposition from local governments and the timber and mining industry ultimately led to the defeat of this legislation. Finally, environmental lobbyists and influential state politicians were instrumental in the inclusion of peatlands under the (state) Wetland Conservation Act of 1991.

The act became a vehicle for the legislature to override local concerns, as it dedicated eighteen peatlands as PSNAs, to be managed by the Minnesota DNR. The management goals for these areas are to 1) restore the hydrologic integrity of the system; 2) reintroduce natural processes such as fire; and 3) preserve these areas as PSNAs. To reach these goals, long-range management plans will be developed for each site. These plans will deal with issues such as the development of winter logging road standards, rerouting of recreational trail systems, ditch systems, and the initiation of prescribed burn management. Monitoring will include hydrologic variables, acid deposition, the impact of fire, and the impact of the use of winter logging roads and recreational trails.

STATUS AND OUTLOOK

Although site-specific management plans are still under development, some strategies have already been implemented successfully. Winter logging road guidelines have been developed and are currently being used. Thus, potential damage from these roads is being limited.

Factors Facilitating Progress

A significant step in the protection of the Minnesota peatlands involved the passage of the 1991 legislation and the inclusion of the peatlands in this legislation. Environmental organizations were instrumental in that accomplishment. Another factor facilitating progress is the low economic value of the peatlands, and the acceptance that the resources forgone from economic utilization of this area are minimal at this time.

Obstacles to Progress

It is unfortunate that several local units of government perceive the protection of the peatlands as meddling of state government in their "back yard," to which they are intensely opposed. Local governments, some foresters, and the timber industry still have little appreciation of peatland ecosystems. In addition, the timber industry does not appreciate the restrictions on the use and location of winter logging roads. Removal of ditches and prescribed burning are

expensive management actions and it is unlikely that the state will make that investment.

Finally, it is expected that the timber industry and local units of government will succeed in having the (state) Wetlands Conservation Act of 1991 amended to allow the use of corridors or disturbance (trails, ditch grades, etc.) for any motorized use (as opposed to allowing only those activities occurring at the time of act's enactment).

Contact information:

Mr. Bob Djupstrom
Scientific and Natural Area
    Supervisor
Minnesota Department of Natural
    Resources
Wildlife - SNA, Box 7
500 Lafayette Road
St. Paul, MN 55155
(612) 297-2357
Fax: (612) 297-4961
E-mail:
    bob.djupstrom@dnr.state.mn.us

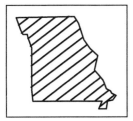

MISSOURI COORDINATED RESOURCE MANAGEMENT

Location:
Missouri

Project size:
45 million acres

Initiators:
Missouri Department of Conservation, U.S.D.A. Forest Service

PROJECT AREA DESCRIPTION

The landscapes of the southern half of the state of Missouri range from rolling prairie-oak woodland landscapes to rugged hills supporting oak and oak-pine forests. The area is primarily used for forest management and hay-pasture land. The Mississippi lowlands of southeastern Missouri consist largely of floodplains and were once covered in swamps, bottomland forests, and wetland ecosystems. The northern part of the state consists of flat to gently rolling plains which used to support prairies and woodlands.

Today, both the Mississippi lowland and the plains are used primarily as cropland and hay-pasture land. The state is home to more than 500 vertebrate animal species and more than 200 species of fish. Of special interest are the breeding populations of migrant neotropical songbirds in the forest interior of the lower Ozarks.

ECOSYSTEM STRESSES

The settlement of the state has led to the conversion of much of the land to agricultural production, especially row crops and pasture. Less than 2 percent of prairie and 10 percent of wetlands remain in the state. In addition, forest lands, while still abundant, have suffered from unregulated timber harvests, grazing, and altered disturbance regimes (especially fire). The big rivers and many other stream systems have suffered the effects of channelization, levees, reservoirs, as well as pollution or siltation from broad land use changes.

PROJECT DESCRIPTION

In response to a growing interest in ecosystem management and biodiversity conservation, the Missouri Department of Conservation (MDOC) and the U.S.D.A. Forest Service (USFS) created a biodiversity task force in the 1980s. In order to address concerns about natural systems and stresses on these systems, the task force recommended a regional planning approach. In response, seven land management agencies came together in 1993. These agencies are the MDOC, Missouri Department of Natural Resources, U.S. Fish and Wildlife Service, National Park Service, USFS, U.S.D.A. Natural Resources Conservation Service, and Army Corps of Engineers.

The goals of the regional planning effort, or "Coordinated Resource Management" (CRM) as it is officially known, are to conserve and restore healthy ecosystems while taking into account the sustainable production of commodities, and while maintaining and creating opportunities for outdoor recreation, education, and interpretation. In order to reach these goals, the state of Missouri has been divided into ten CRM regions. The boundaries of these regions are partially based on ecosystem criteria according to the hierarchical approach to ecosystems developed by USFS and as modified by MDOC.

The boundaries of several CRM regions coincide with major river basin boundaries. A separate management plan will be written for each of these regions, each with its own goals, objectives and strategies. The contents of these plans will depend heavily on the input of the public, which is obtained through public meetings, polls, surveys, and a tool-free number. All plans should be completed by the year 2000. Although the plans will have a fifty-year vision, they may be revised sooner if necessary. After agreeing on the plan, each agency voluntarily determines how it can contribute to the fulfillment of plan goals and objectives. On a voluntary basis, individual landowners are invited to participate as well. Varying strategies may be employed by each plan, for example, reintroduction of fire, erosion control, wetland and prairie restoration, and reforestation. Education of the public and state legislators is an important plan component.

STATUS AND OUTLOOK

All seven agencies have signed a memorandum of understanding, agreeing on the CRM process. Following six public meetings, a draft plan for the Lower Ozark region has been distributed for public review. A final draft will be completed in Summer 1996.

Four public meetings were held for the Grand River region, and work on a draft plan has begun. Planning for the next region, the Springfield Plateau, will be initiated in late 1996.

Factors Facilitating Progress

Several factors have been instrumental in the CRM planning process. They include the national focus on ecosystem management, the support of state administrators, and good interagency cooperation. In addition, many people have attended the public meetings. Since 93 percent of the land in Missouri is in private ownership, citizen participation from the onset of the project is crucial to the success of CRM.

Obstacles to Progress

The biggest challenges to the CRM process revolve around citizen input. How can fear of government and the idea that

CRM is a threat to the state's economic well-being be dispelled? How can the public be assured that their participation in CRM is voluntary, and will not jeopardize private property rights? How can they be assured that CRM will not evolve into a regulatory program? How can rural and urban people be brought together? These questions present hurdles that need to be overcome. In addition, the development and organization of resource information in a comprehensive, compatible format is also a challenge.

Contact information:

Ms. Sara Parker
Policy Analyst
Missouri Department of
  Conservation
P.O. Box 180
Jefferson City, MO 65102-0180
(573) 751-4115 ext. 345
Fax: (573) 526-4495

MISSOURI RIVER MITIGATION PROJECT

PROJECT AREA DESCRIPTION

Before the arrival of European settlers, the Missouri River was a turbid, braided prairie stream. Its width varied from a few hundred yards to half a mile, and it featured both shallow and deep areas. Flow in the river fluctuated greatly from season to season and year to year. Thus, the river supported a large diversity of aquatic habitats. Bottomland hardwoods, wet meadows, and prairies covered the floodplains, providing varied terrestrial habitats.

In the project area, four federally listed threatened and endangered species occur: the bald eagle, pallid sturgeon, least tern, and piping plover. In addition, the project area is home to twenty-five state listed species, including the chestnut lamprey, silverband shiner, and flathead chub.

Although some bottomland hardwood forests and prairies still exist in the riparian area, most of the floodplains have been converted to agricultural use. Large cities, such as Omaha, Kansas City, and St. Louis, are located on the floodplains. The Missouri River itself is used for navigation, commercial and recreational fishing, recreational boating, industrial and household water supply, and dredging of gravel and sand.

ECOSYSTEM STRESSES

Between 1912 and the late 1970s, the entire lower 735 miles of the river was channelized and lined with levees. The river became a 9-foot-deep, 300-foot-wide channel with very little habitat diversity. As a result, many species that historically occurred in the river are now in decline. Channelization has

caused the river to erode its bed and run deeper, and some of the side channels of the river have become perched above the main channel and are no longer connected to it. Many similar hydrologic connections between the river and its wetlands are no longer in place.

Once the river was channelized and the levees were in place, it became feasible to clear the associated floodplains of hardwoods and prairies and to plant crops. Clearing occurred in many cases all the way to the edge of the river. Terrestrial habitat was lost as a result. Non-point source pollutants that find their way into the river, such as sediments, agricultural chemicals, urban runoff, inputs from sewage treatment plants, and effluents from heavy industry pose additional problems.

PROJECT DESCRIPTION

In the mid-1970s the U.S. Army Corps of Engineers (ACOE), the U.S. Fish and Wildlife Service (FWS), and the states of Iowa, Kansas, Missouri, and Nebraska recognized the impact of the channelization of the Missouri River on its fish and wildlife resources. The six agencies subsequently set out to restore approximately 50,000 acres of fish and wildlife habitat in the lower 735 miles of the Missouri River.

The ACOE is the lead agency, with the four states and FWS actively involved in all phases of the project. During the planning phase, specific restoration sites were identified and prioritized, and restoration plans were developed for each site. In 1990, federal funds were made available to ACOE to initiate the implementation phase of the project.

Location:
Lower 735 miles of the Missouri River—southeastern Nebraska, northeastern Kansas, southwestern Iowa, northwestern Missouri

Project size:
50,000 acres

Initiators:
U.S. Army Corps of Engineers, U.S. Fish and Wildlife Service

One major strategy is the restoration of habitat on existing state and federal lands. A second strategy is to acquire land from willing sellers for habitat restoration. In both cases, restoration includes reforestation and the hydrologic reconnection of wetlands and side channels to the Missouri River.

STATUS AND OUTLOOK

It is too early to determine if any of the goals of the project have been realized. Habitat evaluation guidelines have been developed to be used over time to document the benefits of the project. Approximately 8,500 acres of land have been acquired and habitat restoration has started. It is expected that terrestrial habitats will recover more rapidly than aquatic habitats.

Factors Facilitating Progress

So far, excellent cooperation between federal and state agencies has been particularly helpful in the progress of the project. Federal agencies have benefited from the strong and united support of the states and their political representatives in Washington, D.C.

Obstacles to Progress

A major interruption resulted from the floods of 1993, which required the Army Corps of Engineers to direct its attention elsewhere. Continuation of funding is presently a concern. The project is 100 percent federally funded, with annual appropriations required.

Contact information:

Mr. Steve Adams
Natural Resources Coordinator
Kansas Wildlife & Parks
900 SW Jackson, Suite 502
Topeka, KS 66612-1233
(913) 296-2281
Fax: (913) 296-6953

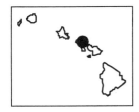

MOLOKAI PRESERVES

Location:
Island of Molokai, Hawaii

Project size:
22,000 acres

Initiator:
The Nature Conservancy

PROJECT AREA DESCRIPTION

The northeastern mountain range of Hawaii's Molokai island is a region characterized by sheer sea cliffs, high narrow mountain ridges, and low steep valley gulches. This area hosts an array of natural communities, including low-land and montane grasslands, dry and moist shrublands, and mesic, tropical, and summit cloud forests. Several rare natural communities are found here, including the Hawaiian Continuous Perennial Stream community, and the 'Ohi'a Mixed Montane Bog and Montane Wet Piping Cave, both found only on Molokai. Many state and federally listed threatened, endangered, and rare species are present as well (mostly plants), including the 'i'iwi, 'o'opu alamo'o, oloma'o, kakawahie, alani, and koloa maoli.

The Pelekunu Valley, one of three major valleys of this mountain system, has one of the last undiverted stream systems on the island. Pelekunu Preserve makes up 89 percent of this valley and is owned by The Nature Conservancy (TNC). Adjacent to Pelekunu Preserve is the 2775-acre Kamakou Preserve, owned by Molokai Ranch with a conservation easement held by TNC since 1983, when TNC first began using adaptive management strategies. Kamakou is one of the primary sources of water for irrigation on the island. Bordering these preserves are the state owned Puu Alii and Olokui Natural Area Reserves, and Kalaupapa National Historical Park (federal). Together, these areas protect more than 22,000 acres of contiguous native ecosystems.

With a population of 7000, Molokai is a traditional, rural, agriculturally-based island. Current land uses consists mostly of grazing and farming. Subsistence and recreational hunting are also prevalent in the region.

ECOSYSTEM STRESSES

Large populations of feral goats, pigs, and axis deer, as well as domestic live-stock, are the primary threat to native vegetation on the island. Because Hawaii has no native terrestrial mammal species, its native plant species are easily destroyed by grazing and trampling. Humans have also introduced fire to these natural systems which encourages invasive, fire-adapted species to establish in the place of natural plant communities. Roads and pine plantations have further impacted native vegetation. Roads serve as invasion routes for non-native species, while conifers encroach on adjacent native communities.

Runoff and siltation from agricultural practices and feral animal damage are degrading the island's reef systems. Furthermore, there is a potential threat of water development for irrigation. The state has plans to draw water from Pelekunu and Wailau valleys, gravely threatening these watersheds. There is also the potential for urban development, which would impact both forest and water systems on the island.

PROJECT DESCRIPTION

TNC's primary goal for this project is to preserve the natural diversity of Molokai. Recognizing that conservation efforts on TNC preserves alone would not achieve this goal, TNC felt a need to affect conservation on an island-wide basis. To foster understanding and support from the community, TNC began training and hiring people from the local community in 1993 to work in preserve management. At this time, TNC also formed an advisory committee made up of respected leaders in the community dedicated to protecting the island's natural heritage.

To address present threats to the ecosystem, TNC works closely with state and federal agencies on adjacent lands, including the Hawaii Division of Forestry and Wildlife, State Historical Preservation Division, National Park Service, and U.S. Fish and Wildlife Service. TNC also works closely with local hunters, assisting them in reaching remote regions in an attempt to control feral mammal populations. Finally, TNC has helped the state develop and receives funding from the Natural Area Partnership Program (NAPP), an program providing funding (two-thirds state, one-third private) for the conservation efforts of private landowners.

STATUS AND OUTLOOK

It appears that native forest vegetation is recovering in some areas, with a decrease in alien mammal and weed species. However, it is too early to detect whether or not the ecosystem is doing better as a whole. TNC feels that continued ecological improvements are dependent on greater community support for conservation efforts.

TNC is now focusing more on outreach, getting involved in community conservation programs inside and outside the preserves, such as volunteer reef protection and weed eradication programs, and educational events that emphasize the ecological, economic, and cultural values of Molokai's natural resources.

Factors Facilitating Progress

Hiring people who are part of the community has helped TNC obtain an image as a local group making local decisions. This has encouraged greater community support and involvement in their conservation initiatives.

Obstacles to Progress

A controversy over TNC's policy on alien species eradication lead to the loss of local support and brought into question TNC's management practices. Animal rights groups and local hunters alike raised objections to TNC's practice of snaring feral mammals. The animal rights groups raised objections to the inhumane killing of these animals while the hunters felt that snaring threatened traditional subsistence hunting practices. There has also been some opposition to TNC's conservation efforts from supporters of the development of Molokai's land and water resources. Finally, TNC lacks adequate funding, staff, and other resources to fully carry out its conservation goals.

Contact information:

Mr. Ed Misaki
Director of Programs
The Nature Conservancy of
  Hawaii
Molokai Preserves
PO Box 220
Kualapuu, HI 96757
(808) 553-5236

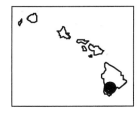

NATURAL RESOURCE ROUNDTABLE

Location:
Kona and Kohala, Island of Hawaii

Project size:
Kona—100,000 acres; Kohala— 22,000 acres

Initiators:
Hawaii State Office of Planning, Hawaii State Department of Land and Natural Resources, Hawaii County

PROJECT AREA DESCRIPTION

The roundtable covers two Natural Resource and Watershed Areas in Hawaii. The larger Kona project site is located on the slopes of Mauna Loa and Hualalai mountains, and ranges from gently sloping to steep lands. The land is characterized by a thin layer of soil over basalt and lies over a high-quality aquifer, with rainfall and fog drip serving as the aquifer's recharge source. The vegetative cover ranges from dense stands of relatively intact native forest to scattered native-exotic mixes, grasslands with sparse tree coverage, and open lava fields in the process of re-colonization by native plant species.

The smaller Kohala site, on the flank of the Kohala mountain range, is considered an important water recharge area. It contains native forests ('ohi'a primarily) and pasture lands. Both areas contain critical habitat for native honeycreepers (birds) and 'alala (Hawaiian crow), all of which are listed as threatened or endangered by the federal and state governments. The lands are predominantly in private ownership and are used for pasture (cattle primarily) and forestry.

ECOSYSTEM STRESSES

The most significant stresses are conversion of the land for suburban subdivisions and non-point source pollution due to erosion from logging, grazing, and roads. The native koa tree is a highly-valued, high grade timber and their removal degrades native forest bird habitat. Exotic species (pigs, feral sheep, goats, non-native birds, mosquitoes, non-native grasses) and alteration of hydrology (due to removal of trees essential to capturing fog drip) are other

notable stresses. Less significant stresses include livestock grazing and road construction associated with development.

PROJECT DESCRIPTION

In 1992, the State Land Use District Boundary Review, conducted every five years, identified the Kona and Kohala Natural Resource and Watershed Areas as areas of critical concern. Subsequently, the Hawaii state legislature requested the State Department of Land and Natural Resources (DLNR) and Hawaii County to conduct roundtable discussions to address management and protection concerns of the area. The Center for Alternative Dispute Resolution of the Judiciary of Hawaii, created by state law to assist in out-of-court settlements of complex litigation and public policy disputes, was requested by the agencies to facilitate the roundtable discussions.

Starting in 1992, a roundtable of landowners, developers, environmental and community groups, and government agencies met over the next two years, initially to determine the various interests of the region and its users, and later to frame a common set of principles, leading finally to the development of a set of recommendations. The group sponsored a series of three informational meetings in 1993 to bring all stakeholders up to the same level of scientific and regulatory knowledge about the area. All of the roundtable group's decisions were reached by consensus.

STATUS AND OUTLOOK

Proceedings of the 1993 meeting were published that year. Officially, the roundtable ended in mid-1995, with a report outlining the principles and recommendations of the roundtable published in July 1995. The most important outcomes are described as better awareness and recognition of the importance of the natural resources and water recharge characteristics of the area. The roundtable also proposed legislation designed to provide a source of funds accessible to both private and government landowners to optimize natural resource/watershed management on sensitive lands. This legislation was introduced in the 1996 state legislative session.

Other important outcomes include better communication, understanding, and cooperation between stakeholders; enhanced public awareness; and, indirectly, changes in the state DLNR so that sustainability is the primary goal of all of the department's activities. Finally, there was greater recognition that property rights, property owner cooperation, and economic factors needed to be in balance in any recommendations. Another roundtable is being considered to address more specific 'next step' implementation issues identified in the 1995 report.

Factors Facilitating Progress

Strong political support by the former governor and administrative support by the state natural resource agency have been essential to the effort's progress. A willingness on the part of private landowners, despite significant misgivings, to recognize the importance of the issues and work with all parties toward creative solutions, clarity of project goals, and leadership by the State Office of Planning and roundtable participants were also described as benefits to the process. Finally, increased scientific understanding of the area has become a priority and the general public appears to be in support of the effort.

Obstacles to Progress

While many stakeholders have participated in the roundtable discussions, landowner and developer resistance to the effort has been significant, based on their past experience with government regulations and their perception that the state's five-year boundary review was excessively aggressive in its efforts to apply broad-brush regulatory downzoning as a catch-all solution to a far more complex set of issues. Thus, not all stakeholders are in agreement as to a common vision for the area's future, and stakeholder conflict has yet to be reduced uniformly. Support from federal and local public agencies has not been as high as from state agencies.

A new state governor and a state budget crisis have clouded the effort's future. The new governor is attempting to remove regulatory controls for natural resource management in favor of free-market forces. His focus is on economic development strategies to jump-start the local economy.

The budget crisis has resulted in significant downsizing of the state planning and natural resource agencies, in both personnel and appropriations. Those landowners who became engaged in the process out of fear that the state would impose regulatory restrictions no longer seem as interested in the process; they feel the regulatory "teeth" may no longer be present. Nevertheless, the majority of landowners and public are still committed to finding ways to produce better resource management strategies by partnering with the government.

In any case, funding for the effort, for future roundtables and public and private resource management incentives, had never been guaranteed beyond yearly appropriations by the state legislature and agencies.

Contact information:

Mr. Scott A.K. Derrickson
Hawaii Office of State Planning
PO Box 3540
Honolulu, HI 96811-3540
(808) 587-2805
E-mail: sderric@pixi.com

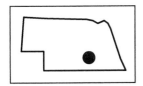

NEBRASKA SANDHILLS ECOSYSTEM

Location:
North-central Nebraska

Project size:
12.5 million acres

Initiator:
U.S. Fish and Wildlife Service

PROJECT AREA DESCRIPTION

The dunes which form Nebraska's sandhills were created by wind-blown sands. These sands are now held in place by mixed grass prairie vegetation. The groundwater table beneath the dunes is very high. Groundwater discharge gives rise to spring-fed streams and various types of wetlands (1.3 million acres in total) in the valleys between the dunes. Deciduous trees, such as willows and cottonwoods, can be found along some of the streams.

Twelve federally listed threatened and endangered species can be found in this ecosystem. Some examples include Hayden's penstemon, the prairie-fringed orchid, and American burrowing beetle. In addition, six state listed species occur here, four of which are fish living in the spring-fed streams. Currently, over 95 percent of the sandhills are covered by native grassland. The lands are primarily in private ownership and used for grazing. In addition, limited row crop agriculture takes place. However, due to the poor, sandy soils, crop agriculture has not been successful in the past.

ECOSYSTEM STRESSES

In order to maintain a year-round ranching operation, cattle need to be fed hay during the winter. Wetland areas are capable of producing much of this hay. To obtain this hay, ranchers drained many wetlands, starting in the early 1900s. As the wetlands were drained, their buffering function was lost. Water that used to be absorbed by wetlands ended up in streams. As a result, stream bed erosion accelerated and the groundwater table lowered. This has led to a subsequent thinning of vegetation, and, in turn, to a higher susceptibility of the soil to erosion.

In addition to causing drainage of wetlands, agriculture has led to desertification in small areas within and adjacent to the ecosystem, resulting from exposure of bare soils. However, in some areas, irrigation has led to creation of wet areas, through the pumping of groundwater to the surface, although groundwater has also been polluted with nutrients and herbicides.

PROJECT DESCRIPTION

In recognition of the uniqueness of the ecosystem, the U.S. Fish and Wildlife Service (FWS) drafted a program in 1980 designed to purchase easements on the wetlands to protect them from drainage. After vehement opposition of local landowners, the plan was discontinued. In 1990, a different approach was taken, when a sandhills coordinator was hired whose main task was to develop a program with the support of the landowners. As a result, the Sandhills Task Force was formed. The majority of its members are local landowners, many of whom were selected by locally active organizations such as the Nebraska Cattlemen, Upper Loup Natural Resources District, and the Nebraska Association of County Officials.

After much initial distrust between the landowners and agency representatives, participants discovered commonalities between landowner and wildlife needs. Subsequently, the task force has developed into a team of people who trust and respect each other, and who make decisions by consensus. The task force generated a management plan, which, after incorporation of public comments, was signed in September 1993. The plan has been implemented since then. The task force members decided to remain active as a group to help carry out the objectives of the plan. In January 1995, task force membership was extended to a representative of The Nature Conservancy.

The overall goal of the plan is "to enhance the sandhill wetland-grassland ecosystem in a way that sustains profitable private ranching, wildlife and vegetative diversity, and associated water supplies." In order to reach that goal, a variety of strategies are employed. Since implementation began, several educational projects have been organized, including for instance, a two-day workshop on ranching and environmental issues. Another strategy involves the provision of technical assistance to landowners in wetland restoration and erosion control. The plan also recommends land acquisition or land easements if deemed appropriate. However,

the sandhills management plan states that land acquisition would only "be a last alternative to ensure that unique ecosystems will remain."

STATUS AND OUTLOOK

The new management approach has already proven fruitful. Successful projects have been completed throughout the region. Not only are wetlands being restored, but the relationship between local landowners and FWS has greatly improved. Congressional support was voiced to the secretary of the U.S. Department of Interior commending this program. In addition, the management plan has gained the support of nongovernmental organizations, agencies, and the governor. This has encouraged others to participate.

Factors Facilitating Progress

Open communication and the discovery of common ground between landowners and FWS has greatly benefited the project. The involvement and support of

local landowners from the onset of the project has been invaluable. Support of elected officials has also been beneficial.

Obstacles to Progress

Currently, work carried out by agencies such as the U.S.D.A. Natural Resources Conservation Service, the Nebraska Game and Parks Commission, and the FWS is coordinated by the sandhills coordinator. Unfortunately, the coordinator does not have any staff, which is greatly needed. Funding is lacking for additional positions. In addition, future local opposition is anticipated, for instance, if the task force should support land acquisition by The Nature Conservancy.

Contact information:

Mr. Gene Mack
Sandhills Coordinator
U.S. Fish & Wildlife Service
Kearney Field Office
PO Box 1686
Kearney, NE 68848
(308) 236-5015
Fax: (308) 237-3899

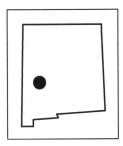

NEGRITO PROJECT

Location:
Southwestern New Mexico

Project size:
120,000 acres

Initiators:
Private citizens, U.S.D.A. Forest Service

PROJECT AREA DESCRIPTION

The Negrito Creek watershed in New Mexico's Gila National Forest has a varied landscape, ranging from river bottom to the upper Sonoran Desert and sub-alpine mountain slopes. The watershed is 80 percent forested, including ponderosa pine, mixed conifers, and pinyon-juniper woodlands. The remaining 20 percent consists of grasslands and cienega-type meadows.

The watershed provides habitat for the Mexican spotted owl and Gooddings onion, both federally listed threatened species. It is also home to the Apache northern goshawk, classified by the U.S.D.A. Forest Service (USFS) as a sensitive species, and the spotted bat, a USFS species of concern. The region also supports a very large elk herd. Due to protection efforts for the spotted owl, restrictions on logging have drastically reduced timber harvest from the watershed. As a result, small sawmills have been forced to close and local communities have experienced high unemployment and depressed economies.

ECOSYSTEM STRESSES

The most profound stress to the watershed has been the effects of fire exclusion, which began in the early 1900s. In the absence of fire, a "vast sea" of abnormally dense, even-aged ponderosa pine stands have developed, while pinyon and juniper encroach on meadows and grasslands. Another problem in the watershed has been erosion damage to riparian areas, caused by year-round flooding and grazing of livestock and elk. The relationship between past logging activities and landscape fragmentation is a current source of debate.

PROJECT DESCRIPTION

Under the 1992 USFS New Perspectives program, a coalition of citizens proposed an ecologically-based pilot project to the Gila National Forest supervisor. The proposal was adopted and a working group was formed, composed of a diverse array of dedicated and experienced individuals from five agencies, two conservation groups, two industries, two educational institutions, and the local community.

The Negrito watershed was chosen for the project for four major reasons: 1) the USFS had good existing field data on the watershed as well as geographic information systems (GIS) hardware and expertise; 2) there were few scheduled management activities that would affect project recommendations; 3) Negrito Creek is a critical watershed to the Gila River system; and 4) severe county-wide economic stress has not been successfully relieved by traditional resource management practices.

This project is unique in that most of the work has been carried out by non-agency personnel; this is a community-based project where all planning and decision-making efforts are carried out by consensus through facilitated meetings. The overriding goal of the project is to have a watershed that sustains over time its ecological functions while providing for the human community. The project addresses economic concerns, with a goal of maintaining resource-based employment (i.e., supporting surviving local, small-scale sawmill operations, and maintaining grazing through range improvement projects and new grazing systems).

STATUS AND OUTLOOK

A series of small-scale watershed improvement projects developed by the working group are being funded and implemented in a prioritized fashion. An information needs assessment has been completed and extensive field data was collected during Summer 1995. Currently, the field data is being entered into a database for GIS analysis. By the end of September 1995, preliminary statements of desired future conditions had been developed based on the data, as was a list of proposed projects.

The working group is trying to set up a nonprofit community watershed organization to serve as a focal point for broad-scale planning, fundraising, and other activities separate from USFS.

Factors Facilitating Progress

The commitment from members of the working group to a consensus-based process, the emphasis on keeping the project community based, and keeping human needs in the equation have all lent to the progress of the project thus far. The project has also received strong support from the USFS forest regional supervisor and funding from USFS economic diversification and ecosystem management programs. Finally, having GIS in place has made many broad-scale planning efforts possible.

Obstacles to Progress

The Federal Advisory Committee Act (FACA) may limit this effort, as there are non-federal members in the working group. There are continuing differences between how the public and USFS define "success" and what desired future conditions should be. For example, some opponents from both environmental and commodity interests, whose agendas are too restrictive to participate in this collaborative process, have attempted to derail planning efforts.

The federal budget stalemate in early 1996 was beginning to cause concerns for the project as budgets became tighter. Also, more forest districts are adopting ecosystem management activities, causing greater internal competition for fewer and fewer administrative resources.

Contact information:

Mr. Don Weaver
USDA Forest Service
Gila National Forest
Reserve Ranger Disitrict
PO Box 170
Reserve, NM 87830
(505) 533-6231

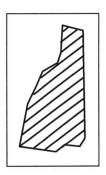

NEW HAMPSHIRE FOREST RESOURCES PLAN

Location:
New Hampshire

Project size:
5 million acres

Initiator:
New Hampshire Division of Forests and Lands

PROJECT AREA DESCRIPTION

Because this project is a New Hampshire state agency policy, the project area is defined by the state's boundaries. The landform varies from mountainous terrain in the northern portion of the state to the Connecticut River Valley and the small seacoast area to the south. The state is 85 percent forested, with spruce fir in the high elevations, predominantly northern hardwoods in the north, and a mix of white pine and red oak in the south. Federally listed threatened and endangered species include the small whorled pogonia. Significant landholdings are in public ownership, including the White Mountain National Forest at 740,000 acres. The predominant land use in the northern part of the state is forest products, agriculture in the Connecticut River Valley, and manufacturing and service industries in the southern portion.

ECOSYSTEM STRESSES

The biggest threat to the area is the conversion of forest to non-forest uses, occurring most heavily in southern New Hampshire due to the urban sprawl northward from Boston. Technology has also made it easier for people to live further from their office. The result is an increase in development and fragmentation from large to small blocks. These blocks have no management strategies. Roads often run through traditional wildlife habitat.

PROJECT DESCRIPTION

Forest resource plans are legislated policy documents that direct state forest policy for a period of ten years. The plan currently being drafted is the fourth forest plan in New Hampshire; the first was in 1952 and the most recent in 1980. In the past, these plans tended to be issue-oriented, looking at forest resources as a commodity and focusing on the forest industry. The current plan is quite different, reflecting the change in knowledge and scientific base with regard to forest resources. This plan has a more ecologically-based approach and is vision driven instead of issue driven. This new direction of forest planning comes on the heels of the work of the Northern Forest Lands Council (see page 215), an effort focused on maintaining large forest tracts in the region.

A steering committee of twenty-eight people from diverse backgrounds was brought together by the state forester in April 1994 to guide the forest resources plan. This group comprises landowners, forest industry, state resource agencies, and property rights and environmental groups. The steering committee outlined a vision for the desired future landscape condition for the next 50 to 100 years and recorded thirteen of the biggest challenges in reaching this vision. In order to assess the current condition of the forest, a group of forty-five additional people were gathered into three assessment groups: economic issues, ecological resources, and human and social values.

STATUS AND OUTLOOK

A 250-page assessment was presented to the steering committee by the assessment groups. The committee will develop a plan that will meet the challenges based on the information from the assessment, relying on their own knowledge and expertise as well as on people from the forest resources community. The plan is due to be completed in 1996.

Factors Facilitating Progress

New Hampshire has a history of cooperative policy development. The forest resources and environmental communities are small: most of the people have worked together previously on other projects. This familiarity aided during plan development.

Obstacles to Progress

The project follows an ambitious schedule due to legislated deadlines. Since much of this work is being carried out for the first time, working out details may take longer than anticipated, making deadlines more challenging to meet.

Contact information:

Ms. Susan Francher
New Hampshire Division of
*   Forests and Lands*
PO Box 1856
Concord, NH 03302-1856
(603) 271-2214

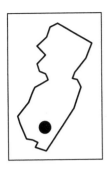

NEW JERSEY PINELANDS

Location:
Southern New Jersey

Project size:
1.1 million acres

Initiators:
Environmental organizations, Private citizens

PROJECT AREA DESCRIPTION

Located in the midst of the densely-populated northeast corridor, the New Jersey Pine Barren ecosystem is characterized by droughty, sandy soils underlain by the 17 trillion gallon Cohansey Aquifer, one of the largest sources of water in the world, and vital to the 30 million nearby residents. With nearly flat topography, cedar and red maple dominate the region's lowlands, while scrubby pine and oak forests, with an understory of plants of *Ericacea* (heath family), are typical of the highlands. A significant concentration of rare and unusual plants and animals can be found in the Pinelands, including the federally listed swamp pink. In addition to undeveloped forests, some in preserves, other land uses include commercial and industrial development, sand and gravel mining, glass manufacturing, farming, housing, forestry, and recreation.

ECOSYSTEM STRESSES

Disruption of the fire regime is one of the greatest stresses to this fire-dependent ecosystem. Non-point source pollution from agricultural fertilizers and pesticides, as well as from septic systems and landfill leachates, is a significant stress to water quality, because of the sandy substrate that easily drains to the aquifer. Additional stresses, such as rural and urban development, mining, and recreation, result in habitat destruction and fragmentation.

PROJECT DESCRIPTION

A proposal for a major new airport in the area in the 1970s was the catalyst for current protection of the Pinelands. The support of then-Governor Brendon Byrne and the strong voice of environmentalists were vital in gaining protection for the region under the National Parks Act of 1978 (federal) and the New Jersey Pineland Protection Act of 1979 (state). Shortly thereafter, the Pinelands Commission was formed, with seven appointees of the governor, seven appointees of the stakeholder counties, and one appointee from the federal government.

In 1981, a comprehensive management plan (CMP) was published, focusing on the ecological significance of the Pine Barrens and on land use planning. The CMP is coordinated at the state level, with local government responsibility focused on compliance plans and only minimal federal involvement. The CMP operates through prescriptive zoning designations and strict performance standards, with mandatory local compliance and regional oversight.

The CMP divides the region into two main areas based on different development allowances. The core Preservation Area of 300,000 acres has the most restrictive guidelines. The Protection Area contains the remainder of the acreage and is further delineated into six areas, each with a different focus and growth allowance. These allowances include the following: regional development and rural development areas, agricultural production areas, forest areas, military areas, and towns and villages. The overall goals of the two main areas include 1) preserving the essential character of the Pinelands environment, including plant and animal species; 2) promoting the continuation and expansion of horticulture and agriculture; 3) discouraging piecemeal development; 4) and protecting the quality of surface and groundwater.

STATUS AND OUTLOOK

Since 1981, the commission has set environmental and development standards for fifty-three municipalities and seven counties. Over the past fifteen years, 93 percent of new homes in the Pinelands have been constructed in the four management areas designated for growth, areas representing 25 percent of the Pinelands. Only 5 percent of the new homes have been in two management areas designated for greater protection (57 percent of the Pinelands). The two agricultural management areas (12 percent) have received only 1 percent of the new homes.

This project, created before ecosystem-scale management was seriously considered, has realized its principal objectives: preventing development in critical habitat and maintaining water quality. The current focus is to continue to manage development and monitor ecological and economic factors as an indicator of the project's success.

Factors Facilitating Progress

The initial support from Governor Byrne and the continued support from the environmental community and state residents has been key to progress. Strong national and state legislative protection has also been critical for progress, as well as the abundance of regional natural history data on the region.

Community involvement in the CMP drafting, and eventual economic support programs, helped offset local opposition with regard to the plan's early negative economic consequences on some individuals and communities. Finally, an abundance of natural history data on the region was instrumental, especially in the early phases of the effort.

Obstacles to Progress

The commission does not have enforcement authority similar to other regulatory agencies, which can be a problem. Continued pressure from developers and private property rights advocates and the initial exclusion of one-fifth of the ecosystem in the legislative management area delineations, are additional concerns. However, legislative efforts to enact "takings" legislation and even eliminate the Pinelands Commission have received little or no support. Conversely, the environmental community would like to see stronger controls set in place.

Contact information:

Mr. Don Kirchhoffer
Project Manager
Pinelands Preservation Alliance
114 Hanover Street
Pemberton, NJ 08068
(609) 894-8000
Fax: (609) 894-9455
E-mail: DKirk100@aol.com

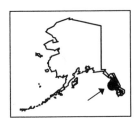

NORTHEAST CHICHAGOF ISLAND

Location:
Southeastern Alaska

Project size:
275,000 acres

Initiators:
*U.S.D.A. Forest Service,
Alaska Department of Fish and Game*

PROJECT AREA DESCRIPTION

Part of the Alexander Archipelago, Chichagof Island lies across the straits from Glacier Bay National Park, about fifty miles southwest of Juneau. A 300 yard wide land bridge connects the northeast sector to the rest of the island. Anadromous fish streams run to the island's bays and estuaries. The limestone ridge that splits the island is a demonstration of the underlying karst topography.

Hemlock, Sitka spruce, muskeg, and blueberrry cover the island. The subalpine forests consist of yellow cedar and shore pine (a dwarf type of lodgepole pine). Over 233,000 acres of northeast Chichagof are Tongass National Forest lands. Timber harvesting has occurred on 23,000 acres. Patented claims for gypsum and silver mining exist, but none are currently active. The north end of the island contains the bulk of native and native over-selection lands (42,000 acres).

ECOSYSTEM STRESSES

Timber harvesting is the most prominent stress on the ecosystem. Native corporations, not subject to the same laws and restrictions as the U.S.D.A. Forest Service (USFS), have been logging more heavily than the USFS, resulting in unsustainable harvesting levels on some native lands. All eighteen of the island's watersheds have been entered for logging. Although heavy equipment was driven up some streams to access timber in the Tongass, this did not occur on northeast Chichagof Island. Yet, 250 kilometers of roads have been built in the region since 1980. In some areas, road layout, poor drainage control, undersized and unaligned culverts, and the lack of a realistic road maintenance program have contributed to fish passage problems, sedimentation, and altered hydrologic regimes. Road building associated with timber harvesting has resulted in increased access to previously inaccessible areas, leading to overhunting of brown bears.

PROJECT DESCRIPTION

This initiative is a joint research project of the Alaska Department of Fish and Game (DFG) and the USFS. Alaska DFG has provided input to USFS on timber projects since harvesting began in the mid-1980s. The realization that logging was occurring at an unsustainable rate and threatening wildlife populations sparked the formation of a four-member team of DFG and USFS personnel in 1993.

Northeast Chichagof Island is being used as a demonstration project for ecosystem-based approaches that can be exported to other parts of Tongass National Forest. The goal of the project is to provide information needed to make landscape-scale predictions concerning silvicultural practices and wildlife habitat needs over the length of the timber rotation period. Previously, all timber projects were analyzed individually; cumulative effects were not assessed.

The team is conducting a woodpile analysis to determine sustainable yields for northeast Chichagof Island. The analysis begins with the 200,000 acres of USFS lands. Buffer zones for fish streams, areas with hazardous soils, and areas where timber is broken up by muskeg are subtracted to determine the total area where harvesting could occur.

Habitat Conservation Areas (HCAs), designated to support minimum viable populations of species with the largest ranges, are also withdrawn so that harvesting activities do not negatively impact wildlife. With growth rates for the forest in hand, the project team can determine the sustainable yield for 100-year rotations.

STATUS AND OUTLOOK

In January 1995, the Tongass forest supervisor unilaterally disbanded the eco-team assigned to the project before the final season of data analysis and some on-the-ground implementation of the findings could occur. However in early 1996, USFS decided to spend a limited amount of time to prepare a final report. The report, likely to be published in mid-1996, will not include all components of the project due to time and data restrictions; it will focus instead on three or four components with reportable accomplishments.

Factors Facilitating Progress

The public's acceptance of eco-system-based approaches was cited as a factor facilitating progress, although the public has not been involved directly as a partner. The expertise and technical capability of the eighty-five resource specialists within the Alaska DFG, USFS, and other state and federal agencies participating in various review roles were invaluable assets to the project. Geographic information systems data allowed information to be communicated more quickly than with past projects.

Obstacles to Progress

Gathering information has been expensive due to inter-island travel. Technical concerns include not having a control area for the project due to the previous disturbances to all of the island's watersheds. Some scientists are concerned that basing HCAs on minimum viable populations is risky to the long-term survival of certain species.

Because the eco-team was disbanded before its work was completed, the findings in the final report may never be applied to management of the Tongass, as the report expected to reflect the findings only partially, and there is no planned follow-up for implementing recommendations. Finally, the planning process has not included the general public at any step of the project's development. As a result, there are no advocates for the project aside from the agency personnel involved.

Contact information:

Mr. Phil Mooney
Habitat Biologist
Alaska Department of Fish & Game
Habitat and Restoration Division
304 Lake Street
Room 103
Sitka, AK 99835-7563
(907) 747-5828
Fax: (907) 747-6239

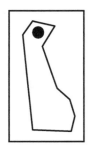

NORTHERN DELAWARE WETLANDS REHABILITATION PROGRAM

Location:
Northern Delaware

Project size:
10,000 acres

Initiator:
Delaware Department of Natural Resources and Environmental Control

PROJECT AREA DESCRIPTION

The Northern Delaware Wetlands Rehabilitation Program is focused on an area that includes 10,000 acres of tidal freshwater and brackish wetlands. The project is located along the urban corridor of the Christina and Delaware rivers in New Castle County. The ecosystem is dominated by wetland vegetation including red maple and cattail. Federally listed threatened and endangered species include the bald eagle and short-nosed sturgeon. Land usage is dominated by residential, urban, and industrial development.

ECOSYSTEM STRESSES

The wetlands in this area have been subject to a long history of impoundment for farming purposes, starting as early as the 1600s. Many dikes have been maintained for 300 years. As a result, the marshes have been excluded from tidal exchange and cut off from the natural system. Many wetlands in the area have also been drained and filled for a variety of reasons, including agriculture, landfills, and industrial development. Degradation of wetlands was lessened with the passage of state and federal wetland acts. Highly-polluted Superfund sites, resulting from a large port in the area serving the chemical industry, continue to threaten the wetlands. Non-point source pollution and the nuisance plant species *Phragmites* are also current stresses.

PROJECT DESCRIPTION

The Northern Delaware Wetlands Rehabilitation Program is a regional, non-regulatory restoration program. The program was initiated by the Delaware Department of Natural Resources and Environmental Control in conjunction with civic and business leaders, scientists, resource managers, and property owners. An inventory of wetlands was conducted and thirty-one wetland sites have been identified as needing rehabilitation. A two-tiered organizational approach was established with a steering committee and an adjunct committee. A multi-agency approach is encouraged, thus replacing the traditional individual agency approach. In order to increase public awareness to better ensure long-term stewardship, the community is actively involved in management of the site.

Wetland rehabilitation plans are developed for each site, constructed around scientifically-based biological inventories and ecological evaluations of the site area. The rehabilitation plans call for site-specific goals and action steps with measurable evaluative criteria. Goals include reestablishing tidal exchange and *Phragmites* control plans.

A systematic management procedure was developed, which provides a framework to apply across numerous sites. This planning document is expected to serve as a model for future coastal and wetland restoration programs.

STATUS AND OUTLOOK

Currently, eleven of the wetland sites are in the planning and restoration phases. Planning includes preparing for restoration activities, putting partnerships together, and trying to secure funding for restoration. Five of those sites are undergoing restoration, with two of these essentially completed as of early 1996.

Factors Facilitating Progress

The responsiveness of the wetlands to rehabilitation efforts has been a benefit to this effort's progress. A high level of cooperation from landowners and support from the community have been crucial. Another benefit has been regional federal and state funding opportunities that were recently realized.

Obstacles to Progress

Several marshes, initially expected to be easy to rehabilitate, turned out to be far more difficult to restore, because Superfund-class contaminants were found on the sites.

The logistics of organizing and maintaining communication channels between many groups was a second obstacle. Site-specific operation, maintenance, and monitoring plans are also required for each site. Ensuring long-term future funding for these efforts is an ongoing challenge .

Contact information:

Mr. Robert Hossler
Delaware Division of Fish &
   Wildlife
250 Bear/Christiance Road
Bear, DE 19701-1041
(302) 323-4492
Fax: (302) 323-5314
E-mail: rhossler@dnrec.state.de.us

NORTHERN FOREST LANDS COUNCIL

Location:
Maine, New Hampshire, Vermont, New York

Project size:
26 million acres

Initiators:
Governors Task Force, Congress

PROJECT AREA DESCRIPTION

The study area is most closely defined by the ecological boundary of the spruce-fir zone in the northeastern United States, extending over several states (but not into Canada). Landforms in this area range from lowland swamps to high mountain ranges. The vegetative cover, depending on the topography, is predominantly a mix of spruce-fir and northern hardwoods. Federally listed threatened and endangered species in the area include the bald eagle, eastern mountain lion, and osprey. The most common land use is private timber production; a small amount of agriculture is present. A portion of land is maintained in private reserves, as well as in state forests and parks.

ECOSYSTEM STRESSES

Threats to the ecosystem include suburbanization and road development northward from Boston. Development is especially drawn to waterfront areas of lakes and major rivers, and can harm water quality. Population growth in the Northeast and increased recreation use in highly scenic or fragile areas also stresses the ecosystem. Intensive timber production can impair wildlife habitat and water quality. Periodic spruce budworm epidemics have affected the ecosystem as well.

PROJECT DESCRIPTION

The project was motivated by public concern over large-scale changes in landownership patterns and traditional uses. Due to increased land values in the 1980s, many lands were transferred from regional companies to multinational corporations with little allegiance to the region. The public was concerned that these companies were looking at timber as an undervalued asset without considering the environmental or social implications.

In 1988, Congress funded the Northern Forest Lands Study. A four-state governors task force guided the study, and recommended formation of the Northern Forest Lands Council. The council, funded by Congress for four years beginning in 1990, was comprised of four members from each state and one from the U.S.D.A. Forest Service. Their goal was to make recommendations to both Congress and the governors on what could be done to maintain the large forested tracts of lands from being converted to nonforest use. The council submitted a final report of thirty-seven recommendations in September 1994. Many of these recommendations are not specifically ecosystem focused.

The recommendations fall into the following categories: foster stewardship of private lands, protect exceptional resources, strengthen the economies of rural communities, and promote more informed decisions. An example of one of the specific strategies designed to reach the stewardship goal is tax equity—taxing forest land at its use value instead of its development value. Changing current tax policies aims to prevent driving land toward subdivision and development.

STATUS AND OUTLOOK

The council dissolved in September 1994 and their report is now being implemented at the state and federal levels, albeit selectively since the recommendations are voluntary. This is a long-term effort and many recommendations, such as policy changes to tax codes, will not be realized immediately.

Factors Facilitating Progress

The willingness of the seventeen council members to work together was extremely important to progress, especially since their interests were so diverse. The financial backing of Congress and the support of the four governors were also essential. The common

perception of a threat in the region helped the project to maintain momentum, and the residents of the region were involved at all stages of the council's work.

Obstacles to Progress

Coordination of four states, each with a different view and way of doing business, was a challenge throughout the process. In addition, getting groups with different philosophies and regional perspectives to agree on

a middle ground demanded innovative approaches.

Contact information:

Mr. Charles Johnson
Vermont Department of Forests,
*   Parks & Recreation*
103 S. Main St., 8 South
Waterbury, VT 05671-0601
(802) 241-3652
Fax: (802) 244-1481
E-mail:
*   cjohnson@fr.anr.state.vt.us*

NORTHERN LOWER MICHIGAN ECOSYSTEM MANAGEMENT PROJECT

Location:
Northern lower Michigan

Project size:
19 million acres

Initiators:
Michigan Department of Natural Resources, U.S.D.A. Forest Service

PROJECT AREA DESCRIPTION

The landscape of Michigan's northern lower peninsula (NLP) is dominated by glacially-formed features. The sandy high plain in the interior is surrounded by lower elevation lake plain, ground moraine, and outwash. Upland pre-settlement vegetation consisted of northern hardwood forests on moister sites, and oak-pine or pine forests on droughty sites. Swamp and bog communities were common. Twelve federally listed threatened and endangered species occur in this area, including Kirtland's warbler, Michigan monkey-flower, and dwarf lake iris. A variety of state listed species and candidate species for federal listing are also present.

The Huron-Manistee National Forest extends over one million acres; three state forests spread out over an additional two million acres. Both public and private lands are used for recreation, timber and wildlife management, and oil and gas production. Public lands also provide for protection of critical species and communities. Uses on privately-owned lands include hunting, residential (second) homes, commercial and industrial development, and agriculture.

ECOSYSTEM STRESSES

The logging history of the state and the conversion of land to agriculture have changed forest composition dramatically. Early successional deciduous species have increased greatly in abundance, whereas the occurrence of some conifer species has been reduced.

Currently, the most significant stresses to the ecosystem are land conversion for expanding communities, resort development, home and cabin sites, additional roads or road improvements, and natural gas development. These lead to land fragmentation, disruption of habitat, introduction of exotic species, displacement of natural processes, and noise pollution.

PROJECT DESCRIPTION

In 1992, the Michigan Relative Risk Analysis Project, administered by the Michigan Department of Natural Resources (MDNR), concluded that one of the state's top environmental concerns was the lack of integrated land use planning. In response, Governor John Engler requested that the MDNR address this problem. This request, in conjunction with the realization that the MDNR and the Huron-Manistee National Forest often manage the same ecosystems and in some cases share boundaries, led to the formation of a joint ecosystem team in 1994. The Bureau of Land Management (BLM) joined the team shortly afterwards.

The overall goal of the team and the project is to address land and resource management issues that span the entire NLP. An important objective is to coordinate public land management in the NLP through the development of a guiding document, known as the Resource Conservation Guidelines. This document will be ecosystem based, include public wishes and concerns, and be developed using an adaptive planning process. It will provide an ecosystem-based umbrella for such efforts as the Huron-Manistee National Forest plan revision, joint planning for oil and gas development, and the update of the Kirtland's Warbler Recovery Plan. In a memorandum of understanding signed in November 1994, the three agencies represented in the team, as well as the U.S. Fish and Wildlife Service and the National Park Service, agreed to cooperate in the development of this guiding document. The ecosystem team has also established partnerships with local agencies and organizations, who provide advice on the appropriate public participation process.

STATUS AND OUTLOOK

The most important outcomes to date are the increased communication within and between agencies and disciplines. In addition, a one-year grant has been secured from the federal Coastal Zone Management Program to initiate this regional ecosystem planning effort. A public participation strategy has been developed, as well as an educational slide program.

Factors Facilitating Progress

Strong support from leadership and staff of the MDNR has been very helpful. Individuals with a strong interest in the project and a willingness to explore new ideas have been an important factor facilitating project progress. A willingness to cooperate and to share information has also been beneficial. Early federal funding has been helpful as well.

Obstacles to Progress

An overall shortage of funding has been problematic. Funding is needed for analysis of information, for delineation and characterization of land-type associations, and for analysis of social aspects of the project. Uncertainty regarding the planning process has also been a barrier. In addition, working with many people is at times cumbersome and challenging for decision making.

Contact information:

Mr. Michael T. Mang
Michigan Department of Natural Resources
PO Box 667
Gaylord, MI 49735
(517) 732-3541 ext. 5042
Fax: (517) 732-0794

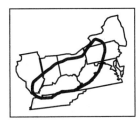

OHIO RIVER VALLEY ECOSYSTEM

Location:
Ohio River Valley through ten northeastern and midwestern states (IL, IN, OH, KY, WV, PA, and small portions of MD, VA, NY, and TN)

Project size:
92.2 million acres

Initiator:
U.S. Fish and Wildlife Service

PROJECT AREA DESCRIPTION

The boundaries of this project, set by the U.S. Fish and Wildlife Service (FWS), are based on the Ohio River Valley drainage area, crossing three FWS Regions and encompassing a portion of ten states. The upper reaches of the drainage basin extend into New York State, the lower reaches into Kentucky. A broad range of topography is present, including high Appalachian mountains, lower foothills, and floodplains. The vegetative cover in the upper reaches is conifer, while the lower reaches are deciduous and the floodplains mostly mature bottomland forests. Federally listed threatened and endangered species within the watershed include the peregrine falcon, Indiana bat, and various freshwater mussels. Land uses in the Ohio River Valley include navigational travel and heavy industrial uses. Row crop agriculture is present in the floodplains.

ECOSYSTEM STRESSES

Abandoned mines are a major stress to the ecosystem due to the devastating effects of acid mine drainage and sediment runoff on water quality and wildlife habitat. There are also several current proposals for new strip mine operations in the river valley, as well as a proposal for the largest pulp wood mill in the world. These could have significant impacts on the health of the river due to a decrease in vegetative coverage, an increase in sedimentation, and point source contaminant loading. Additional threats to the water quality of the system are posed by agriculture and timber operations, and industries such as steel mills along the river corridor. The exotic zebra mussel is also in the early stages of colonization,

which could devastate native populations of mussels.

PROJECT DESCRIPTION

This project is part of the national FWS shift toward ecosystem management. The Ohio River Valley Ecosystem Team was established in May 1994 and includes thirty-five representatives from FWS field stations within regions 3, 4, and 5. Members include personnel from hatcheries, refuges, ecological services offices, and law enforcement divisions.

The team's first goal was to define the resource priorities for the watershed and to outline the associated action strategies to meet these priorities. Examples of the seven priorities that were defined include protecting karst cave habitats, freshwater mussel communities, and wetlands. Action strategies for the resource priority on freshwater mussel communities will likely include activities such as monitoring the rate of zebra mussel infestation and compiling a database of information to help guide management decisions. Some action strategies within the plan are a continuation of existing duties, while others, such as protecting karst cave habitats are newer, more broad-based activities.

STATUS AND OUTLOOK

The team is focusing on refining a list of twenty-seven draft action strategies that will help them meet the resource priorities of the watershed. Outside partners, including other federal agencies, nonprofit organizations, state agencies, universities, and water sanitation divisions, will be brought in to work on the specific action strategies. The team is also in the process of compiling a geographic information systems (GIS) baseline map for the watershed.

Factors Facilitating Progress

Although the project is in its early stages, an immediate benefit to the ecosystem approach has been increased communication across the different agency functions. This has been helpful in using scarce resources more effectively. The dedication of the team and the field personnel has greatly facilitated the process.

Obstacles to Progress

Since the project area crosses three FWS regions, coordinating activities takes much time. Also, the initial guidance from the national level was not fine tuned and lacked consistency, which made management at the regional level difficult. Finally, funding for the resource priorities has yet to be secured.

Contact information:
Ms. Kari Duncan
Team Leader, Ohio River
* Ecosystem*
U.S. Fish & Wildlife Service
White Sulphur Springs National
* Fish Hatchery*
400 East Main Street
White Sulphur Springs, WV 24986
(304) 536-1361
Fax: (304) 422-0754

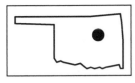

OKLAHOMA TALLGRASS PRAIRIE PRESERVE

Location:
Northeastern Oklahoma

Project size:
37,000 acres

Initiator:
The Nature Conservancy

PROJECT AREA DESCRIPTION

In presettlement times, the tallgrass prairie spanned 142 million acres from Canada to the Gulf of Mexico. It evolved under the forces of weather, bison grazing, and fire. Starting with European settlement, over 90 percent of the tallgrass prairie has been converted to agriculture.

The Oklahoma Tallgrass Prairie Preserve is located in the Osage Hills. The rolling hills are covered by tallgrass prairie interspersed with streams and post oak-blackjack oak savannas. The preserve encompasses most of the upper portion of the Sand Creek watershed, and is buffered by adjacent privately-owned cattle ranches. Some 500 to 700 plant species can be found in the preserve.

The preserve is used for recreational activities such as hiking, photography, and nature observation. In addition, 107 oil and gas wells are located on the preserve. The wells are independently owned, and produce oil and gas in compliance with a contract with the owners of the mineral rights, the Osage native American tribe.

ECOSYSTEM STRESSES

Fire suppression and prescribed burns that only poorly mimic natural fires may lead to woody encroachment or shifts in species composition. Cattle grazing, as opposed to bison grazing, may also lead to a shift in species composition. Both stresses used to impact the tallgrass prairie on the preserve, and still affect neighboring properties. However, the natural fire-bison regime is currently being restored on the preserve. The threat of an accidental discharge of oil or salt water from oil wells poses a potential threat to the preserve.

PROJECT DESCRIPTION

As early as the 1930s, the National Park Service recognized that tallgrass prairies were not protected in a large enough area to recreate a functioning tallgrass ecosystem. Since the only sizable tracts of tallgrass prairie could be found in the Flint and Osage hills of Kansas and Oklahoma, subsequent efforts by the National Park Service and conservation organizations focused on these states. These efforts collapsed with the failure of a bill proposing a Tallgrass National Preserve in Osage County, Oklahoma.

In consultation with an interdisciplinary team of experts, The Nature Conservancy (TNC) realized that a tallgrass preserve area should encompass a watershed, and should be large enough to support a genetically viable bison herd. In Fall 1989, this realization led to the purchase of the Barnard Ranch (a 29,000-acre parcel) by TNC, which gained possession of the ranch in January 1991. Additional acquisitions have since enlarged the preserve to 37,000 acres.

The prairie was allowed to rest for two years, during which time buildings were restored and an ecological inventory completed with the Oklahoma Natural Heritage Program, using satellite imagery of vegetation. The overall management goal is to restore the full complement of ecological processes on the prairie. This involves the recreation of the fire-bison interaction, resulting in a dynamic landscape patch mosaic. Prescribed burns were begun in

1993 and conducted in a manner intended to mimic presettlement burn patterns. Presettlement grazing patterns have been replicated through the reintroduction of an American bison herd on an initial 5,000 acres of the preserve. While the herd builds itself up, additional grazing pressure is exerted by cattle on 24,000 acres initially.

TNC has worked diligently to gain the trust of local oil producers, the Osage Agency, the Bureau of Indian Affairs, and ranchers. Non-interference with oil production, cooperation in fire management and wildfire suppression, and the maintenance of a disease-free bison herd have all contributed to the development of this trust.

STATUS AND OUTLOOK

Although fire and bison have been reintroduced, only parts of the preserve are currently subjected to these natural forces. These will be restored to the entire preserve over the next ten years.

Approximately 400 bison now occupy 5,760 acres of the preserve; an additional 1,200 acres were to be added in Spring 1996 with a corresponding

decrease in cattle grazing. More than a dozen research projects have been completed or are under way on the preserve. For instance, a behavioral study of the bison is under way, as is a bird study in conjunction with the Southern Aviary Research Center and monitoring of bison and cattle impacts on vegetation.

Factors Facilitating Progress

In this area of the country, large ranches do not come up for sale often. Finding a suitable tract of high-quality tallgrass prairie for sale was a major factor in the success of this project. Also very helpful was the national attention that the project received, leading to the raising of $15 million in private funds. According to TNC, after the controversial and unsuccessful efforts of the federal government to create a National Preserve, only a private fundraising effort by an organization like TNC was likely to succeed. The commitment—financial and otherwise—of many partners from assorted public and private communities has contributed greatly to this effort's progress.

Furthermore, the expertise of the interdisciplinary team, the invol-

vement of local people, and the tremendous dedication of many individuals has been very important.

Obstacles to Progress

Although the $15 million fundraising goal has been reached through pledges, arriving at that goal was a challenge, and it will be several years before all of the funds are received. Furthermore, TNC now recognizes that the $3 million endowment for management expenses will probably not be sufficient to achieve all of its goals. Restoring and managing a large-scale fire-bison regime is a very expensive enterprise, requiring perpetual institutional support. So far, it has involved procuring equipment and training for prescribed burns, as well as obtaining a bison herd and building a bison corral and fences.

Contact information:

Mr. Harvey Payne
Director
Tallgrass Prairie Preserve
The Nature Conservancy
PO Box 458
Pawhuska, OK 74056
(918) 287-4803; (918) 287-1290
Fax: (918) 287-1296

OUACHITA NATIONAL FOREST

Location:
West-central Arkansas, eastern Oklahoma

Project size:
1.6 million acres

Initiator:
U.S.D.A.. Forest Service

PROJECT AREA DESCRIPTION

The Ouachita ecosystem consists of a series of east-west ridges. The project area (the Ouachita National Forest, or ONF) is primarily mountainous, and lowland valleys are mostly in private ownership. The region is a transition zone between southeastern evergreen forest type (shortleaf pine) and eastern deciduous forest type (oak-hickory). Inholdings in the national forest account for approximately 30 percent of the land within its boundaries. The area includes ten federally listed threatened and endangered species, including the red-cockaded woodpecker, leopard darter, two mussels, and one plant.

Timber was and still is the primary extractive use of the ONF. Other uses of the ONF include recreation (motorized and non-motorized), historic grazing, historic lead and zinc mining, and limited crystal mining. Over 600,000 acres of the ONF are off-limits to timber production, because they are in wilderness and wild and scenic rivers, or are too steep for logging. In the valleys, agriculture (poultry) predominates.

ECOSYSTEM STRESSES

Historically, timber harvesting, over-harvesting of animals, and conversion to agriculture (in the valleys primarily) were the primary stresses. Today, roads, mostly unpaved, stress the system by increasing sediment loads into streams and breaking up natural fire patterns. Other stresses include localized water development, conversion to urban land uses, and barriers to fish movement.

ONF decision makers and U.S.D.A. Forest Service (USFS) scientists feel that fire suppression associated with traditional timber harvesting has been a stress, and so the USFS has accelerated ONF's prescribed burning program. Critics from the environmental community argue that fires were never a significant natural process (due to high rainfall and topography) and that prescribed burns are now stressing the system. Critics also allege that ONF is further stressing the system by converting mixed stands to pine stands as part of its efforts.

PROJECT DESCRIPTION

The supplemental planning process for the ONF, beginning in 1987, was the first step toward ecosystem management. Clear-cutting on the Ouachita had brought much public and political pressure, including a lawsuit by the Sierra Club in early 1990 when the amended plan was published. In August 1990, USFS Chief Dale Robertson met with Senator David Pryor (D-AR), which led to a halt in the clear-cutting and the declaration of the ONF as a pilot site for the USFS's recently-announced New Perspectives program.

The New Perspectives ONF project became the ecosystem management program in 1992, and has three components. The first component consists of showcase projects (e.g., on old-growth and on the shortleaf pine-bluestem-red cockaded woodpecker), which incorporated increased ecological sensitivity and enhanced public involvement. The second component is a three-phase interdisciplinary research effort, considered of critical importance in establishing the scientific basis for ecosystem management: a) a series of demonstrations of clear-cutting alternatives; b) a ten-year formal study of alternative harvesting methods; and c) research at the landscape level. The third component is a congressionally- authorized ecosystem management advisory committee of thirteen professionals, including foresters, a rural sociologist, an ecologist, a recreation specialist, and landscape architects. The committee's composition is intended to maintain a broad spectrum of professional experiences and backgrounds. Meeting three or four times a year, the committee reaches decisions through consensus and makes recommendations to the forest supervisor.

Various stakeholders have been involved in the effort, particularly in research endeavors, including state agencies, other federal agencies, environmental/ land trust groups, Weyerhauser, and universities. A major ONF project partner is USFS's Southern Research Station, which developed the three-phase research program with cooperators from universities and industry.

The goals of the effort are 1) sustaining and, where appropriate, restoring biodiversity and ecological integrity; 2) integrating human uses with environmental values; 3) maximizing options for future generations; and 4) emphasizing natural qualities and social values. To meet these goals, six strategic objectives with action items have been drawn up by the committee.

ONF amended its forest plan in 1994 to place more emphasis on monitoring in the following areas: forest health; impacts of forest management on the

environment and social conditions; public demands; and management efficiency.

STATUS AND OUTLOOK

The level of conflict between the USFS and the public, while still significant, has been reduced in part due to the advisory committee's work. The first phase of research is complete, the second is in its fourth year, and the third was recently initiated. More ecosystem monitoring is occurring than existed before. Clear-cutting has been effectively eliminated on the Ouachita, leading to visual and habitat improvements. Herbicide use has been reduced, prescribed burning is increasing, and restoration of native forest composition is underway (although there is debate between the USFS and its critics on these issues).

Factors Facilitating Progress

The research effort, a committed public, USFS leadership, and adequate funding have been instrumental. Principal research scientists are located on-site, optimizing coordination of the research with the operational ONF program and creating a robust scientific basis for effort. The advisory committee has been instrumental, for example, in providing a forum in which issues can be discussed openly. Support from the congressional delegation has been important.

Obstacles to Progress

Initially, setting up data collection was delayed due to difficulties in achieving appropriate scientific methodology. Using ecosystem management without a clear definition of the paradigm has created difficulties, particu-

larly for external overview of ONF operations, although ONF does not feel that a precise definition is needed to proceed. Implementation of ecosystem management schemes has been limited at times by conflicting USFS goals.

Some skepticism by the public remains. Critics claim that the Ouachita is using ecosystem management to justify traditional forest management. Opposition by local and national environmental organizations is still significant. ONF was featured in a *Sierra* magazine article criticizing the USFS's ecosystem management efforts, specifically on the prescribed burning and forest composition issues.

Future funding is a concern, especially for research and for justifying the effort's funding level. Being able to demonstrate that ecosystem management "works" is a another key concern.

Additional training of personnel is needed, as is greater commitment to effective interdisciplinary teamwork by ONF specialists. Having adequate geographic information systems (GIS) capabilities would be beneficial. Finally, open and positive relations with all parties is essential, according to various stakeholders.

Contact information:
Mr. Bill Pell
Ecosystem Management
*   Coordinator*
USDA Forest Service
Ouachita National Forest
P.O. Box 1270
Hot Springs, AR 71902
(501) 321-5202

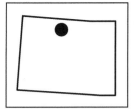

OWL MOUNTAIN PARTNERSHIP

PROJECT AREA DESCRIPTION

Extending northwest from the Continental Divide and south from the Wyoming border lies the North Park region of the Colorado Rocky Mountains. This region is bounded by high mountain ridges and includes the Arapaho National Wildlife Refuge, serving as a major wintering area and migration route for elk. The North Park region is characterized by coniferous forests, rolling sagebrush uplands, and extensive pasture lands and hay fields. There are approximately 700 square miles of intermittent and perennial streams within its watersheds, and many neotropical birds summer in this region.

The drainages of the Michigan and Illinois rivers define the project boundaries of the Owl Mountain Partnership, which includes 67 percent public lands and 33 percent private lands. Agriculture (primarily livestock grazing), logging, and recreation provide the economic foundation of this rural area, the least-populated area of Colorado.

ECOSYSTEM STRESSES

Extensive logging has occurred in North Park, particularly in the Owl Mountain area. A large network of logging roads and extensive clearcutting are the legacy of past forestry practices on private, state, Bureau of Land Management (BLM), and U.S.D.A. Forest Service (USFS) lands. These practices have affected wildlife migration, and increased wildlife harassment and mortality. As a result, greater numbers of wildlife, particularly elk, have retreated to private lands at lower elevations. Furthermore, severe drought conditions in 1994 have exa-

cerbated degraded range conditions from past grazing practices and have stressed the region's water resources. The resulting loss of forage has caused a decline in sage grouse and deer populations.

PROJECT DESCRIPTION

The project originated with the Colorado Division of Wildlife Habitat Partnership Program (HPP) in which private landowners and land management agencies work together to resolve local conflicts over forage consumption of elk and livestock on public and private rangelands. In 1992, the HPP committee received a grant from the Seeking Common Ground Work Group for a cooperative ecosystem management demonstration project.

In 1993, the Owl Mountain Partnership was born, with a steering committee established to identify priority issues and problems to be addressed by the project. The steering committee includes representatives from the Colorado Division of Wildlife, Colorado State Forest Service, Colorado State Land Board, BLM, U.S.D.A. Natural Resources Conservation Service, USFS, U.S. Fish and Wildlife Service, Colorado State University, several ranchers and other private landowners, an environmental representative, and the North Park community. All planning and management decisions for the project area are made by the committee, using a consensus-based process.

The intent of the Owl Mountain Partnership is to meet the economic, social, and cultural needs of the North Park community while developing adaptive long-term landscape management pro-

Location:
North-central Colorado

Project size:
246,000 acres

Initiators:
Private citizens and landowners

grams to ensure ecosystem sustainability. To attain these goals, the project focuses on creating partnerships to resolve resource conflicts, developing and implementing a management plan across political, administrative, and ownership boundaries, and promoting communication and education efforts for the partners and the community.

STATUS AND OUTLOOK

So far, a comprehensive plan has not been developed for the project area. However, extensive inventory and monitoring work on vegetation, wildlife, and aquatic systems are underway. As of early 1996, several surveys had been completed, for instance, on neotropical birds and mycorrhizae soil ecology. A water monitoring study funded by the U.S. Environmental Protection Agency has also been completed. Meanwhile, smaller projects have been implemented to address immediate concerns of water scarcity and fence damage.

Eight individual partnership plans (for ranchers) have been prepared.

By direction of the steering committee, one method for vegetative inventory has been developed for all land management agencies and participating private landowners so that data can be consistent and easily-shared among partners. Despite the wide range of interests and the difficult hurdles, the committee has been very successful in maintaining cooperation in addressing resource issues.

Factors Facilitating Progress

To address the initial distrust and misunderstanding of the project's intent, a great deal of outreach to local ranchers clarified these issues and achieved the support of most ranchers in the project area. The committee has also become more sophisticated in seeking funding through grants. Finally, the project is driven by a locally-based group and has full-time

agency staff who work only on these ecosystem planning efforts.

Obstacles to Progress

Proposals for ski area developments within the project area have split the community; some want to protect the rural agrarian lifestyle and the wildlife habitat the area provides, while others, such as county commissioners, see the development as a greatly-needed economic opportunity for the North Park community. The committee hopes that through collaborative planning at the community level, long-term social, economic, and ecological sustainability can be achieved.

Contact information:

Mr. Jerry Jack
Project Manager
Bureau of Land Management
Kremmling Resource Area
1116 Park Avenue
P.O. Box 68
Kremmling, CO 80459
(970) 724-3437
Fax: (970) 724-9590

PARTNERS FOR PRAIRIE WILDLIFE

Location:
West-central and southwestern Missouri

Project size:
49,920 acres

Initiator:
Missouri Department of Conservation

PROJECT AREA DESCRIPTION

The project area is separated into two target zones: a northern zone encompassing 24,320 acres in west-central Missouri south of Sedalia, and a southern zone encompassing 25,600 acres south of the town of Nevada in southwest Missouri. Both areas are former tallgrass prairies on hilltops or side slopes, giving way to woodlands in bottomlands. Once, the prairie was dominated by warm-season grasses such as big bluestem, Indian grass, and little bluestem. Today, much of the prairie has been converted to exotic cool-season grasses for grazing and haying and crop land.

Several threatened and endangered species are known to exist in the project area, including Henslow's sparrow, Mead's milkweed, upland sandpiper, and the prairie-chicken.

Most of the land is privately owned with core areas owned by the Missouri Department of Conservation (DOC) for conservation and recreation (hunting, fishing, viewing, hiking).

ECOSYSTEM STRESSES

Fire suppression has allowed woody invasions in fence rows and the upper reaches of some draws to fragment the prairie landscape. Conversion of the prairie to exotic vegetation for pasture and row cropping has decreased plant diversity, and year-long grazing does not provide suitable structure for nesting and brood rearing by a variety of birds, including the prairie chicken.

Finally, agricultural non-point source pollution and timber and forest management on side slopes and bottomlands are present but less significant as stresses.

PROJECT DESCRIPTION

The prairie chicken is considered an indicator of overall prairie biodiversity, and although the chicken's populations had stabilized on some public lands, the overall population was continuing to decline according to a 1992 report published by the DOC. Recovery of the species, and by extension the landscape as a whole, was not considered possible relying only on public land. The decline of the prairie chicken, which is promoted as a flagship species because of its high visibility and the concern it garners from landowners, stimulated the program's development.

Following the 1992 report, the DOC decided to develop a ten-year pilot program coordinating conservation efforts on public and private lands, the first of its kind in the state. In 1992 and 1993, DOC biologists selected and inventoried the two project target zones. Public-private resource management strategies to improve prairie habitat were developed in 1994. County Soil and Water Conservation District boards (SWCD) representing landowner interests, were asked to comment on the approach, with their feedback incorporated into revisions of the final program.

The program is designed to increase grassland habitat by encouraging landowners to change management to better favor prairie wildlife, with a goal of developing suitable prairie wildlife habitat on 40 percent of each target area. To that end, the program has three objectives: 1) enhance grassland diversity and structure to improve wildlife nesting and brood-reading cover; 2) reduce fragmentation of prairie landscapes by removing invading trees; and 3) demonstrate that agricultural production and prairie wildlife habitat improvements can be compatible.

The program is operated in co-operation with the Missouri Department of Natural Resources Soil and Water Conservation Districts Program and the country SWCDs, with additional funding from the U.S. Fish and Wildlife Service and Monsanto Agricultural Group. Monitoring will be based on annual prairie-chicken census routes, annual breeding bird surveys, habitat inventory to monitor change, and surveys of landowner participation and perceptions.

Ten different incentive options, based on 50 to 75 percent of average conversion costs to farmers, range from converting fescue to native grasses, to removing invasive trees from fence rows. Landowners submit applications to the local SWCD; the applications are then evaluated for acceptance into the program. Technical assistance is provided to landowners by DOC biologists. Finally, active public education, recognition and outreach efforts will be developed as important components of this program.

STATUS AND OUTLOOK

The program is operational, with the first sign-up having occurred in Spring 1995 and a second in Fall 1995. Two to three sign-up periods per year are envisioned. An internal program plan was developed; individual management plans are developed for each landowner.

It is too early to assess the effectiveness of on-the-ground improvements, although cooperation and communication between stakeholders and agencies has occurred, and the program appears to be well received by participating landowners. However, as of early 1996, 75 percent of the tree removal goal was near completion, at or just slightly ahead of schedule.

Factors Facilitating Progress

The Missouri DOC's move toward ecosystem management, or Coordinated Resource Management (see page 195) as it is officially known, has been an important benefit to this effort. Enthusiasm on the part of other agencies and private sector groups has also helped the program. In the south zone, participation has been particularly good, thanks to a letter of support sent by the local Soil and Water Conservation District to all landowners in the area.

Funding of this effort, for monetary incentives and program administration, appears to be assured. Finally, several other factors have been cited as positive factors, including a favorable level of scientific knowledge (inventory, understanding of processes), and the clarity and measurability of goals.

Obstacles to Progress

Convincing some traditional constituent groups that trees that fragment former prairies should be removed to benefit wildlife has been a challenge at times, in part because of DOC's earlier success at promoting tree planting as a general way to improve wildlife habitat. DOC is addressing those concerns as they arise, largely through public education efforts.

Contact information:

Mr. William D. McGuire
Private Land Coordinator
Missouri Department of
  Conservation
Wildlife Division
P.O. Box 180
Jefferson City, MO 65102-0180
(314) 751-4115 ext. 148
Fax: (314) 526-4663

PATRICK MARSH WETLAND MITIGATION BANK SITE

Location:
Southern Wisconsin

Project size:
270 acres

Initiators:
*Wisconsin
Departments of
Transportation and
Natural Resources*

PROJECT AREA DESCRIPTION

Formed during the last ice age, Patrick Marsh is a 160-acre restored wetland within a larger project site, with gently-rolling topography, typical of this gla-cially-influenced landscape. Located in what is today an agricultural and urban fringe region near Sun Prairie, this site once supported a prairie-oak savanna complex. Today, predominant uses of the surrounding privately-owned lands are row crop agriculture, dairy opera-tors, and rapidly growing residential and light-industrial development. The project area is owned entirely by the Wisconsin Department of Natural Resources (DNR).

Today, a rich variety of species of plants and animals are recolonizing the site, including forty-two species of birds in the marsh and surrounding uplands, frogs and other amphibians, and native aquatic and upland vegetation such as aquatic sedges, burr oak, and various prairie forbs and grasses. It contains no endangered species.

ECOSYSTEM STRESSES

There are no direct significant stresses on the project site, since it is protected in public ownership. The greatest threats, however, are an inadequate buffer zone around the wetland and urban development on adjacent up-lands. Non-point source pollution from highway runoff, seed corn operations, and future urban stormwater could be significant stresses.

PROJECT DESCRIPTION

As a result of the expansion of State Highway 151 and the loss of twenty-six acres of wetlands in another location,

the Wisconsin Department of Transportation (DOT) was required to replace the lost wetlands. In consultation with the Wisconsin DNR, the DOT purchased the Patrick Marsh property in 1991 (at that time, it was a cornfield), and transferred ownership to the DNR. Soon after, a pumping station which had kept the field dry was removed, and the marsh began to refill with water. By the following Spring, the area had reached its present size, with the water level stabilizing by 1994.

Only two agencies, the DOT and DNR, have been involved in this effort. While the public has not been involved in decisions regarding the marsh, DOT purchased the land through voluntary agreements (as opposed to condem-nation). An informal management plan was developed by DNR, initially to be used for DOT negotiations with land-owners, but is now used to guide resto-ration and management efforts.

Project literature describing the site including its history, goals, and recov-ery—has been developed. The effort's goal is "to recreate ... a large, thriving wetland community surrounded on the uplands by oak openings and tall-grass prairie." Five specific objectives are listed: 1) establishment of a wildlife area for visitation and education; 2) reconstruction of mesic and wet-mesic prairie communities on uplands; 3) restoration of remnant oak openings through weed tree removal, prescribed burns, and reseeding; 4) public educa-tion on the function and value of res-tored wetland and prairie communities; and 5) evaluation of wetland restoration through monitoring.

STATUS AND OUTLOOK

Because the restored marsh is larger than the wetlands lost as part of the original highway expansion, the extra wetlands 'created' as a result of this effort can be used to mitigate against future wetland losses from other DOT projects. Thus, Patrick Marsh is a bank from which wetland credits can be used. This is the first such wetlands mitigation bank site in Wisconsin, and as such, is considered an important pilot for wetland mitigation banking in the state.

Cooperative efforts with local schools and interpretative activities have also been developed. Monitoring by graduate students and the DNR, for example, of breeding birds and aquatic vegetation, is occuring on a limited basis.

Factors Facilitating Progress

Interagency cooperation between the DOT and DNR, as well as the two agencies' support for this effort, were cited as benefits to the process. Visitation, particularly by school groups, has helped create local support for the marsh.

Obstacles to Progress

The availability of funds to purchase additional uplands and for interpretive activities was cited as an obstacle to this effort's progress. The DNR project manager, who has other responsibilities, is able to devote only a very small percentage of his time to the marsh. Nearby landowners have been unwilling to sell lands which could serve as buffers or native uplands, because of skyrocketing land values which make their lands more valuable for development. Ironically, the restoration of the marsh has made the surrounding uplands even more appealing for housing development. More DNR outreach with the community and local officials may have aided with buffer land acquisitions. Finally, stresses outside the project area which impact the marsh and uplands are not being reduced and may become more significant with increased housing development.

Contact information:

Mr. Alan Crossley
Wildlife Biologist
Wisconsin Department of Natural
 Resources
3911 Fish Hatchery Road
Fitchburg, WI 53711
(608) 275-3242
Fax: (608) 275-3338

PHALEN CHAIN OF LAKES WATERSHED PROJECT

Location:
South-central Minnesota

Project size:
14,790 acres

Initiators:
Ramsey-Washington Metro Watershed District, eight local governments, University of Minnesota Department of Landscape Architecture, Minnesota Department of Natural Resources

PROJECT AREA DESCRIPTION

During the past 35,000 years, the landscape of this watershed has been worked and reworked by glaciers. The glaciers left a landscape of rolling, well-drained uplands dotted with lakes and wetlands in low areas. The chain of lakes lie along an old river valley of the St. Croix River. During the last glaciation, gravel and soil were deposited in the valley, and large chunks of ice were left in low areas, forming the chain of lakes.

At the time of the original land survey (1845-47), most of the watershed was covered by oak woodland or oak savanna, with scattered groves of northern pin oak, bur oak, white oak, and aspen trees. Only scattered clusters of large, presettlement oaks remain today in some older neighborhoods and parks, and a few acres of remnant prairies remain along railroad tracks, near wetlands, and in other undeveloped areas. Today, the watershed is dominated by low-density residential land use, with commercial/industrial areas near freeways and major streets, and parks and open spaces along lakes, creeks, and wetlands. Among the area's state listed endangered species are three prairie plants and Blanding's turtle.

ECOSYSTEM STRESSES

Urbanization has fragmented the landscape significantly and affected the ecological communities and systems of the watershed in many ways. Settlement has nearly eliminated the native vegetation of the area, replacing it with impervious surfaces. Habitat diversity and the varieties of birds and animals that inhabited the area have been much reduced. Exotic species have gained a foothold in the watershed and have become a problem.

Hydrologic alteration has also been extensive. Urban development has increased soil compaction, creeks have been paved and channelized, and over half of the wetlands in the watershed have been eliminated. These changes cause more water to run off the land with each storm at a faster rate, creating flood and erosion problems.

PROJECT DESCRIPTION

The Phalen Chain of Lakes Watershed Project is a partnership among the Ramsey-Washington Metro Watershed District, eight local governments, the University of Minnesota Department of Landscape Architecture, and the Minnesota Department of Natural Resources. The partnership was formed in 1992 when it applied successfully for foundation funding to hire a project coordinator. The purpose of the project partners was to develop and implement a comprehensive, locally-based plan to protect, restore, and manage land and water resources while accommodating continued urban growth in the watershed over the next fifty years.

The project partners organized into a steering committee and began meeting in May 1993. The committee included representatives from city councils and planning commissions, the county board, the watershed district, local businesses and developers, environmental organizations, lake associations, private citizens, and agencies.

At the beginning of the steering committee's planning process, members received a broad education on watershed history, resource conditions, and trends. As a result of the education process, the committee articulated a powerful and collective vision for the watershed's future. Seven goals supported by recommendations and action steps were also developed.

STATUS AND OUTLOOK

The watershed plan was completed in April 1994. Steering committee members brought the plan to their community constituencies for approval and implementation. The plan has been formally endorsed by local governments, organizations, and others working in the watershed. The project coordinator is now working to implement the plan. Citizens, businesses, local governments, and agencies are now undertaking specific actions to improve and manage natural resources in the watershed.

To date, the partners have raised over $500,000 to support demonstration projects such as an innovative stormwater management project and an eight-acre wetland restoration effort on an old shopping center site; completed demonstration plantings to restore native wetland, upland and aquatic plant communities; and are working with cities to develop model environmental ordinances and projects.

Factors Facilitating Progress

The commitment of four dedicated people allowed the project to progress during its early stages. Members of the steering committee represent the diversity of local and community interests. Furthermore, steering committee members have a mutual respect for one another and have committed themselves to work together in new ways. The common vision shared by committee members is also an asset, as is the creativity used to act against threats to the watershed. Other major positive influences are the level of political support the project receives, the project's administrative structure, and the capability of project leadership.

Obstacles to Progress

Conflicts within and between agencies were identified as obstacles to progress. Although agency support is another factor facilitating progress, agency conflicts have required the project administration to undertake a 'balancing act.' Constraints have also been imposed by an insufficient inventory of the ecosystem's biophysical components. Furthermore, there is a low degree of certainty as to how these components interact with one another.

Contact information:

Ms. Sherri A. Buss
Phalen Watershed Project
*   Coordinator*
Ramsey-Washington Metro
*   Watershed District*
1902 East County Road 13
Maplewood, MN 55109
(612) 777-3665

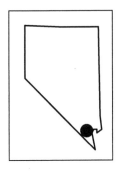

PIUTE/EL DORADO DESERT WILDLIFE MANAGEMENT AREA

Location:
Southern Nevada

Project size:
531,000 acres

Initiator:
Clark County, Nevada

PROJECT AREA DESCRIPTION

Thirty miles south of Las Vegas, the Piute/El Dorado Desert Wildlife Management Area (DWMA) encompasses the northeastern portion of the Mojave desert and the McCollough mountain range watershed eastward to the Colorado River. Four major mountain ranges border this region, which is comprised mostly of creosote scrublands, Joshua Tree woodlands, and scattered blackbrush, yucca, and cacti communities. The DWMA is also dissected by washes (periodically-flowing draining channels) with unique vegetation such as acacia and mesquite. Federal and state listed threatened or endangered species in the area include the desert tortoise, peregrine falcon, gila monster, chuckwalla, and phainopepla.

Current land uses permitted in the DWMA include mining and recreation. Eighty percent of the project area is owned by the Bureau of Land Management (BLM), with most of the remaining lands controlled by the National Park Service (NPS).

ECOSYSTEM STRESSES

Hard rock mining has led to the loss of desert wildlife habitat. Roads and power line corridors provide unsupervised access into the interior regions of this management area, fragmenting habitat and increasing species collection, harassment, and mortality. Unregulated commercial and illegal collecting of reptiles and plants has caused a decrease in certain lizard and snake populations and yucca and cacti communities. Exotic grasses and other annual plant species introduced for grazing have replaced native plant communities and has affected soil characteristics. These threats to native species have been exacerbated by a long period of drought in the region, increased human activity on the landscape due to the expansion of urban areas and small communities, and increased recreational pressures in southern Nevada. The species most threatened by all of these stresses is the desert tortoise.

PROJECT DESCRIPTION

The project was initiated as a result of the federal listing of the desert tortoise in 1990. In response, Clark County, and in particular Las Vegas, sought to develop a Habitat Conservation Plan (HCP). As the primary facilitator of this process, The Nature Conservancy (TNC) helped delineate the DWMA area for tortoise recovery and began purchasing grazing allotments on public lands in 1991. A Desert Tortoise HCP steering committee was established with representatives from Clark County; Nevada Divisions of Wildlife, Agriculture, and Transportation; BLM; NPS; U.S. Fish and Wildlife Service (FWS); Southern Nevada Home Builders Association; Las Vegas Water District; Southern Nevada Off-Road Enthusiasts; mining operators; the Tortoise Group; and TNC. The HCP was the result of the committee's recommendations developed through consensus and proposed to the Clark County Commission.

STATUS AND OUTLOOK

Although the original intent of the DWMA was the protection of the desert tortoise, the committee now takes a multiple species and ecosystem approach in making management recommendations. However, funding restrictions limit monitoring capabilities, so that the tortoise continues to serve as the indicator species for the entire Mojave desert ecosystem.

The committee produced a thirty-year plan which was under review by FWS in 1994. In the meantime, several committee management recommendations have been implemented. BLM has closed roads, restricting access to the DWMA, and increased law enforcement, deterring illegal collection, hunting, off-road vehicle use, and other violations. The impacts of highways in the project area are proposed to be mitigated through fencing and underpasses. Since the removal

of livestock grazing, vegetation is persisting longer. The committee is currently working with a local community on a conservation easement for 85,000 acres of land in the northern end of the planning area, and has set aside 531,000 acres of tortoise habitat.

Factors Facilitating Progress

The involvement of all interest groups from the effort's inception has been the key to this project's accomplishments.

Obstacles to Progress

There has been some negative sentiment, however, from local communities and local land users who feel that the DWMA impedes their traditional/cultural uses of the land and the ability of

small communities to grow. Some backlash has occurred, such as the deliberate killing of tortoises and illegal ORV travel. Public education efforts, including billboards, radio spots, and school curricula, are being used to abate these negative sentiments.

Contact information:

Mr. Jim Moore
Field Representative
The Nature Conservancy
Nevada Field Office
1771 East Flamingo Road
Suite 111B
Las Vegas, NV 89119
(702) 737-8744
Fax: (702) 737-5787

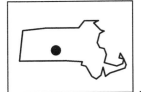

PLAINFIELD PROJECT

Location:
Central Massachusetts

Project size:
13,632 acres

Initiator:
U.S.D.A. Forest Service

PROJECT AREA DESCRIPTION

The project boundaries are those of the town of Plainfield, Massachusetts. The area is approximately 86 percent forested and is dominated by mature forests of northern and central hardwood species. Apart from two large tracts of land owned by Massachusetts Audubon and the state, 78 percent of the forests are in small private ownerships. The project area is largely undeveloped and rural, with limited agriculture and industry.

ECOSYSTEM STRESSES

Development poses the largest threat to the natural system. Land conversion to urban uses and development of roads and infrastructure would cause fragmentation and loss of open space, and would threaten the area's rural character with which it is identified.

PROJECT DESCRIPTION

This effort is a pilot project in the U.S.D.A. Forest Service's (USFS) Forest Stewardship Program. The program is run by the Division of Forests and Parks of the Massachusetts Department of Environmental Management. The overall goal is to increase the awareness of private landowners beyond their property boundaries and to motivate a shift in values toward a landscape-level perspective. To reach this goal, the program is creating incentives and building mechanisms that encourage cooperative projects by adjacent landowners.

In order to be considered for the cost-sharing incentive that the project offers, landowners initially were required first to have a ten-year forest stewardship plan developed by a natural resource professional. This plan has to include a forest inventory, statement of objec-

tives, and schedule of activities. Then, three or more neighboring properties could receive cost-sharing dollars to hire a consultant to review landowner plans and to look at opportunities that would be enhanced through cooperative efforts.

Examples of projects that cross boundaries include locating projects to enhance or protect a wildlife corridor, sharing timber access to minimize costs and environmental impacts, or extending a trail network. Cooperative efforts are made as suggestions and are known as "neighborhood checklists."

Planning for this project began in Spring 1994. The active project members are the stewardship coordinator for Massachusetts and the extension forester at the University of Massachusetts at Amherst. The Plainfield Conservation Commission, a community organization, has also been involved. A community volunteer in Plainfield, who received training under the COVERTS program, has spearheaded the program locally. (COVERTS is an educational project developed at the Universities of Vermont and Connecticut, teaching non-industrial private forest owners how to improve management of private woodlands.)

In Fall 1994, the Forest Stewardship Program held a public workshop and used geographic information systems (GIS) maps as an educational tool to communicate a landscape perspective to private landowners. An emphasis was placed on wildlife habitat. This workshop also highlighted the pilot project and explained to landowners what it would mean for them to get involved. Although it is difficult to

measure an increase in awareness and a shift in values, the
project team held a roundtable
discussion in Summer 1995 to
survey landowners and gauge
success.

STATUS AND OUTLOOK

This pilot project is coming to a
close. The next step for the
coordinators is to look back at the
process, make adjustments, and
expand it to a different area. The
process will be simplified in
order to gain participation from
more landowners; for example,
the requirement for landowners
to have a forest management
plan prior to receiving assistance
from the program will be eliminated. The coordinators expect to
be more actively involved in
working directly with landowners, as opposed to relying so
much on consultant foresters. An
outreach effort to landowners
was anticipated for Spring 1996.
Finally, this project was featured

in a *Journal of Forestry* article in
early 1996.

Factors Facilitating Progress
This project is a non-regulatory
program and therefore is less
threatening to landowners. Since
the project is assistance based,
the overall tone has been very
positive and the project has been
well-received. The enthusiasm of
the COVERTS cooperator has
also been an important
component of this effort.

Obstacles to Progress
Generally speaking, changing the
values and mentality of how
people view the land is a difficult
process which will not happen
overnight. Trying to affect
landowners within a non-
regulatory setting is a challenge
that demands creative and innovative solutions. Logistically, it is
also difficult to work with the
mosaic of landowners. It is important to respect individual in-

terests while also weighing in the
overall interest.

Adding an ecosystem approach
to the Forest Stewardship
Program without making the
process cumbersome and bureaucratic is another challenge. More
specifically to this program,
additional follow-up with
landowners by consultant
foresters would have benefited
this pilot project. This will be
addressed in subsequent efforts,
with the coordinators becoming
more directly involved in the
process.

Contact information:

Ms. Susan Campbell
Stewardship Coordinator
Massachusetts Department of
  Environmental Management
463 West Street
Amherst, MA 01002
(413) 256-1201

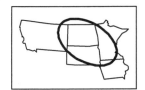

PRAIRIE POTHOLE JOINT VENTURE

Location:
Iowa, Minnesota, Montana, North Dakota, South Dakota

Project size:
64 million acres

Initiator:
U.S. Fish and Wildlife Service

PROJECT AREA DESCRIPTION

The Prairie Pothole region extends over a large area in both the United States and Canada. The region is characterized by gently-rolling knob and kettle terrain. In presettlement times, it supported millions of small wetlands surrounded by mid-grass prairie. Currently, most of the area is used for agriculture and grazing. In the United States, 95 percent of the area is privately owned.

During spring and fall migration, almost all waterfowl crossing the central United States and Canada pass through the Prairie Pothole region. Examples include snow geese and sandhill cranes. The region is also important breeding habitat for waterfowl. The area supports many federally listed threatened and endangered species, such as the piping plover, least tern, whooping crane, pallid sturgeon, and western prairie fringed orchid.

ECOSYSTEM STRESSES

Many of the prairies and wetlands of presettlement times have been converted to crop lands, thus resulting in extreme loss of waterfowl habitat, as well as degradation of adjacent habitat. Although some native grassland remains (e.g., 10 million acres in North Dakota), the structure of the prairie plant communities has changed due to overgrazing. As a result, these grasslands are no longer suitable as waterfowl habitat. Furthermore, almost all rivers and streams in the region have been altered as a result of the construction of dams and reservoirs. For example, in North Dakota only sixty out of 450 miles of the Missouri River are still free-flowing.

PROJECT DESCRIPTION

In response to declining duck populations, the North American Waterfowl Management Plan (NAWMP) was signed in 1986 by the United States and Canada. Implementation of the NAWMP is the responsibility of "joint ventures," which are cooperative efforts between governments, private organizations, and individuals. These joint ventures are active in important waterfowl habitat areas, as identified by the NAWMP. One of these areas is the United States portion of the Prairie Pothole region. In 1987, the U.S. Fish and Wildlife Service (FWS) organized the Prairie Pothole Joint Venture (PPJV) Steering Committee, consisting of FWS, state wildlife agencies in Iowa, Montana, Minnesota, North and South Dakota, and five conservation organizations. A plan previously developed by FWS ("Concept Plan for Waterfowl Habitat Protection") became the basic document for framing the overall design of the PPJV plan. The latter was approved in 1989.

The goal of the PPJV is "to involve the public in a broad-scale unified effort to increase waterfowl populations by preserving, restoring, creating, and enhancing wildlife habitat in the Prairie Pothole region of the United States." The objective is to maintain an average breeding population of 6.8 million ducks, and 13.6 million ducks in the fall flight by the year 2000. The PPJV Steering Committee attempts to ensure that this goal and objective are reached by providing a framework within which individual projects can be developed and implemented in each state. These projects will be developed and implemented by the individual participating agencies and organizations. The steering committee coordinates their efforts. In addition, it mediates conflicts between participants, reviews recommendations, determines policy, and guides implementation of PPJV projects. Such projects involve management of both existing and newly-acquired public lands for increased waterfowl production, as well as the development and maintenance of habitat on private lands through consecutive short-term agreements with landowners.

Cooperation with private landowners through the provision of financial and technical assistance is a major strategy. On public as well as private lands, a large number of individual management techniques may be applied. These include wetland restoration, delayed haying agreements, seeding cropland back to grassland, and provision of nesting structures.

STATUS AND OUTLOOK

Technical analyses of the factors inhibiting achievement of desirable wildlife populations have been conducted and have been used to determine the needed quantity and quality of ecosystems. Cooperation and communication among conservation partners has greatly increased. In both 1994 and 1995, a tremendous recovery of waterfowl numbers was observed.

Pilot projects have demonstrated successful techniques and practices and established coordination networks among stakeholders. Furthermore, goals of many individual projects have been realized. For example, one of the projects that was developed under the PPJV umbrella was the Chase Lake project. This project developed an action plan with thirty-eight action points, including acquisition, and public relations, among others. A number of these points have been completed, whereas others are partially completed or have yet to be initiated.

Factors Facilitating Progress

Having clearly-identified goals for major waterfowl species has allowed a clear definition of the desired type of habitat or ecosystem restoration. In addition, the positive attitude and cooperation of agency personnel and landowners has been very instrumental in the success of the project. Recognition that ecosystem management is a useful analytical tool has also helped.

Obstacles to Progress

Funding is a major problem. Only approximately 10 percent of the estimated cost of the NAWMP has actually been appropriated. As a result, the project has progressed much more slowly than anticipated. With the reorganization and streamlining of federal government, the restriction of funds and personnel is expected to remain problematic and may work counter to ecosystem management.

Contact information:

Mr. Mike McEnroe
Ecosystem Team Leader
U.S. Fish & Wildlife Service
Region 6
1500 E. Capitol Avenue
Bismarck, ND 58501
(701) 250-4418
Fax: (701) 250-4412

PRINCE WILLIAM SOUND– COPPER RIVER ECOSYSTEM INITIATIVE

Location:
South-central Alaska

Project size:
30 million acres

Initiators:
*U.S. Department of Interior,
National Biological Service*

PROJECT AREA DESCRIPTION

Diverse landforms comprise this vast project area on the Gulf of Alaska. For administrative purposes, the ecosystem has been subclassified into four resource areas: the Copper River Delta, Prince William Sound, Copper River Basin corridor, and the continental shelf. The peaks in Wrangell–St. Elias National Park and Preserve which serve as the headwaters for the Copper and Chitina rivers reach up to 16,000 feet above the coastal and ocean components of the ecosystem.

The climatic and elevational diversity accounts for the broad range of vegetative types. The ecosystem's deciduous forests include aspen, cottonwood, and birch. Mixed conifer and deciduous forests and black and white spruce forests cover broad stretches of the mountain and coastal areas. Cranberries and blueberries are widely found. Many of the wetlands contain cotton grass. Other ecosystem components include eel grass beds on mud flats, kelp beds in intertidal pools, and spectacular black spruce bogs in low, wet areas.

All types of motorized and non-motorized recreation occur. Commercial and sportfishing are prominent uses of the inland, coastal, and ocean waters. The U.S.D.A. Forest Service, Bureau of Land Management, National Park Service, state of Alaska, and native corporations are the principal landowners in the region.

ECOSYSTEM STRESSES

Timber harvesting is among the most significant stresses affecting the region.

Runoff from clear-cuts has led to the sedimentation of streams. Native corporations are concerned about trash and garbage being left behind by an increasing number of recreational users. Treatment of village garbage and sewage is also an important concern.

PROJECT DESCRIPTION

This project is one of twelve ecosystem initiatives undertaken by the National Biological Service (NBS). Many of the area's residents are dependent on the area's natural resources for their livelihoods. The Exxon Valdez oil spill was a significant impetus for this effort, with a partnership established among the region's stakeholders to avoid future "train wrecks."

The goal of the project is to promote a multi-lateral exchange of information among resource managers in the Prince William Sound–Copper River region. NBS hopes that natural resource managers will make decisions that incorporate the concerns and interests of other landowners once the impacts of their own activities on their neighbors are understood.

The project has two primary components: partnerships and information management. NBS is responsible for initiating 1) an agreement with agencies, private landowners, and other individuals with resource management roles to identify their science and data needs for this area; and 2) an agreement with agencies, groups, and individuals conducting research and collecting data in this area. Given the recent establishment of the agreements, partners are still determining which agencies will interact and how. NBS is also developing an information management plan focused on integrating the dispersed information sources on Prince William Sound and the Copper River. Included in the information management plan is a bibliography that references all work completed within the region, a current research profile, a directory of natural resource managers, and a Metadata Catalog of Spatial Data.

It is important to note that NBS is not conducting original research through this initiative, although it does have other active research projects in the region. Rather, the agency is responsible for gathering the information that exists, identifying information gaps, and facilitating the sharing of information among the region's natural resource managers.

STATUS AND OUTLOOK
The cooperative agreement is ready to be finalized and was expected to be signed by more than a dozen partners in early 1996. Resource managers who have not spoken in over a decade despite working close proximity

of each other are communicating about the resources in the sound.

Meanwhile, related activities are beginning to occur between partners. For example, one of the partners is spearheading a workshop on a geographic information systems (GIS) database which was developed on the region. The database has been distributed to more than 700 people, mostly researchers, in the U.S. and abroad. But the partnership's intent is to get this information to natural resource managers. The workshop will bring together these managers, as well as cooperators outside the partnership, for training on the database.

Factors Facilitating Progress
The support and experience of the project supervisor has been cited as the primary reason for progress to date. Having been on the ground for two similar efforts in the lower forty-eight states, the project supervisor has seen how well such partnerships can work.

Obstacles to Progress
Suspicions of the federal government by native corporations and "turf" wars

among government agencies were cited as obstacles to progress. Keeping the partners interested and demonstrating that the project is working for them will determine future success.

Continued Congressional funding for the effort is a great concern. Current funding has been delayed due to the 1995-1996 federal budget impasse, which slowed or prevented the NBS from conducting routine project business, such as purchasing equipment and travelling to meetings with partners. Future funding is even less certain.

Contact information:

Ms. Lisa Thomas
Fish and Wildlife Biologist
National Biological Service
Alaska Science Center
1011 East Tudor
Anchorage, AK 99503
(907) 786-3685
Fax: (907) 786-3636

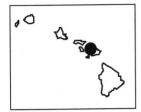

Pu'u Kukui Watershed Management Area

Location:
West Maui, Hawaii

Project Size:
8,600 acres

Initiator:
Maui Land and Pineapple Company

PROJECT AREA DESCRIPTION

The project area consists of private forest preserve lands owned by the Maui Land and Pineapple Company (MLPC), extending from Honokowai and Honokohau valleys to the north, to the 5,788-foot summit of Pu'u Kukui to the south. The watershed is remote and extremely rugged, with a mix of vegetation: lowland communities include native koa and 'ohi'a wet and mesic forests as well as uluhe wet and shrubland forests; montane communities include mixed fern and shrub cliffs, and various bogs and wet shrublands, some of which are considered rare natural communities. In general, the vegetation protects the fragile mountain soil by absorbing water from heavy rainfalls. Water is then gradually released into streams and groundwater aquifers. Several plants and snails, some of which are endemic to west Maui, are candidates for the federal threatened and endangered species list.

The project area is zoned as a conservation area by the state of Hawaii Land Use Commission. Lands on either side of the watershed are part of the state's West Maui Natural Area Reserve or are privately-owned with a conservation easement held by The Nature Conservancy of Hawaii (TNCH). Total size of the protected areas is over 13,000 acres.

ECOSYSTEM STRESSES

Exotic plant, bird, and invertebrate species compete with native species for food, shelter, and other resources, thereby destroying and altering habitat and natural processes. Feral pigs are the greatest threat to the watershed, as they disrupt the ground cover and soil through foraging and rooting, allowing

exotic plants, which cannot root in undisturbed soil, to readily become established. Also, it is believed that pig activity results in erosion and changes in plant communities, which affect infiltration rates of water into the soil and thus the area's hydrology. Finally, grazing and recreation (trespassing hikers, mountain bikes, motorcycles) are additional, less significant stresses.

PROJECT DESCRIPTION

In 1988, the MLPC asked TNCH for assistance in creating a private preserve in lieu of giving TNCH a conservation easement to the property and having TNCH manage the area. At the same time, MLPC hired a part-time manager for the watershed. In 1993, MLPC joined the state of Hawaii's Natural Area Partnership (NAP) program to obtain funding to expand the program.

TNCH developed a long-range management plan, whose broad goal is "to maintain the best possible watershed through protection of native ecosystems," especially the remaining rare natural communities of west Maui and the species that comprise them, with a focus on controlling feral ungulates (through snaring, hunting, and fencing) and exotic plants (with mechanical, chemical, and biological methods).

The watershed was divided into ten management units based on topographical and biological features. The upper elevation bogs and forests were designated a Special Ecological Area (SEA), with management beginning there and continuing downslope. Priority was given to keeping the SEA free of exotic ungulates and plants. Access to the watershed is restricted to

MLPC personnel, volunteers, and researchers. Finally, a monitoring system of feral ungulate and exotic plant populations and native plant and animal population dynamics was established. The plan also called for coordinating management with adjacent landowners and state conservation efforts.

STATUS AND OUTLOOK

Currently, the MLPC is in its third year of the six-year NAP program. Some of the major program goals have been met: the pig control program is nearly complete and feral pig populations have been virtually eliminated in otherwise pristine areas, with the subsequent recovery of native flora. Significant efforts still need be devoted to combating exotic plant species in the remaining predominantly native cloudforests. The MLPC hopes to be able to renew the six-year NAP program commitment.

Factors Facilitating Progress

Strong personal ties of the MLPC President/CEO to The Nature Conservancy (he was a founding member of TNCH) was described as essential to this effort's inception and subsequent progress. TNCH's efforts to secure passage of the Natural Areas Partnership Program by the Hawaii state legislature also benefited this project.

Obstacles to Progress

Although the MLPC is technically guaranteed funding into 1998 through the state NAP program, there is concern that the current fiscal crisis in Hawaii may affect the state's ability to provide program funding for this effort. Similarly, renewal of the NAP status in 1998 could be affected by the fiscal situation.

Contact information:

Mr. Randal T. Bartlett
Watershed Supervisor
Maui Pineapple Company, Ltd.
4900 Honoapi'ilani Highway
Lahaina, HI 96761
(808) 669-5439
Fax: (808) 669-7089

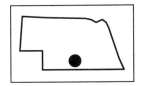

RAINWATER BASIN JOINT VENTURE

Location:
South-central Nebraska

Project size:
2.7 million acres

Initiators:
Nebraska Game and Parks Commission, U.S. Fish and Wildlife Service, Ducks Unlimited, The Nature Conservancy, and others

PROJECT AREA DESCRIPTION

The Rainwater Basin can best be described as a flat loess plain, pock-marked with many deep wind-blown depressions. These depressions can vary in size from one to 2,000 acres. In presettlement times, some 4,000 major wetland basins could be found in these depressions, surrounded by native prairie. Currently, fewer than 400 wetlands remain, and the prairie has been replaced with croplands.

The Rainwater Basin is located in the center of the Central Flyway, a major migration route of migratory birds; the basin is thus critical to the preservation of healthy populations of migratory waterfowl on the North American continent. For instance, more than 90 percent of the mid-continental population of white-fronted geese use the Rainwater Basin as a staging area during migration. There the geese gain strength before continuing on to their breeding grounds. In addition to migratory waterfowl, the basin is frequented by shorebirds and by several federally listed threatened and endangered species, such as the whooping crane, peregrine falcon, bald eagle, and least tern.

ECOSYSTEM STRESSES

The conversion of land use to crop agriculture has had many direct and indirect impacts on the ecosystem. The disappearance of the prairie and the drainage of many wetlands has significantly reduced wildlife habitat. Furthermore, the recharge of the remaining wetlands has been reduced as a result of irrigation tailwater recovery pits which collect water from an irrigated field for later redistribution. These pits not only collect irrigation water, but

natural runoff as well. Land leveling and manipulation of water by the county road and ditch drainage system has also reduced wetland recharge.

PROJECT DESCRIPTION

Most wetlands in Iowa and eastern Nebraska have disappeared, and the Rainwater Basin is the last remaining migratory bird staging area in the region. Recognizing the importance of the basin, the Nebraska Game and Parks Commission, the U.S. Fish and Wildlife Service, Ducks Unlimited, The Nature Conservancy, and others developed a concept plan and subsequently applied for joint venture status under the North American Waterfowl Management Plan (NAWMP) in 1990. (The NAWMP was signed in 1986 by the United States and Canada in response to declining duck populations. Implementation of the NAWMP is the responsibility of "joint ventures," which are cooperative efforts between governments, private organizations and individuals.) After the concept plan was approved, joint venture status was granted in 1991. Following public input, the concept plan was expanded to become an implementation plan, which was finalized in 1992. This plan lists goals, objectives, strategies, and tasks.

The overall goal of the Rainwater Basin Joint Venture is to "restore and maintain sufficient wetland habitat in the Rainwater Basin area of Nebraska to assist in meeting [waterfowl] population objectives identified in the North American Waterfowl Management Plan." This goal will be realized through the protection, restoration, and creation of 25,000 acres of wetlands and 25,000 acres of associated uplands, including the safeguarding of the water supply to these wetlands; and optimization of wetland values to waterfowl. These objectives will be accomplished through the coordination of non-regulatory activities in the area, such as the provision of technical and financial assistance to private landowners attempting to restore migratory waterfowl habitat.

The joint venture hopes to provide structure and focus for state, local, and federal agencies, private landowners, and corporations who want to contribute to a unified wetland restoration effort. Organizations that have become involved include the U.S.D.A. Natural Resources Conservation Service, National Audubon Society, Nebraska Environmental Trust, natural resource districts, and many others.

STATUS AND OUTLOOK

Since the completion of the plan in 1992, a joint venture coordinator has been hired; a private lands work group has been established; a private lands-protection program has been developed; and existing wetlands have been identified and prioritized. Some wetlands have been acquired. Also, several wetland restoration projects on private lands have been carried out.

Factors Facilitating Progress

A recognition among government agencies that the needs of private landowners must be recognized; that no single agency can accomplish the task; and strong administrative support for ecosystem-based assistance have all helped this project to progress. The partnership of all involved agencies and the commitment of its personnel have been very important. In addition, national support through the North American Waterfowl Management Plan has been crucial for the success of this effort. The protection of the wetlands greatly benefits from the development of the implementation plan which streamlines the efforts of individual agencies and landowners.

Obstacles to Progress

Governmental programs offered through the U.S. Department of Agriculture (USDA) tend to provide incentives to keep even extremely poor cropland in production (e.g., through farm programs and set-aside provisions). Thus, landowners are not inclined to look at alternative uses for the land. However, until recently it had been difficult to involve USDA in the effort. Prior to the employment of a joint venture coordinator, the partnership tended to focus on agencies normally working with wildlife.

An additional obstacle to progress has been the lack of both local and agency awareness of the resource, as well as landowner mistrust of government.

Contact information:

Mr. Steve Moran
Coordinator
Rainwater Basin Joint Venture
1233 North Webb Rd., Suite 100
Grand Island, NE 68803
(308) 385-6465
Fax: (308) 385-6469

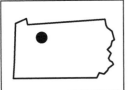

ROBBIE RUN STUDY AREA

Location:
Northwestern Pennsylvania

Project size:
60 acres

Initiator:
U.S.D.A. Forest Service

PROJECT AREA DESCRIPTION

The project area is a part of a 82,000 acre landscape corridor defined by the U.S.D.A. Forest Service (USFS), consisting of seven core areas in the Allegheny National Forest. The purpose of the landscape corridor is to connect existing islands of late-successional and old-growth forests and to protect riparian values. The area is managed by the USFS for old-growth values. It is characterized by sweet birch, black cherry, maple, and beech. Primary use of the area is forest and timber management.

ECOSYSTEM STRESSES

Stresses to the ecosystem include insect infestation and diseases such as beech bark disease and sugar maple decline. High deer populations are a stress to the biodiversity of the ecosystem, because of overcrowding and competition for food.

PROJECT DESCRIPTION

The 1986 Allegheny National Forest plan mandated that the USFS manage for old growth. The impetus for this landscape corridor designation in the Bradford Ranger District is to fulfill this mandate. Adaptive management and various studies, such as the Robbie Run project, are being used to assess the range of management options available within the corridor and to determine the extent that the corridor is achieving the designated objectives. The project was initiated in Spring 1994 and cooperators include, in addition to the national forest, the USFS's Northeastern Forest Experiment Station.

The goal of the project is to develop guidelines for managing snags and logs for wildlife benefits, and to document the impact of such management on wildlife communities within a land-scape perspective. Specific objectives to meet this goal include determining 1) how to create snags and logs; 2) when cultured snags become useful to wildlife, by what wildlife, and for how long; and 3) if different initial starting conditions elicit different responses in wildlife communities.

STATUS AND OUTLOOK

The project area is currently undergoing pretreatments with final treatments planned for 1996. Treatments include girdling live trees to create snags and felling trees as a source of logs. Species richness and abundance of songbirds, reptiles, amphibians, and small mammals will be sampled every other year for nine years. Vegetation will be sampled every three years.

Factors Facilitating Progress

The 1986 forest plan brought people to the table to discuss landscape-level values. Without this impetus, a study project like Robbie Run may not have been conceived.

Contact information:

Chris Nowak or Dave deCalesta
USDA Forest Service
Forestry Sciences Laboratory
P.O. Box 928
Warren, PA 16365
(814) 563-1040
Fax: (814) 563-1048

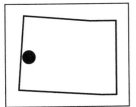

RUBY CANYON & BLACK RIDGE ECOSYSTEM MANAGEMENT PLAN

Location:
Western Colorado

Project size:
118,700 acres

Initiator:
Bureau of Land Management

PROJECT AREA DESCRIPTION

The Ruby Canyon and Black Ridge ecosystem consists of Ruby Canyon and the areas immediately north and south of it. The canyon, characterized by steep red sandstone walls, is an approximately twenty-five mile long section of the Colorado River corridor in western Colorado (Mesa County). The area north of the canyon consists of rolling high desert grass-shrub lands.

The 74,000-acre Black Ridge Wilderness Study Area is located south of the canyon. This area is characterized by sheer-sided, red-rock canyons, sandstone arches, caves, and granite outcrops with spectacular waterfalls and pools. Located between the canyons are rolling sagebrush pinyon-juniper mesas. Several federally listed threatened and endangered species occur in the project area, including the bonytail chub, humpback chub, Colorado squawfish, razorback sucker, and American peregrine falcon. In addition, much waterfowl populates the river, particularly during the Spring and Fall.

Ninety percent of the planning area is public land administered by the Bureau of Land Management (BLM). The other 10 percent consists of private inholdings. The dominant land uses are recreation, cattle grazing (on a limited number of parcels), and education.

ECOSYSTEM STRESSES

Fire, a natural disturbance factor, has been suppressed, leading to increasingly dense pinyon-juniper stands. Survival of desert bighorn sheep may be reduced, because of their limited ability to traverse the pinyon-juniper areas. Exotic species such as salt cedar, knapweed, and cheat grass have altered plant community composition by replacing native species. In addition, increased recreational use has had impacts on vegetation and wildlife, as well as on the experiences of visitors.

PROJECT DESCRIPTION

Increasing recreational pressures, in combination with BLM's shift in management focus from multiple use to ecosystem management, led to this effort. Development of the Ruby Canyon and Black Ridge ecosystem management plan began in 1994 with the publication of a management summary, which illustrated how different aspects of previous management of Ruby Canyon were related. This publication was followed by the formation of an ad-hoc committee consisting of representatives of user groups, local, state, and federal agencies, grazing interests, environmental groups, and community representatives. In addition, technical committees were formed, addressing either recreational aspects or vegetation management.

Based on the findings of the technical committees, the ad-hoc committee helped the BLM refine and develop possible management strategies for the area. It also developed a vision document, expressing the overall goal of the effort: "The Ruby Canyon–Black Ridge area will continue to contribute to the current quality of life for the Grand Valley and will be managed for an ideal balance of use and preservation."

The plan will incorporate a "Benefits-Based Management" approach to recreation management, which is based on a visitor study conducted from 1992 to 1994. It ensures that BLM will provide those experiences most sought by visitors, rather than unwanted services or contrary settings. Vegetation management will be based on an ecological site inventory which was conducted in 1993.

Strategies include the development of partnerships, managing recreational pressures, and vegetation management. In addition, some of the private inholdings will be acquired by the BLM either through land exchanges or purchases. Monitoring will focus on measurable management objectives (e.g., wildlife, vegetative response, visitor satisfaction). The management plan will be periodically revised based on monitoring results and public involvement.

STATUS AND OUTLOOK

Some of the most important outcomes of the project to date include the beginning of collaborative public land management, as well as the realization that public lands play a part in a community's quality of life. BLM has received valuable support for many management objectives. Some issues, however, have not been resolved to everyone's satisfaction, including mountain biking in and access routes to the Black Ridge Wilderness Study Area, group size limitations for river floating in Ruby Canyon, and restrictions on motorized use of the river in the canyon.

Factors Facilitating Progress

Several factors helped this project proceed, including a good ecological site inventory and visitor use survey, community-based partnerships, willing partners, and a professional facilitator.

Obstacles to Progress

The potential threat of a lawsuit under the Federal Advisory Committee Act (FACA) has been a problem. Scheduling a meeting so that all partners can attend has been difficult. Obtaining consensus from a large number of partners has not been easy. Different entities were not always willing to compromise on their use of the natural resources. Until a facilitator was brought in, little progress was made.

Contact information:

Mr. Harley Metz
Grand Junction Resource Area
2815 H Road
Grand Junction, CO 81506
(970) 244-3076
Fax: (970) 244-3083

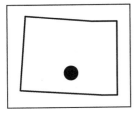

SAN LUIS VALLEY COMPREHENSIVE ECOSYSTEM MANAGEMENT PLAN

Location:
Southern Colorado

Project size:
2.56 million acres

Initiator:
U.S. Fish and Wildlife Service

PROJECT AREA DESCRIPTION

The San Luis Valley ecosystem is a high mountain desert valley, surrounded by the Sangre de Cristo and San Juan mountain ranges. This area makes up the Upper Rio Grande ecosystem, consisting primarily of salt desert shrub and wet meadow riparian habitat on the valley bottom, and pinyon-juniper and spruce-fir habitats at higher elevations.

Since the early 1900s, most of the valley bottom has been irrigated for agriculture, converting salt desert shrub to wet meadow habitat. This altered habitat provides important migration and wintering grounds for many waterfowl species. Valley wetlands also support several federally listed threatened, endangered, and candidate species, including the whooping crane, bald eagle, peregrine falcon, and snowy plover. State listed species include the white-faced ibis, sandhill crane, and slender spider flower.

Current human uses in the valley consist mainly of agriculture, ranching, and mining. There are a few dams present in the Upper Rio Grande ecosystem.

ECOSYSTEM STRESSES

Irrigation has had a positive impact for migratory bird and other wildlife species, but has contributed to a decrease in groundwater levels, adversely affecting some wetland areas. Furthermore, structures associated with irrigation (i.e., power lines) have caused high mortality rates for birds. Another serious threat to the ecosystem is the Summitville Mine, a nearby Superfund site that has contaminated the Alamosa River. Population growth in the state of Colorado also impacts the San Luis Valley. Increased water demand and consumption, road building, the loss of wildlife habitat due to subdivision development, and the change in rural character of the valley are all consequences of this growth. Finally, the introduction of an extremely aggressive invasive plant species, tall white-top, has had significant impact on wet meadow riparian habitat.

PROJECT DESCRIPTION

Recognizing that long-term sustainability requires an ecosystem approach to management, the U.S. Fish and Wildlife Service (FWS) began building partnerships with private landowners for wetland development and habitat management through FWS's Partners for Wildlife Program in 1990. At the same time, public agencies in the valley began discussing ways in which they could work together to manage the valley as a whole. This cooperative spirit grew out of an earlier valley-wide effort to stop a major water development project which planned to pump water from San Luis Valley to Colorado's rapidly growing front range.

As a group, local, state and federal agencies, private landowners, and private organizations from both environmental and commodity interests began to work together on a common goal: to ensure the ecological, social and economic sustainability of the San Luis Valley by protecting, restoring, and maintaining viable levels of biotic diversity.

STATUS AND OUTLOOK

FWS has formed 110 partnerships with private landowners. So far, only one private landowner has left the Partners for Wildlife Program, and the demand for contracts from private landowners far exceeds the FWS's financial resources. In only four years, these and other partnerships have increased wetlands in the valley by 12,000 acres. Overall wildlife diversity is increasing, particularly wetland-dependent species, and a major outbreak of avian cholera has been eliminated. The group is now discussing a proactive initiative to develop a hatchery for sensitive wetland species to prevent further endangered species listings.

Factors Facilitating Progress

Funding and support provided by organizations such as Ducks Unlimited and the Colorado Division of Wildlife have made many of the project's programs possible. Perhaps more important, however, has been the attitude and support of the residents of the San Luis Valley of good land stewardship; the belief that biological diversity leads to ecological, social, and economic sustainability has fostered a strong desire to work together and resolve issues through common goals. To promote such efforts, an education center has been created to increase educational and communication opportunities for valley residents.

Obstacles to Progress

To date, the project has run smoothly, but two factors threaten the future of the project. First, a lack of funding is the major limiting factor for the Partners for Wildlife Program. Second, increased water demand that accompanies the increased population growth of Colorado threatens the water resources of rural areas like the San Luis Valley.

Contact information:

Refuge Manager
U.S. Fish & Wildlife Service
Alamosa/Monte Vista National
  Wildlife Refuge
9383 El Rancho Lane
Alamosa, CO 81101
(719) 589-4021
Fax: (719) 589-9184

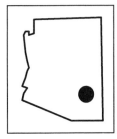

SAN PEDRO RIVER

Location:
Southeastern Arizona

Project size:
512,000 acres

Initiators:
Bureau of Land Management, The Nature Conservancy, local government, private citizens and landowners

PROJECT AREA DESCRIPTION

Flowing from its headwaters in Mexico to its confluence with the Gila River, the San Pedro River is one of the last un-dammed rivers in the Southwest. Surrounded by the Huachuca and Mule mountain ranges, the San Pedro River Valley sits in a combination of Chihuahuan and Sonoran desert habitat, supporting grassland, mesquite, and riparian communities. The Southwest's most extensive cottonwood-willow riparian forest habitat, a rare and rapidly disappearing forest type in the United States, can be found along the San Pedro River.

This river corridor is prized for its wealth of biological diversity, and renowned by bird-watching enthusiasts. This corridor is an important stop-over for migratory birds and supports over 360 bird species. Major land uses in the upper basin include ranching and suburban development. Mining also occurs within the watershed, both in Mexico and along the lower stretch of the river in the United States.

ECOSYSTEM STRESSES

The most significant threat to the San Pedro River system is past and present groundwater pumping. Hydrologic studies have shown that groundwater pumping for agricultural and domestic uses has created two distinct cones of depression in the upper basin aquifer. Although agricultural pumping has decreased, chances of the aquifer recovering are threatened by rapid growth and increasing subdivision development in the upper San Pedro. In the absence of mitigation measures, increased pumping for domestic uses could dry up the river. The habitat

value of the area has also been impacted by fire exclusion, overgrazing, habitat fragmentation, and introduction of exotic species. Finally, the river is threatened by potential contamination caused by acid discharge from mining operations in Mexico.

PROJECT DESCRIPTION

Approximately six years ago, the Bureau of Land Management (BLM) acquired the 56,000-acre San Pedro National Riparian Conservation Area (SPNRCA) along the San Pedro River. Around the same time, The Nature Conservancy (TNC) and the National Audubon Society identified the river as an important system to protect because of its high species diversity and its riparian habitat. Twenty years earlier, TNC had begun operation of the Ramsey Canyon Preserve, located six miles west of the river.

TNC came to realize that the long-term viability of the San Pedro was intricately tied to water and land use in the region surrounding the river. As a result, TNC began building partnerships with the BLM and other public agencies, citizens, and private landowners in the U.S. and Mexico for the long-term protection of the entire upper San Pedro watershed.

TNC's overall goal for the San Pedro project is to work with local communities to promote support for the long-term protection of the river system.

STATUS AND OUTLOOK

TNC is currently working with the Water Issues Group, a community organization that has brought environmental and economic development interests together to develop water management strategies for the region. TNC also participates in water resource planning through the San Pedro Technical Committee, and assists public and private land managers and local decision-makers in crafting land use and water mitigation strategies for the basin. TNC is also promoting ecotourism to illustrate the economic value of the river in its natural state. Finally, TNC is working with other local conservation organizations such as The Friends of the San Pedro River to increase awareness of the river within local communities.

Factors Facilitating Progress

Efforts between environmental and economic interests in developing a comprehensive water management plan for the entire watershed continue, and initiatives to promote ecotourism have been well received by both the general public and the business community. Furthermore, public education efforts appear to be increasing awareness of the river and promoting community discussion about water management strategies. However, long-term hydrologic protection for the river has yet to be achieved.

Obstacles to Progress

Obstacles to reaching a comprehensive water management plan are many, including inadequate funding for mitigation efforts needed to protect the river, conflicts between development and conservation interests, and a strong mistrust of projects with government involvement. Furthermore, because surrounding communities do not rely on the river economically, there is a lack of interest or support to protect it. Finally, with the large and diverse population base of the project area, it is difficult to craft a solution with so many interests around the table. What is needed is a shared community vision of desired future conditions, and a plan of action that allows for economic viability while providing for the long-term protection of the river.

Contact information:

Mr. Paul Hardy
Program Manager
The Nature Conservancy
27 Ramsey Canyon Road
Hereford, AZ 85615
(602) 378-2785

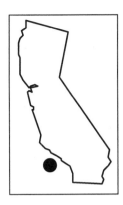

Location:
Southern California

Project size:
42,135 acres

Initiator:
*Santa Catalina
Island Conservancy*

SANTA CATALINA ISLAND ECOLOGICAL RESTORATION PROGRAM

PROJECT AREA DESCRIPTION

Santa Catalina is a rugged, mountainous island characterized by grasslands, chaparral, coastal-sage scrub, oak woodland, dunes, and beaches. The native animals are those that once made it across the channel separating the island from the mainland. These include the Catalina Island fox, an endemic shrew and ground squirrel, several reptiles and amphibians, bats, mice, and many birds. The island is home to many rare and threatened species or communities, including at least one species, on the federal threatened and endangered species list, the bald eagle, and several candidates for the federal list, such as the island fox (also listed by the state), Catalina Island mountain-mahogany, and the Santa Catalina Island monkeyflower (presumed extinct).

Once inhabited by native Americans who lived primarily off of fishing, this isolated island has remained relatively undeveloped through modern times, having generally remained in large tract ownerships. Today, 88 percent of the island is owned by the private, nonprofit Santa Catalina Island Conservancy, which manages it as a nature reserve with low-impact recreation, research, and environmental education. Eleven percent is owned by a tourism-oriented company, Santa Catalina Island Company, with only the remaining 1 percent in private ownership. Southern California Edison owns water rights on the island. Development in the project area is limited to a small ranch that is engaged in limited agriculture and grazing management, facilities in coves along the

coast for lessees, a small airport, three campgrounds, and small native plant memorials and gardens. The ranch also serves as headquarters for conservancy facilities management, while the airport houses educational and interpretive staff. The town of Avalon (not in the project area) has approximately 3,000 year-round residents.

ECOSYSTEM STRESSES

Grazing was introduced to the island by early settlers, thereby disrupting the native plants and damaging shallow soils. As a result of this and later human activities, exotic species were introduced, primarily non-native plants but also goats, cattle, pigs, sheep, mule deer, cats, rats, and a herd of bison. Sheet and gully erosion from livestock overgrazing and trampling have led to excessive stream sedimentation. Hydrologic alteration occurred from impoundments and pumping of water for human and livestock consumption. Today, the conservancy controls goats, pigs, and deer through hunting and trapping. Cattle and sheep are no longer ranched, and the bison herd is managed and limited in size.

PROJECT DESCRIPTION

Conservation practices have been in place on Santa Catalina Island since the middle part of this century, when the Santa Catalina Island Company owned most of the island. The company was controlled by a family with a strong interest in conservation on the island. Members of this family established the conservancy in 1972. Three years later, the company deeded 42,135 acres of the island to the conservancy. This deed included a fifty-year, open-space easement agreement with Los Angeles County, which allows public access to most conservancy land for recreation and education.

The conservancy's legal mandate is to "preserve the island's native plants and animals, its biological communities, and its geological and geographical formations of educational interest." It is also charged with managing Catalina's open space lands for viewing and controlled recreation. The conservancy, which has a paid staff of approximately forty, is administered by a board of directors. The board provides general policy and direction to the conservancy staff, within which projects are developed, with informal consultation from the island's tourism business and recreational users and formal consultation with county and state agencies (in accordance with state and county regulations).

Funding comes from membership dues, large donors, and conservancy revenue-producing operations (such as land leases to various youth groups and recreation clubs). Volunteers are recruited for restoration/conservation

activities, special events, scientific research, and administrative support.

STATUS AND OUTLOOK

The conservancy has an internal long-range plan, with conservation as its highest priority and two additional goals (education, recreation). A more specific and comprehensive resource management plan is being developed and will include a monitoring program.

The most significant outcomes of this effort have been the reduction of feral goats, pigs, and deer, some control of invasive non-native plants, reappearance of rare native plants and animals, and regrowth of vegetation such as coastal sage scrub, oak woodland, chaparral, and native grassland.

As a result, the project area is being restored to a more natural condition. In addition, development of a volunteer program and the development of initial restoration plans are occurring.

Factors Facilitating Progress

Strong volunteer and donor programs have helped this effort progress. Scientific research

(archaeology, floristic studies, wildlife, geology, marine science) is contributing to a better understanding of the island's biophysical components. A strong conservation ethic shown by the conservancy leadership and biological staff has also been credited as a benefit to this effort.

Obstacles to Progress

Several factors are described as specific program concerns or needs, including limited funding for projects and increasing staff; a need to further educate staff about the island's ecology, flora and fauna, and history; and a need to increase island residents' support for and involvement in the conservancy and its activities.

Contact information:

Mr. Allan Fone
Santa Catalina Island
  Conservancy
P.O. Box 2739
Avalon, CA 90704
(310) 510-1299

SANTA MARGARITA RIVER

Location:
Southern California

Project size:
*1 million acres
(540,000 acres in
protective status)*

Initiator:
*The Nature
Conservancy*

PROJECT AREA DESCRIPTION

The Santa Margarita River corridor is the most unique and biologically diverse area in southern California, representing a microcosm of California's natural habitats. Flowing through steep granite canyons, conifer forests, dense coastal willow and cottonwood riparian forests, rare native grasslands, coastal sage scrub, and marsh communities, the Santa Margarita is the last major undammed river in coastal southern California. These diverse habitats support many threatened and endangered species, including the California gnatcatcher, California least tern, Cooper's hawk, least Bell's vireo, and the California mountain lion. The project area abuts Riverside, Orange, and San Diego Counties, and includes the 124,000-acre Camp Pendleton Marine Corps Base.

ECOSYSTEM STRESSES

Fragmentation is a major threat, particularly in the project area's western slope which is experiencing rapid residential development pressures from Orange County. Disruption of the region's natural fire regime including fire suppression and unnatural fire frequencies has affected vegetational composition in the coastal sage scrub and other natural communities. Human uses of the river, such as the development of water supply projects, have lowered the water table, decreased water quality, and disrupted natural flooding patterns, ultimately destroying riparian forest habitat.

PROJECT DESCRIPTION

During a statewide analysis, ecologists identified California's southwest biogeographic region not only as the state's most biologically diverse region, but also as the most threatened region in the state. The Santa Margarita River was identified as the number-one biological hot spot in southern California.

The Santa Margarita has been included in The Nature Conservancy's (TNC) Bioreserve Program, and, since 1982, TNC has placed approximately 17,000 acres under protective management. TNC's conservation plan has focused on the natural processes of the area, specifically fire and flood regimes. As a result, the boundaries of the planning area were based on the size of the area required to maintain those processes.

The project goals include, among others, 1) placing the entire main stem of the Santa Margarita River into conservation ownership; 2) restoring and maintaining the natural occurrences of fire and flooding; 3) creating connecting corridors for wildlife movement from the Santa Margarita River to TNC's Greater Santa Rosa Plateau Preserve (purchased ten years ago), and to other surrounding large conservation ownerships such as U.S.D.A. Forest Service and Bureau of Land Management lands; 4) developing a land management strategy for Camp Pendleton Marine Corps Base that ensures protection of the resources while continuing military activities; and 5) working with private landowners to educate them on the ecological significance of their properties and to enable them to use and develop their properties while meeting Endangered Species Act requirements.

To reach these goals, TNC has continued to build public and private partnerships at the local, state, and federal levels.

STATUS AND OUTLOOK

Ninety-eight percent of the Santa Margarita corridor is presently in conservation ownership, primarily through TNC land acquisitions and state and federal ownership. TNC is starting on the third year of the five to ten-year project. TNC is now working in conjunction with the Natural Community Conservation Program (NCCP) in southern California, an effort to protect coastal sage scrub communities by implementing the program in the project area. Already, several sites have been designated as NCCP sites in the coastal Orange County portion of the project area.

Factors Facilitating Progress

A sound scientific foundation establishing the need for protection, a non-confrontational, team approach, and detailed strategic planning have all contributed to the advancement of this project. The willingness of major landowners to be a part of the process has been cited as particularly helpful.

Obstacles to Progress

Conservation efforts have been disrupted by property rights interests whose explosive and often inaccurate rhetoric has in some cases negatively influenced the landowners with whom TNC is attempting to build partnerships.

Contact information:

Mr. Cameron Barrows
Southern California Area Director
The Nature Conservancy
P.O. Box 188
Thousand Palms, CA 92276
(619) 343-1234
Fax: (619) 343-0393

SIDELING HILL CREEK BIORESERVE

Location:
Western Maryland, southern Pennsylvania

Project size:
66,000 acres

Initiator:
The Nature Conservancy

PROJECT AREA DESCRIPTION

The project area is defined by the watershed of Sideling Hill Creek. The ecosystem has remained fairly pristine due to the low human presence throughout the watershed. Sideling Hill Creek starts in Pennsylvania and flows through the ridge and valley physiographic region in Maryland and finally into the Potomac River. The flow of the creek in the summer can be very low because most rains tend to fall on the Appalachian Plateau to the west before reaching the watershed. During the winter, however, the creek can rise rapidly with intense storm events.

Significant rare communities are present in the watershed, including harperella, freshwater mussels, and shale barrens. The watershed is 75 percent forested, with the predominant vegetative cover in oak-hickory forest. In non-forested regions, land uses include small amounts of agriculture and development.

ECOSYSTEM STRESSES

Roads and major highways are a stress to the ecosystem due to habitat destruction and runoff of salts that threaten the rare mussel populations. Exotic species are also a direct threat due to displacement of native populations within the river, floodplains, and riparian forest lands. These exotic species include zebra mussels, Asiatic grasses, and some insects. Since many of the rare species, such as harperella, are dependent on the flow regime in the creek, anything that disturbs the hydrology threatens their existence.

PROJECT DESCRIPTION

The Nature Conservancy (TNC) and the Western Pennsylvania Conservancy (WPC) were alerted to the significance of the area by survey work completed by the Maryland and Pennsylvania Heritage Programs. In 1991, TNC began acquiring key parcels of land and increasingly realizing the importance of the watershed. WPC then acquired a key shale barren near the Maryland border. Both groups decided that acquisition of parcels would not provide sufficient protection, and in 1992 began planning for a bioreserve.

The seven goals of the area are to 1) improve decision making through research; 2) protect shale barrens; 3) protect the riparian corridor; 4) control exotic species; 5) maintain water quality; 6) maintain flow regime; and 7) restore riparian forest communities to buffer water quality.

Currently, a large component of the project is research and monitoring. The emphasis is on water quality, habitat requirements, and life history of freshwater mussels. Other strategies designed to meet the goals include 1) continuing protection of shale barrens and riparian areas through acquisition; 2) maintaining buffers; 3) restoring disturbed areas so they do not become avenues for exotic species; and 4) eliminating water quality degradation caused by road salt runoff. These strategies are further outlined in the Sideling Hill Strategic Plan. Additional partners include the Maryland Heritage Program and State Highway Administration.

STATUS AND OUTLOOK

TNC recently purchased a 161-acre property in the Maryland portion of the watershed. In mid-1995, TNC was negotiating with the Maryland State Highway Administration to help purchase two additional parcels of land and to help monitor the impacts of salt and other highway runoff. In Pennsylvania, WPC has acquired three tracts and continues to work on several other parcels. The Pennsylvania Department of Environmental Regulation has recommended that the portion of the stream in Pennsylvania be given Exceptional Value Status, which would provide additional water quality protection. Future plans include expanding the list of federal community partners, and continuing to secure lands to ensure long-term protection.

Factors Facilitating Progress

One factor that facilitated progress from the start was the overall consensus between partners on the ecological importance of the site. In addition, baseline information on Sideling Hill Creek was well documented due to the natural heritage survey, the nature of the threats were easily defined, and the watershed was relatively pristine. These factors facilitated the development of a management plan.

Obstacles to Progress

Securing adequate funding for such a large project areas has been a challenge. These resources are critical in fulfilling research and monitoring goals. It has also been somewhat difficult to obtain the involvement of federal partners. In addition, there was criticism early on from local communities regarding limitations on recreational fishing in the creek.

Contact information:

Mr. Rodney Bartgis
Manage, Sideling Hill Creek
Bioreserve
The Nature Conservancy
2995 Grade Road
Martinsburg, WV 25401
(304) 754-6709

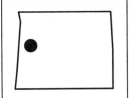

SNAKE RIVER CORRIDOR PROJECT

Location:
Northwestern Wyoming

Project size:
69 river miles

Initiator:
Teton County, Wyoming

PROJECT AREA DESCRIPTION

The project site is located within the Greater Yellowstone Ecosystem and focuses on the segment of the Snake River between the outlet of Jackson Lake Dam to the Palisades Reservoir. The river flows under the shadow of the Teton Range as it cuts across the Yellowstone Plateau. It contains one of the most floated stretches of whitewater canyon in the country, although the river is known primarily for its excellent fishing opportunities.

The river is classified as a braided system with complex wetland, riverine, palustrine, and riparian areas throughout. Narrowleaf cottonwood and willow inhabit riparian areas while Douglas-fir and several species of pines cover the uplands. Moose, buffalo, trumpeter swan, osprey, mule deer, sandhill crane, grizzly bear, and other species occupy the corridor. The region is one of the most productive nesting sites for bald eagles. Although still a predominant land use, ranching is gradually giving way to high-end residential development along certain segments of the river.

ECOSYSTEM STRESSES

Construction of a forty-two mile levee system by the Army Corps of Engineers in the 1950s and 1960s has negatively impacted the river's fisheries and riparian lands. Changes in the composition of vegetation in the riparian zone are attributed to the levees' disruption of the flood regime. Introduction of the levee system in the heart of Jackson Hole allowed for wetland areas to be considered upland areas for development purposes. Multi-million dollar homes now occupy the

cottonwood forest that had been located in the riparian zone.

PROJECT DESCRIPTION

In Fall 1993, Teton County applied for a technical assistance grant from the National Park Service's River and Trails Conservation Assistance Program. This program provided the county with professional planning assistance for the Snake River corridor.

The project's narrow initial scope on recreational planning has been expanded. Currently, the project's goals are to 1) preserve and enhance the natural character of the Snake River; 2) provide improved recreational opportunities within the corridor, consistent with minimum impact on river resources, adjacent private lands, and quality of experience; and 3) create a system of cooperative planning for river management among local, state, and federal agencies, and community organizations.

Future desired conditions for the river are being developed with input from numerous organizations. Input is gathered by river stretch rather than by agency jurisdiction, demonstrating the project's focus on ecological boundaries. Among the participants are the U.S.D.A. Forest Service, Bureau of Land Management, National Park Service, Army Corps of Engineers, Trout Unlimited, Jackson Hole Alliance for Responsible Planning, Lower Valley Power and Light Company, over a half dozen state and local government entities, and many citizens and landowners.

These organizations are not trying to develop a management plan for agencies to follow. Instead, they participate in roundtable discussions to resolve disputes, streamline management, and improve stewardship of the river resources. Geographic information systems mapping will allow tracking of changes and progress in the corridor. Specific management activities are handled by individual agencies or organizations according to their mandates or charters.

STATUS AND OUTLOOK

Public workshops have been held in Wyoming, Idaho, and Utah in order to define the project's geographic area and identify public concerns. The project is expected to last until 1996-97, although the project may continue if funding to support a local staff position becomes available. Although organizations active within the

corridor share their proposed management activities, sharing proposed plans has done little to promote coordination.

Factors Facilitating Progress

The project has been assisted by the strong support of agency management in Grand Teton National Park and Bridger-Teton National Forest. The federal agencies have encouraged the county to take a leadership role in conserving the river. Continued commitment of staff time and funding are required for future progress.

Obstacles to Progress

To explain the project and the benefits it can provide to the public and to government agencies has been challenging. It has been incumbent upon the project facilitator to demonstrate that

cooperative roundtable meetings serve a useful role in conserving the river's resources. Keeping a sense of momentum and support among Teton County commissioners may be difficult. Wavering county support, lack of landowner support for increased use or regulation, and federal agency management mandates which do not permit compromise threaten the project's future.

Contact information:

Mr. Tim Young
Project Facilitator
Snake River Corridor Project
Teton County
PO Box 1727
Jackson, WY 83001
(307) 733-8225
Fax: (307) 733-8034
E-mail: tyoung@wyoming.com

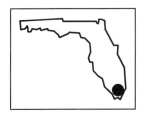

SOUTH FLORIDA/EVERGLADES ECOSYSTEM RESTORATION INITIATIVE

Location:
Southern Florida

Project size:
11.5+ million acres

Initiator:
U.S. Department of Interior

PROJECT AREA DESCRIPTION

The south Florida watershed extends from Orlando, the headwaters of the Kissimmee River, through Lake Okeechobee and the Everglades into Florida Bay, the Florida Keys, and the near coastal waters beyond. The entire area is a flat wetland, underlain by a limestone aquifer, with semi-tropical and tropical vegetation. The Everglades itself is actually a 150-mile long, 50-mile wide river, extending from Lake Okeechobee, the source of its water, to Florida Bay, with the elevation declining only 20 feet over that distance. On its year-long passage through the Everglades, water passes through vast expanses of sawgrass (unique to the area), small hummocks of hardwoods, and mangrove swamps. The entire area is rich in all types of plant and animal life, including fifty-six federally listed threatened and endangered species such as the snail kite, Florida panther, and wood stork, and many state listed species.

Much of the lower part of the system is in public ownership. Federal lands include Everglades National Park, Loxahatchee National Wildlife Refuge, Big Cypress National Preserve, and the Florida Panther National Wildlife Refuge to the west. Extensive state conservation lands include areas recently acquired for Everglades recovery efforts. The Big Cypress Seminole Indian and Miccosukee Indian Reservations are located within the system. The remainder of the land is in row crop farming (primarily sugar cane and vegetables), some grazing and citrus, and large urban centers to the east.

ECOSYSTEM STRESSES

Severe hydrologic alteration is the predominant stress to the region. Much of the area south of Lake Okeechobee and north of Everglades National Park has been drained for agriculture and flood control, through 1500 of miles of canals built after World War II as part of the Central and South Florida (C&SF) Project. The drainage system no longer supports natural plant communities, and the disruption of the complex water flow patterns into the Everglades is responsible for a host of problems on land and in Florida Bay.

Water quality has also suffered, due largely to agricultural runoff. Mercury contaminants, whose sources are unclear, have bio-accumulated in fish. Land sprawl from the ever-growing MiamiñFt. Lauderdale metropolis is consuming areas essential for natural system function or restoration. Exotic species, especially the plant melaleuca, are invading vast expanses of the Everglades, thriving on the excessive agricultural nutrient runoff and reducing biodiversity.

PROJECT DESCRIPTION

The current initiative is the latest chapter in various efforts to restore the south Florida/Everglades ecosystem, whose decline has been well documented and has gained public attention in recent years. In 1984, Governor Bob Graham initiated the state's "Save Our Everglades" program, involving a number of projects in the area. In 1989, Governor Bob Martinez initiated efforts to restore the Kissimmee River, once a meandering river that was channelized and straightened. In the same year, Congress authorized the expansion of Everglades National Park by over 107,000 acres to include lands essential to the Everglades' restoration.

In 1988, an unusual lawsuit was brought by the U.S. District Attorney against Florida, charging the state with failing to implement its own water quality laws, specifically with regard to agricultural runoff from the Everglades Agricultural Area. After the state fought the suit for five years, newly-elected Governor Lawton Chiles abruptly admitted the state's culpability and began cooperating with the federal government, largely as a result of public opinion and the high cost of fighting the suit.

The current initiative is being coordinated by the South Florida Ecosystem Restoration Task Force, created by U.S. Department of Interior Secretary Bruce Babbitt in 1993 and chaired by Interior Assistant Secretary George T. Frampton, Jr. The task force seeks to coordinate activities of federal and state governments and includes federal representatives from the Departments of Interior, Commerce, Justice; the Environmental Protection Agency; the Army Corps of Engineers (ACOE); and others. The initiative's goal is to restore the lost natural function in the Everglades and for human uses in south Florida to be sustainable.

Also in 1993, at the direction of Congress, the ACOE began a "Restudy" of the C&SF project, which the Corps had largely been responsible for building. The

study will be used to investigate ways to modify the C&SF Project to restore the Everglades while still providing water-related services to residents and industry. Finally, in 1994, Governor Chiles created the Governor's Commission for a Sustainable South Florida, whose task was to develop and implement options for restoring the ecosystem in cooperation with all stakeholders.

The list of stakeholders is long, including many public agencies at all levels, native American tribes (Miccosukee, Seminole), agricultural interests (sugar cane, vegetable), commercial and recreational fishers, environmental groups, developers, other private landowners, the business community, and the general public.

STATUS AND OUTLOOK

Much of the task force's efforts have been in setting up collaborative structures and agreements, including funding schemes, coordinating research, and developing plans for sustainable activities in south Florida. Selected restoration efforts have begun. For example, twenty-six miles of the Kissimmee River are being allowed to return to their original meandering state. Efforts to mimic the hydrology of the Everglades are being designed and attempted, through interagency cooperation and a broader systems approach to restoration. Nearly two billion dollars in public projects (one-third federal, two-thirds non-federal) have been proposed to restructure the region's water control infrastructure (although many projects have yet to be fully funded). In addition, improved communi-

cation and cooperation between stakeholders, particularly public agencies, has been cited as another positive outcome.

Factors Facilitating Progress
The high visibility of the ecological problems in south Florida has brought public and political support for this effort. It is the largest and most complex of four federal restoration initiatives, an indication in itself of the level of commitment from the federal government. The national and state environmental communities, under the umbrella of the Everglades Coalition, are aggressively promoting this effort as well.

Obstacles to Progress
Operating at such a large scale, maintaining individual and agency energy for the long term is a challenge. Continued state and federal funding—for infrastructure, personnel, inventories, and monitoring, among others—is far from certain. While monitoring is occurring on a smaller scale, it has yet to be instituted effectively at a landscape level. There is insufficient scientific understanding or inventories of the ecosystem's biophysical components. While the restoration effort enjoys broad bipartisan political support, opposition to the effort is still significant, particularly from a very powerful sugar lobby.

Contact information:

Col. Terrence Salt
Director, South Florida Ecosystem
Restoration Task Force
Florida International University
OE Building, Room 148
University Park Campus
Miami, FL 33199
(305) 348-4095

ST. MARYS RIVER REMEDIAL ACTION PLAN

Location:
Northern Michigan, western Ontario

Project size:
75 river miles

Initiators:
International Joint Commission, Ontario Ministry of Natural Resources, Michigan Department of Natural Resources, U.S. Environmental Protection Agency, Environment Canada, U.S., Canadian, and tribal fish and wildlife agencies

PROJECT AREA DESCRIPTION

The St. Marys River connects Lake Superior and Lake Huron, and forms one of the borders between the United States and Canada. The river is fast moving; rapids near Sault Ste. Marie required the building of the famous Soo Locks for navigation. The land surrounding the river is rolling, formed by bedrock overlain by a relatively thin layer of glacial material. Vegetative communities include upland Great Lakes hardwood forests, mixed hardwood-conifer boreal forest, wet forests, and wetlands. Federally listed threatened species in the area are the bald eagle and the piping plover.

Most of the land immediately adjacent to the river is privately owned. Eighty-three percent of the land within three and one-half miles of the river is undeveloped forest or wetlands. Ten percent is used for farmland (mostly hay), and 5 percent is urban (Sault Ste. Marie). Industries in the Sault Ste. Marie area include hydroelectric power production, steel and paper mills, and shipping.

ECOSYSTEM STRESSES

The hydrology of the river in the Sault Ste. Marie area has been altered substantially to support navigation and power generation, resulting in significant habitat loss. Land conversion for urban development and roads have also led to habitat loss. An additional stress is posed by non-point and point source pollution. Significant point sources of pollutants include paper and steel industries and sewage treatment works. Chromium contaminants have entered the river from a former tannery nearby, now a Superfund site. The sea lamprey,

an exotic species, is also a major problem. The rapids are prime breeding grounds for this species, which parasitizes and often kills large game fish.

PROJECT DESCRIPTION

In 1985, the St. Marys River was identified by the International Joint Commission as an Area of Concern (AOC) based on the extensive environmental problems of the river. In response, state, provincial, and federal governments in the U.S. and Canada initiated a Remedial Action Plan (RAP) process. This process is characterized by three stages. In stage I, problems and their causes were identified by a RAP team, consisting of Ontario Ministry of Natural Resources (OMEE), Michigan Department of Natural Resources (MDNR) U.S. Environmental Protection Agency (EPA), Environment Canada, and U.S., Canadian, and tribal fish and wildlife agencies. The RAP team was advised by members of the Binational Public Advisory Committee (BPAC), consisting of U.S. and Canadian citizens, environmental and industrial interests, and academicians. The Stage I document was published by OMEE in 1992.

During Stage II, currently underway, remedial actions are identified and implemented. In order to prepare a Stage II document, facilitated task groups have been created. These groups consist of both RAP team members and BPAC members and focus on specific issues such as habitat and point source pollution. The Stage II document will suggest remedial actions, which may include increased treatment of sewage and industrial waste, sediment remediation, and habitat improvement. The implementing agency will vary depending on the problem to be addressed. During Stage III, monitoring will take place and reasons for delisting the area as an AOC will be demonstrated.

In addition, the project attempts to educate the general public about the river, its problems, and actions undertaken to correct these problems. Education takes place through river appreciation days, booths at meetings and shows, and BPAC.

STATUS AND OUTLOOK

Many activities are already underway. They include sewer separation in the city of Sault Ste. Marie, Michigan; improved waste treatment by a local steel mill resulting in significant reductions in loadings; ongoing water column and sediment monitoring; Superfund remediation work; and several pilot-scale sediment remediation projects. In addition, the project has resulted in a greater public awareness of the issues. A decrease in the number of overlapping efforts by governmental land management agencies is another result. Fish and wildlife management agencies in both the U.S. and Canada are cooperating.

Factors Facilitating Progress

A team-based approach to the identification of problems and solutions has been helpful. Also beneficial has been the participation of individuals who care about the St. Marys River and are willing to contribute their time to the effort.

Obstacles to Progress

Obtaining balanced representation of all stakeholders in the BPAC has been difficult. The confrontational approach of some interests on the BPAC may have kept other stakeholders from participating. Staff turnover at the Ontario Ministry of Natural Resources has created vacancies, thus slowing the effort's progress. Political support and leadership are concerns for completing the process.

Contact information:

Ms. Susan Stoddart
Coordinator
Ontario Ministry of Environment
  and Energy
747 Queen Street East
Sault Ste. Marie, Ontario,
  CANADA P6A 2A8
(705) 949-4640

STATE LINE SERPENTINE BARRENS

Location:
Southern Pennsylvania, northern Maryland

Project size:
2,100 acres primary; 20,000 acres connective and buffer

Initiator:
The Nature Conservancy

PROJECT AREA DESCRIPTION

The State Line Serpentine Barrens are a chain of eight sites spanning thirteen miles along a large serpentinite outcrop straddling the Mason-Dixon line in Pennsylvania and Maryland. Serpentine barrens vegetation is defined as an assemblage of plants with a large proportion of species having high regional fidelity to outcrops of ultramafic rock. Serpentine barrens in the eastern United States include some of the largest patches of truly native prairie in that region.

The eight core sites included in this project comprise the largest area of serpentine barrens vegetation in the eastern temperate zone. This serpentine barren community is exceptionally high in species diversity of herbaceous plants and specialist-feeding insects, with a large proportion of species occurring as disjunct populations from ranges farther south, in the western prairies, and along the Atlantic coastal plain. The core areas are dedicated to reserves and passive recreation while the land uses in the buffer region include agriculture, horse farming, and urban settlement surrounding Baltimore and Philadelphia.

ECOSYSTEM STRESSES

The serpentine ecosystem is fire dependent and as a consequence, the system will cease to exist without the presence of fire. Therefore, anything that disrupts the fire regime threatens the system with succession to mesic species. Proposed development in the buffer areas is a threat to the ability to burn. It is therefore necessary to create large enough buffers for smoke and fire management. Another stress is the intro-duction of invasive species into the barrens (e.g., black locust, tree-of-heaven).

PROJECT DESCRIPTION

In the mid-1980s, this region was identified through the Natural Heritage Inventory System as one of the highest priority areas for conservation in Pennsylvania. The project was formally initiated by the Pennsylvania field office of The Nature Conservancy in 1991; a year later, the ecosystem received a global significance ranking. Several other stakeholders participate in this project, including the Pennsylvania Bureau of Forestry; Lancaster and Chester Counties, Pennsylvania, and Cecil County, Maryland; the University of Pennsylvania and Pennsylvania State University; and The A.W. Mellon Foundation.

The two major project goals are to 1) maintain a viable representation of all priority plant communities within the barrens ecosystem; and 2) maintain the connectivity of the system, which will allow for viable populations of lepidopteran species. Strategies designed to meet these goals include fully protecting the core sites and buffer regions through acquisition and easements, researching how best to manage the ecosystem, and then managing it in perpetuity. These goals and strategies will be included in a formal site conservation plan, currently being written and expected to be completed in 1997.

In order to measure project success, monitoring will be emphasized. One factor that will be measured is the ability of different target communities to persist in a representative composition that is appropriate for the ecosystem.

STATUS AND OUTLOOK

Two research efforts have the goal of defining how ecological transitions occur between communities. One effort was completed in early 1996; the other is still ongoing. So far, about forty burns have been completed as part of the research on the effects of prescribed fire, in order to answer such questions as when and at what severity to burn. This research has already yielded useful results with significant implications for future management.

Factors Facilitating Progress

The rural character of the area has facilitated the project, since the farming community already understands the value of prescribed burning. Also, key local players have been supportive of the project, and funding has been forthcoming.

Obstacles to Progress

Protecting the core areas has been challenging since not all landowners have been supportive of the effort. There are still obstacles to overcome in terms of implementing the prescribed burning. In particular, not all of the land in the area is under conservation protection, thus limiting the ability to burn across large landscapes. Resource constraints have also been a problem. However, there are indications that in the near future, these obstacles will be overcome.

Contact information:

Mr. James Thorne
The Nature Conservancy
1211 Chestnut St.
12th Floor
Philadelphia, PA 19107-4122
(215) 963-1400
Fax: (215) 963-1406
E-mail: jthorne@tnc.org

STEGALL MOUNTAIN NATURAL AREA

Location:
Southeastern Missouri

Project size:
5,387 acres

Initiator:
Missouri Natural Areas Committee

PROJECT AREA DESCRIPTION

Stegall Mountain Natural Area is located in Missouri's lower Ozarks. On top of the 1,300 foot high mountains, a mixture of igneous (rhyolite) glades and oak-pine savannas can be found. Mountain slopes support oak-hickory woodlands, interspersed with oak-pine forests. Bottomland hardwoods can be found along high-quality Ozark headwater streams.

Stegall Mountain Natural Area is home to two species listed by the state as rare, the marsh violet and four-toed salamander. The area also supports the eastern collared lizard, a species on the state's 1992 watch list. In addition, migrant neotropical songbirds, such as the Kentucky, parula, and cerulean warblers, as well as red-eyed vireos and summer tanagers, frequent the area.

Seventy percent of the Natural Area is owned and managed by the Missouri Department of Conservation (DOC), and the remainder by The Nature Conservancy and the National Park Service's Ozark National Scenic Riverways. Stegall Mountain is buffered by surrounding naturally forested lands. The area is used primarily for preservation and recreational activities such as hiking, wildlife watching, and hunting.

ECOSYSTEM STRESSES

Savanna-glade ecosystems are among the most endangered in the United States. The glades as well as the oak savannas found on Stegall Mountain depend on fire to reduce encroachment of woody species and to maintain prairie grasses and forbs. Thus, suppression of fire since the middle of

the century has threatened these ecosystems. Comparison of aerial photos show substantial encroachment of woody species such as post oak and black-jack oak into savannas and glades. Herbivores—elk and bison—are now missing from the system. Exotic species occur in areas adjacent to the creeks and roads.

The state of Missouri provides 80 to 90 percent of the lead supply of the United States. The potential exploitation of lead belts both north and south of Stegall Mountain could lead to air and water pollution, increasing development activity in the area, and could alter the area's aesthetics.

PROJECT DESCRIPTION

The state of Missouri identifies and preserves high-quality natural areas representing its ecosystem diversity. A natural areas inventory conducted in every county supports a statewide natural area classification by the Missouri Natural Areas Committee. This committee is composed of four major land management agencies, including the Missouri DOC, Missouri Department of Natural Resources, National Park Service, and U.S.D.A. Forest Service. Since Stegall Mountain supports representative igneous glade and dry igneous savanna forest ecosystems, it was nominated for official Natural Area status in 1992, which was granted by the committee in 1993. Currently, Stegall Mountain is the largest Natural Area in the state. It will function as a prototype of a landscape scale Natural Area.

As a result of its new status, the management approach of Stegall Mountain

Natural Area has changed dramatically. Whereas the area used to be managed for timber and wildlife, the current management goal is to maintain the area's natural communities in a naturally occurring landscape mosaic.

Management strategies include spring and fall prescribed burns to remove woody encroachment and exotic plants, minimization of human disturbance detrimental to the ecosystem, and research and monitoring. Three research plots and three half-mile monitoring transects are use to track the impacts of the burns on the ecosystem in terms of vegetation diversity, community boundaries, and tree recruitment.

STATUS AND OUTLOOK

The project is in its second year of implementation. Three prescribed burns have been carried out, of which the last (1994) covered 960 acres. Herbaceous plants in the glades are showing an increase in abundance and flowering. After the last burn, for the first time in ten years of monitoring, collared lizards apparently moved between glades. Such movement may be crucial for the long-term survival of populations of these lizards. Another prescribed burn was planned for March 1996. Full "recovery" of these ecosystems is expected to take at least ten to 100 years. Requests for additional funding for research and monitoring will continue.

Factors Facilitating Progress

A long history of successful interagency cooperation between the Missouri DOC and the National Park Service, as well as the support of these agencies for the project, was crucial to the establishment and management of the Stegall Mountain Natural Area. Furthermore, it has been very important that management decisions have been supported by scientific information.

Obstacles to Progress

Other competing programs and agency priorities limit the amount of time and number of staff for this project.

Not all people, including some professionals, are convinced that prescribed burns are an appropriate management technique. Some maintain that timber harvest can stimulate the effects of fire, or that the values enhanced do not warrant the cost. Others argue that nature can take care of itself, and that natural fires will maintain the ecosystems of interest.

Contact information:

Mr. Larry Houf
District Wildlife Supervisor
Missouri Department of
  Conservation
Ozark District Office
Box 138
West Plains, MO 65775
(417) 256-7161

TENSAS RIVER BASIN INITIATIVE

Location:
Northeastern Louisiana

Project size:
750,000 acres

Initiator:
Northeast Delta Resource Conservation and Development District

PROJECT AREA DESCRIPTION

The Tensas River basin is part of the lower Mississippi River alluvial valley (LMRAV) that once supported 25 million acres of forested wetlands. The basin is now only a representative sub-basin of the LMRAV, as over 80 percent of the area has been converted to row-crop agriculture. The economic base today is heavily dependent on agriculture—primarily cotton, soybeans, corn, and rice—a significant portion of which is on marginal land. About 60 percent (80,000 acres) of the basin's remaining bottomland hardwoods are in federal or state ownership.

ECOSYSTEM STRESSES

Habitat destruction is the most significant stress. The remaining habitat is fragmented and the hydrology throughout the basin has been severely altered. Water quality problems have resulted from sediment, pesticide and nutrient runoff. As a result, species abundance and diversity have declined dramatically. Rivers and lakes once were relatively clear and supported an abundant fisheries. Today, the Tensas River is among the most turbid in the state.

Endemic species that are extinct or federally listed as threatened or endangered include the ivory-billed woodpecker, Bachman's warbler, Florida panther, red wolf, and the Louisiana black bear. These and many other species which require large areas of forested habitat have declined, as have those requiring very specialized habitats which have also disappeared.

The socioeconomic condition of the region is depressed. Timber production has dramatically decreased. With the

area now almost totally dependent on a single industry that is depressed (agriculture), job opportunities are seasonal and scarce. The region is characterized by tremendously high unemployment and poverty levels. Population declines have been as high as 60 percent over the last twenty-five years.

PROJECT DESCRIPTION

In the early 1990s, local Conservation Districts requested the U.S.D.A. Soil Conservation Service (SCS; since renamed to Natural Resources Conservation Service or NRCS) to perform a river basin study. Following that request, the nonprofit Northeast Delta Resource Conservation and Development District (NDRCDD) convened a meeting in early 1992 of local, state, and federal officials, landowners, and conservation groups to discuss the study in a larger context, since there appeared to be several focused studies or environmental management efforts proposed for the area.

This core group agreed to establish a model demonstration project, and formed a nineteen-member technical steering committee. The committee is chaired by a local farmer, with representatives from local parishes, many federal and state agencies, a local levee district, The Nature Conservancy, National Fish and Wildlife Foundation, the NDRCDD, and six farmers.

A public advisory committee was set up to assist the steering committee, with representatives including landowners, agricultural chemical and equipment dealers, U.S. Department of Agriculture, and local conservation districts. Through the steering committee's efforts, the SCS study formed the basis for a broader land management plan. This two-part plan consists of 1) an information needs assessment, including determining what baseline data is required for strategy development; and 2) a more comprehensive study identifying stresses and strategies to address them.

The committee operates on a consensus basis and reached early agreement on nearly all components of the plan except for flooding, a sensitive issue in this region given the extensive hydrological modifications of the Mississippi River; a separate flooding study was performed so as not to hold up the broader plan.

STATUS AND OUTLOOK

The conceptual plan has been completed and identifies eight major problems: 1) long-term viability of row crop agriculture; 2) water quality; 3) bottomland hardwood decline; 4) flooding; 5) recreation; 6) reduced fish and wildlife habitat; 7) socioeconomic concerns; and 8) linkage of problems. An implementation strategy is being designed around the study's four areas of recommendations: 1) information/education; 2) reforestation; 3) land treatment practices; and 4) structural/engineering practices. Returning to presettlement conditions is considered unfeasible. The plan also offers to work with local and state officials to develop a detailed socioeconomic recovery plan.

In 1995, the area was designated as one of nine ecosystem-based demonstration projects by NRCS, with several resulting federal grants. The flooding study was completed later and circulated for public comment, which were to be incorporated into the final report for submission to the local levee board in 1996. Approximately 25,000 acres of marginal land have been converted to bottomland hardwoods, and the Louisiana Department of Forestry has placed a forester in the area.

Two other advisory committees have been established to increase stakeholder participation from the agricultural community: 1) the Tensas Watershed Agricultural Council is composed of large-scale farmers; and 2) the Tensas Limited Resource Minority Advisory Council has representatives from this community, who make up 10 percent of the farmers in the project area.

Factors Facilitating Progress

This process would not have progressed without a local, nongovernmental, third-party institution taking the lead (the NDRCDD). The council is not considered an "outside" organization: its representative for the Tensas is from the local community and recognizes the importance of locally-driven processes to gain stakeholder acceptance.

Widespread concern about the region from a broad cross-section of residents and users has been helpful. The federal listing of the Louisiana black bear as a threatened species helped push the effort forward. Despite initial skepticism, the willingness of participants to "come to the table" was instrumental, due in part to an aggressive public outreach effort. The steering committee's consensus approach allowed stakeholders to be equally represented. Communication among stakeholders has improved, particularly between farmers and regulators, such as the state or EPA who previously understood little about each other's motives and operating methods.

Obstacles to Progress

Not all of the stakeholders were involved initially, especially from the agricultural community. While they have agreed to participate in the process, bringing them "up to speed" will take additional efforts.

The poor economic condition of the region may be a major inhibitor of participation in implementation, as only older, established farming families and individuals, who hold the best and most productive land, are able financially to participate. Funding of implementation efforts and incentive programs is unpredictable, especially with the 1995-96 Farm Bill debate in Congress. Already, there have been insufficient funds available to cover all applicants for existing wetland reserve and conservation tillage programs.

Finally, conflict among steering committee members and stakeholders is possible as more precise implementation strategies are developed, when greater compromises will be necessary.

Contact information:

Mr. Mike Adcock
Tensas River Basin Coordinator
Northeast Delta Resource Conservation & Dvlpt. District
PO Box 848
Winnsboro, LA 71295
(318) 435-7328
Fax: (318) 435-7436

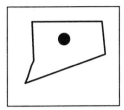

TIDELANDS OF THE CONNECTICUT RIVER

Location:
Central Connecticut

Project size:
2,500 acres

Initiator:
The Nature Conservancy

PROJECT AREA DESCRIPTION

The focus of the project is the riverine system and the connected tidal marshes along the lower Connecticut River. The Nature Conservancy (TNC) has delineated seventeen core sites which are included in the primary boundary of the Tidelands Bioreserve. Sixteen of the seventeen sites include tidal marshes; one is an upland site. The Tidelands project does not include the entire Connecticut River watershed (the "tertiary boundary"—approximately 30,000 acres). The vegetative cover is a wide variety of tidal marsh vegetation and forest lands.

The land use is predominantly residential with some agricultural and urban uses. Landowners include the state of Connecticut, private landowners, TNC, and other private conservation organizations. Federally listed threatened and endangered species include shorebirds, the bald eagle, an insect, a fish, and two wetland plants.

ECOSYSTEM STRESSES

The most serious threat to the area is habitat loss and fragmentation from development, due to significant human presence in the area, a desire to build along the river, and the need of local towns to expand their tax base. The construction and maintenance of docks along the river and its tributaries is also a threat to submerged aquatic vegetation. A second stress to the ecosystem is the introduction and spread of invasive plant species, such as *Phragmites* and purple loosestrife. If uncontrolled, these species tend to dominate the marshes and lessen the habitat value. Other threats include reduction in water quality and barriers preventing water movement or species movement.

PROJECT DESCRIPTION

TNC has been working on Connecticut River issues for many years due to its significance as a hydrologic feature in the Northeast. The seventeen core areas have been defined since the mid-1980s, although the Tidelands project was not formally initiated until 1991. Prior to 1991, the core sites had basic management plans and very little buffer space. Since the initiation of the project, the core sites and buffers have been expanded to take into account larger landscape effects on the ecosystem.

Four priority goals have been developed, which are to protect 1) the tidal marsh system; 2) globally rare species; 3) state listed threatened and endangered species; and 4) species in decline. Strategies designed to meet the goals include 1) an emphasis on outreach and education; 2) traditional land protection through easements and acquisition; 3) control of invasive species; and 4) scientific research. Key partners include the Connecticut Department of Environmental Protection (CTDEP), U.S. Fish and Wildlife Service, Connecticut River Watershed Council, University of Connecticut Cooperative Extension, and fourteen land trusts. The University of New Haven and the University of Connecticut have provided technical support for geographic information systems (GIS) mapping.

STATUS AND OUTLOOK

The project is currently focused on scientific research, including a study of the submerged aquatic vegetation in the project area, and outreach efforts such as developing fact sheets describing the system, individual sites, and various species. In partnership with the University of Connecticut Cooperative Extension Service, TNC is providing residents and decision makers with information and tools that they can use to make better decisions regarding the use and management of their local natural resources, at the level of local watershed subbasins.

Factors Facilitating Progress

One factor that has been helpful to the project is the large amount of conservation work that has already been done in the area. The CTDEP is also progressive and open to innovative solutions.

Furthermore, the populous of the region is both educated and relatively affluent. Therefore, they have been open to learning about the riverine ecosystem and have been willing to provide financial support for conservation efforts.

In addition, partnerships with various organizations, such as local land trusts, Cooperative Extension, and the Connecticut River Watershed Council, are multiplying the benefits of what any one group could accomplish.

Obstacles to Progress

Although approximately 30 percent of TNC's efforts in Connecticut are focused on the Tidelands project, funding is still considered insufficient for such a large project. There are eighteen towns to work with in the project area, a challenge further complicated by the frequent turnover of members on town boards. Also, land is very expensive to acquire in this area and not all residents are open to preservation concepts.

Contact information:

Dr. Juliana Barrett
Tidelands Program Director
The Nature Conservancy
Connecticut Chapter
55 High Street
Middletown, CT 06457
(860) 344-0716
Fax: (860) 344-1334

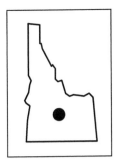

TRAIL CREEK ECOSYSTEM ANALYSIS

Location:
South-central Idaho

Project size:
20,000 acres

Initiator:
U.S.D.A. Forest Service

PROJECT AREA DESCRIPTION

In this mountainous region, Hyndman Peak towers 12,000 feet over Sawtooth National Forest. Aspen, Douglas-fir, limber pine, and spruce grow along the mountains' north slopes. Sagebrush and forbs dominate the south slopes. The region is suited for gray wolf habitat and is a wintering area for bald eagles. Sheep grazing, timber harvesting, and various types of recreation take place in and around the National Forest. Part of the National Forest is designated as a Research Natural Area. The city of Sun Valley lies at the site's western boundary.

ECOSYSTEM STRESSES

Land conversion to residential development and associated road building are primary stresses on the ecosystem. These activities are closely linked to the heavy recreational use the region is experiencing. Automobiles driven alongside Trail Creek have created roads. The cars have thrashed vegetation in riparian zones, led to changes in species composition, and compacted soils. Sediment loading in downstream riverbeds has occurred. The potential exists for stream dislocation or filling of downstream reservoirs with sediment.

PROJECT DESCRIPTION

In 1989, the Ketchum Ranger District started a Total Quality Management process that changed the way the staff looked at the interrelationship of the district's natural resources and their "customers." Following a survey of the district's internal and external customers, a strategic process was developed to generate a list of site-specific actions that address issues and problems concerning five distinct topographic

drainages. The district used a five-year planning horizon for its decisions.

Broad participation by regional stakeholders forms the crux of the "ecosystem analysis decision process." The Trail Creek Ecosystem Analysis was the first to be completed. The document satisfied National Environmental Protection Act (NEPA) guidelines, was not appealed upon its completion, and was signed by the district ranger in 1991. The district staff have been implementing site-specific improvements since then.

The district's philosophy is that enhanced communication of customer needs and ecosystem capacity forges a deeper understanding between the district, local residents, and the surrounding natural systems. The district has aggressively sought the input and participation of the following groups: internal personnel, Idaho Department of Fish and Game, Idaho Department of Water Resources, the administrator of the city of Sun Valley, and numerous recreation and environmental groups. The district remains committed to those beneficiaries who do not vote or have a voice in the process: the wildlife and vegetation species that live on the Ketchum Ranger District as well as the future generations that will be able to use and enjoy a healthy and intact ecosystem.

STATUS AND OUTLOOK

Among the activities completed to date are road closings; opening up compacted soils and seeding them with natural riparian vegetation; stabilizing stream banks with rocks; massive replanting of willow species; and installing water guzzlers for sheep so the animals do not have to come to the stream to drink. The community has been involved in implementing ecosystem improvement activities during specially-designated work days. Riparian areas are already recovering.

Current and future visitors to the Sun Valley area will be the beneficiaries of this analysis process. Local residents have found a successful way of working together and anticipating and solving management challenges before they occur.

Factors Facilitating Progress

Supportive superiors and forest supervisors allowed the experimentation with the innovative quality approaches to take place.

Public involvement up front and public support throughout the decision process is a primary reason for progress. Continually adjusting the process for changing social and ecological conditions is required for future progress.

Obstacles to Progress

As Trail Creek was the first ecosystem analysis decision, the process was very time consuming. Time constraints of the team leader led to long gaps in between meetings of the customers. Moving ahead without all customers present at group meetings led to some backtracking in the planning process.

Securing funding for the project was also difficult. The current guidelines and funding mecha-

nisms do not always support or recognize the innovative approaches to land management being applied by the Ketchum Ranger District. The district frequently must educate governmental and nongovernmental funding sources as to the approaches and the benefits of the Total Quality Management framework for ecosystem decisions.

Contact information:

Mr. Alan Pinkerton
USDA Forest Service
Sawtooth National Forest
Ketchum Ranger District
PO Box 2356
Ketchum, ID 83340
(208) 622-5371

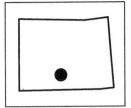

TROUT MOUNTAIN ROADLESS AREA

Location:
Southern Colorado

Project size:
32,000 acres

Initiator:
U.S.D.A. Forest Service

PROJECT AREA DESCRIPTION

The Trout Mountain roadless area in Colorado's Rio Grande National Forest is located in the south-central highlands of the Rocky Mountains. Within this temperate steppe region, the project area consists of two watershed drainages, bordered to the north by the Weminuche Wilderness area and to the south by Highway 160 which runs parallel to the South Fork of the Rio Grande River. Despite historical timber and mining in the area, the majority of Trout Mountain has remained roadless old-growth forest. Present-day use consists primarily of recreation. There are no known threatened or endangered species in the area at present.

ECOSYSTEM STRESSES

Recent timber sales using shelterwood harvest treatments which result in clearcuts are the most serious threats to the region. Motorized vehicle use on a forest system gravel road which lies between Trout Mountain and the Weminuche Wilderness has increased threats to wildlife. Finally, historic mining and timber use have fragmented portions of the project area.

PROJECT DESCRIPTION

The project began as a traditional U.S.D.A. Forest Service (USFS) timber sale proposal. Signed in 1985, the Rio Grande forest plan proposed a set of timber sales which would result in the removal of 9 million board feet of timber from the Trout Mountain roadless area. With growing public concern for old growth and wildlife, the 1985 forest plan became very controversial. The plan is currently under revision to address issues raised by the public concerning the effects of timber harvest and road construction on the unroaded old-growth character of the Trout Mountain area.

The stated goal of the USFS is to remove timber in a manner that does not sacrifice the ecological and biological needs of the project area; the amount of timber removed from the area is secondary to those needs. USFS feels that ecological concerns have driven the planning process.

The project is attempting to look beyond the two drainages and National Forest boundaries to consider adjacent ecological areas that encompass most of southern Colorado and portions of northern New Mexico. Habitat capabilities and relationships between adjacent areas such as the Weminuche Wilderness area are important considerations in the plan. The planning team identified an area they felt might function as a connective corridor to meet the requirements for species migration. However, environmentalists have dubbed it the "corridor to extinction," as the proposed area passes over a major highway. The forest plan also proposes to build twelve miles of temporary roads. Once the timber is removed, some roads would be closed, recontoured, and revegetated. Finally, the plan calls for uneven-aged management which would leave 70 percent or more of the existing vegetation standing, including trees in the old-growth age class.

The project planning process has remained an internal effort by a USFS interdisciplinary planning team. However, between 1990 and 1993, the planning team worked closely with a public working group made up of twelve to fourteen people representing various interests in the region. This group participated in the development of project alternatives which were ultimately selected by the planning team. Six members of the working group were directly employed by industry (sawmill or loggers); another six members represented multiple use interests (recreation, hunting, outfitters, etc.); and only one member from the Colorado Environmental Coalition (CEC) represented environmental interests.

The planning team actively solicited other individuals with environmental interests because of this low representation. The planning team has also worked very closely with Colorado's Division of Wildlife in ongoing and extensive monitoring of species.

STATUS AND OUTLOOK

After the Final Environmental Impact Statement (FEIS) was published in 1994, the project was appealed by CEC to the USFS regional forester who upheld the decision. As of early 1996, the project was in federal district court. Meanwhile, CEC is working with USFS on another revision of the forest plan and has submitted a management alternative to be analyzed in the FEIS for the revised plan. Furthermore, species monitoring has continued in the Trout Mountain region.

Factors Facilitating Progress

USFS feels very good about the project's nontraditional decision-making process (i.e., increased stakeholder participation, level of communication with public), the level of analysis, and how they are proposing to do the work.

USFS feels that these elements are significantly different than how things have been done in the past.

Obstacles to Progress

As evidenced by significant public opposition and the forest plan being contested in court, much controversy remains over the plan. In particular, the issues are focussed on old growth—how it is defined, and whether or not a roadless area should be opened up to timber harvesting.

Contact information:

Mr. Ron Pugh
Forest Planner
USDA Forest Service
Rio Grande National Forest
1803 West Highway 160
Monta Vista, CO 81144
(719) 852-5941

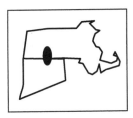

UPPER FARMINGTON RIVER MANAGEMENT PLAN

PROJECT AREA DESCRIPTION

Situated in the foothills of the Berkshires in the northeastern highland ecotone, the project area encompasses a fourteen-mile reach of the west branch of the Farmington River, extending from the Hogsback Reservoir downstream to an area approximately one mile upstream of the confluence with the Nepanug River. The vegetative coverage is mixed deciduous and coniferous forest. Up until 1994, the only mating pair of bald eagles known in Connecticut lived in the project area. Rural development is the predominant land use, with some of the upper watershed in agriculture. Landholdings upstream of the project area are dominated by a governmental water supply.

Location:
Northern Connecticut, southern Massachusetts

Project size:
14 river miles

Initiators:
Private citizens, U.S. Congress

ECOSYSTEM STRESSES

Water quantity, water diversions, and in-stream flow management are primary concerns in the watershed, a fact reflected in the project's management plan. Another threat is increased recreational use of the river by anglers and boaters, especially as its Wild and Scenic River (WSR) status becomes more widely known.

PROJECT DESCRIPTION

In response to public concern over water diversion, federal legislation was proposed in 1986 to designate this area under the Wild and Scenic Rivers Act. At that time, a study committee was set up by Congress to verify that the river met the appropriate qualifications for WSR designation. At the same time, an investigation was initiated to see if all stakeholders in the region supported this designation (e.g., state regulatory agencies, citizens, Farmington Watershed Association, Metropolitan District

Commission, and the town of Farmington).

In order to help these groups reach consensus on designation, a management plan was developed by the study committee. The plan was to be used regardless of designation and contains the following goals: 1) conserve and enhance important land-based natural resources; 2) encourage effective management of river-related growth; 3) balance the legitimate demands on the river water supply; and 4) manage river recreation.

In 1993, the study document concluded that stakeholders agreed that the upper Farmington River should received WSR designation, which occurred the following year. This process was unique because management plans are not usually considered during the designation phase. Since this area is dominated by private landholdings instead of public land, this management plan was a key step in gaining buy-in from stakeholders.

STATUS AND OUTLOOK

Currently, the project is moving into the implementation phase of the plan. In August 1994, Congress passed legislation which was signed by President Clinton giving official WSR designation to the west branch of the river. As a result, the Farmington River Coordinating Committee was formed in 1995, consisting of eight member organizations (four towns, the Metropolitan District Commission, Farmington Watershed Association, National Park Service, and the state of Connecticut). Each organization designates a committee representative and alternate.

Among the committee's responsibilities will be monitoring and ensuring that the protection goals are being met.

The coordinating committee became a legal entity when its members adopted bylaws in February 1996. A memorandum of understanding outlining specific member roles and responsibilities was expected to be signed by committee members in March 1996. Meanwhile, the towns along the designated river reach have enacted special rules regarding use and development along the river, such as new zoning designations. The state has specifically identified the river as a "unique resource deserving special consideration."

Factors Facilitating Progress
The National Park Service and the Metropolitan District Commission provided funds for an in-stream flow study while the Connecticut Department of Environmental Protection managed the project and provided technical guidance. The 1992 finding that the instream flow was sufficient for all management objectives was a significant force allowing the project to progress; any other finding would have ended the effort. The interdisciplinary and cooperative spirit of all those involved has also helped the project advance. An immediate result has been an increase in communication between the citizens, the Metropolitan District Commission, and the state regulatory agencies.

Obstacles to Progress
There were initial philosophical differences between some of the managing entities. These have been addressed, however, as communication between parties has increased.

Contact information:

Mr. Thomas Stanton
Chair, Farmington River
  Coordinating Committee
c/o Town of Colebrook
119 Beech Hill Road
Winsted, CT 06098
(860) 379-8704

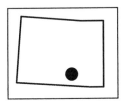

UPPER HUERFANO ECOSYSTEM

Location:
South-central Colorado

Project size:
250,000 acres

Initiator:
Colorado Division of Wildlife

PROJECT AREA DESCRIPTION

The headwaters of the Huerfano River drainage are fed from the west and east by the Sangre de Cristo and Wet mountains. The Sangre de Cristo mountain range boasts many 14,000 foot peaks, including the project area's Blanca Peak, where one of the country's southernmost glaciers is found. Within the Huerfano River drainage, the dominant vegetative types include spruce-fir, aspen, ponderosa pine, pinyon-juniper, and grassland communities. Cotton-wood-willow riparian forests run along the main stem of the Huerfano River. The river supports the greenback cut-throat trout, a federally listed threatened species and one of three pure native genetic strains in Colorado.

Half of the upper Huerfano region is in private ownership, made up of small ranching and farming communities. The remaining lands are both state and federally owned. These public lands provide summer range for elk, while private lands provide most of the winter range. In 1994, much of Colorado's Sangre de Cristo mountain range was designated as wilderness. There is a great deal of recreational activity in this region, including fishing, camping, off-road vehicle use, hiking, hunting, and climbing of "14ers" (mountains exceeding 14,000 feet in elevation).

ECOSYSTEM STRESSES

In the last thirty years, elk populations have increased substantially in the region. Overgrazing of both livestock and elk has impacted grassland and riparian areas on both public and private lands. Furthermore, an eighty-year history of fire suppression has altered the structure and composition of fire-dependent ponderosa pine, aspen, and pinyon-juniper communities, and could potentially result in catastrophic fires. Finally, increased recreational activities have resulted in concentrated use and erosion damage in the San Isabel National Forest and wilderness areas. The combination of uses and impacts over the last 120 years has significantly altered the structure and pattern of vegetative communities in the region.

PROJECT DESCRIPTION

In 1991, the Colorado Division of Wildlife (CDOW) initiated the Sangre de Cristo Habitat Partnership Program (HPP) for the Huerfano region. A committee was established to address conflicts concerning overgrazing and elk use on public and private lands. The HPP committee includes representatives from the CDOW, U.S.D.A. Forest Service (USFS), U.S. Bureau of Land Management (BLM), private landowners, Farm Bureau, and sportsmen's groups. The committee has identified the need to work cooperatively to address issues within the entire Huerfano region, thus adopting an ecosystem approach.

The goals of this project are two-fold: 1) to assess the existing ecological conditions and the ecological potential of the Huerfano ecosystem; and 2) to incorporate the social and economic needs of local communities into the management of the area's natural resources.

STATUS AND OUTLOOK

Desired future conditions for the region are being developed jointly by the HPP committee, local government, local landowners, and nongovernmental organizations including the Sangre de Cristo Mountain Council, a coalition of environmental, economic, and recreation groups. Inventory and analysis has been a joint effort between all four public agencies and was to have been completed in 1995.

USFS, BLM, and the U.S.D.A. Natural Resources Conservation Service are looking at management opportunities across administrative boundaries. The intent of this joint effort is to better understand resource conditions and relationships at larger scales for management issues regarding water quality, recreation, vegetative condition, biodiversity, and elk distribution.

Factors Facilitating Progress

The level of cooperation and communication, particularly between the major landowners in the region (USFS, BLM, private), has provided more flexibility in decision making and greater opportunities to meet the needs of all interests. The Sangre de Cristo Mountain Council and the HPP committee have played vital roles in bringing parties together and building local support for this project through outreach. Also, a great deal of sharing of information and expertise has occurred between groups involved in the project.

Obstacles to Progress

Limited funding has restricted the inventory work. The reluctance of some private landowners to participate, based on government distrust and land use interests that conflict with elk usage has impeded the project's progress to some degree.

Contact information:

Ms. Nancy Ryke
Wildlife Biologist
USDA Forest Service
San Isabel National Forest
San Carlos Ranger District
326 Dozier Avenue
Canon City, CO 81212
(719) 275-4119

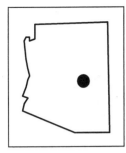

VERDE RIVER GREENWAY

PROJECT AREA DESCRIPTION

Located in central Arizona, the 180-mile Verde River is one of only a few rivers in Arizona that are free flowing for most of their journey. Approximately twenty-five miles downstream from the river's headwaters in Arizona's central highlands, the Verde River Greenway is a designated six-mile stretch within the river's 100-year floodplain in the Verde Valley. The Fremont cottonwood/Goodding willow riparian gallery forest, the dominant forest community type within the greenway, is one of five such stands remaining in Arizona, and one of less than twenty in the world.

This forest community supports an abundance of wildlife diversity, including state and federally listed threatened and endangered species such as the razor-back sucker, southwestern willow flycatcher, gray hawk, and common black hawk. Landownership within the greenway is a mix of state, federal, municipal, and private. Land use is equally diverse, including agricultural, grazing, industrial, commercial, residential, and recreational uses.

ECOSYSTEM STRESSES

The primary impact to the river in the Verde Valley is water diversion. During the summer months, up to 90 percent of the water in the channel is diverted, mainly for agricultural use. Dramatic population growth in the Verde Valley has led to rapidly-expanding residential and commercial development, which has increased water demands and sedimentation of the river. Grazing on national forest lands in the upper watershed have also led to high amounts of erosion and sedimentation down-

Location:
Central Arizona

Project size:
410 acres over six river miles

Initiators:
Governor Bruce Babbitt,
Arizona State Parks

stream. Furthermore, runoff from tailings left over from a historic copper smelter operation may impact water quality along the greenway. Finally, historic sand and gravel mining has channelized parts of the river, altering its natural flow, increasing erosion, and decreasing bank stability and regeneration of native vegetation. The U.S. Environmental Protection Agency has placed a moratorium on mining activities until mitigation efforts are in place.

PROJECT DESCRIPTION

The Verde River Greenway project was initiated by then-Governor Bruce Babbitt in 1986 after the project area was identified as a critical stretch of the river, based on its rare riparian forest community type, its rich cultural resources, and a growing demand for recreational opportunities. State legislation was then passed and funds appropriated authorizing Arizona State Parks to acquire property along the greenway. In 1990, the Arizona Heritage Fund initiative was enacted. Supported by state lottery revenues, this fund provided additional funding for acquisition of lands, establishment of an ongoing ecosystem monitoring program, and development of a management plan.

Arizona State Parks realized from the greenway's inception that to successfully manage this riparian resource, a coordinated management system was required, and had to include both public and private landowners along the greenway and in the surrounding communities. As a result, the following project goals were established based on issues raised by landowners and the public in the region: 1) conserve, protect, and enhance the ecological resources of the Verde River Greenway; 2) preserve cultural resources, including native American historical sites; 3) develop recreational opportunities that are compatible with the conservation goals of the greenway; and 4) build partnerships. A steering committee was set up to advise Arizona State Parks during the management planning process. A public survey was conducted and several public meetings were held throughout the duration of the planning effort.

STATUS AND OUTLOOK

The Verde River Greenway management plan outlines management strategies for greenway lands owned by Arizona State Parks, but also offers recommendations for all landowners within the 100-year floodplain. Implementation of the plan is now under way. A full-time coordinator oversees management on the greenway, and the community continues to strongly support Arizona State Parks' protection efforts along the greenway. Interpretive educational programs, riparian restoration efforts, and ecosystem monitoring have begun. Cooperative relationships between landowners are also beginning to develop.

Factors Facilitating Progress

The initial funding and political support to protect this very significant resource have been the major factors allowing this project to move forward. Also, local communities have supported Arizona State Parks in land acquisition and planning for the greenway.

Obstacles to Progress

There has been some opposition toward the plan from private landowners concerning the amount of public access allowed along the greenway on private lands. Conflict has arisen over what should be considered appropriate recreational use within the greenway. Finally, some elements of the plan may be difficult to implement if the state legislature reduces Arizona Heritage funds in the future.

Contact information:

Arizona State Parks
1300 West Washington Ave.
Phoenix, AZ 85007
(602) 542-4174
Fax: (602) 542-4180

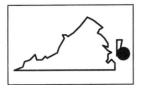

VIRGINIA COAST RESERVE

Location:
Eastern Virginia

Project size:
45,000 acres

Initiator:
The Nature Conservancy

PROJECT AREA DESCRIPTION

The Virginia Eastern Shore is a narrow peninsula of land attached to southern Maryland, with the Atlantic Ocean to the east and the Chesapeake Bay to the west. On the Atlantic ocean side is a string of barrier islands, coastal bays, and extensive salt marshes. The Nature Conservancy's (TNC) Virginia Coast Reserve encompasses 45,000 acres of this coastal area. Ecosystems range from salt marsh habitats to upland wooded areas. The vegetative cover on the islands includes dune vegetation and maritime forest, while the mainland is predominantly farmland and hardwood forest. Federally listed threatened and endangered species include the Delmarva fox squirrel and numerous shore birds such as the piping plover. Farming and seafood harvesting are traditional industries.

ECOSYSTEM STRESSES

The barrier islands have been substantially protected, but the mainland buffer area faces threats due to development. Water quality and quantity are concerns for the sole-source, deep confined aquifer of the Eastern Shore. Development could increase the need for water while upsetting the recharge balance of the aquifer by creating impermeable surfaces. Another major stress is degradation of lagoons, which adversely impacts the shellfish communities. Recently, fecal coliform has been found at high levels due to raccoon scat. Overfishing is also a concern.

PROJECT DESCRIPTION

This conservation project began in the 1970s when TNC purchased the three southernmost islands from a New York corporation, which originally had plans to build a resort. TNC continued assembling the Virginia Coast Reserve and today owns and manages 45,000 acres, including fourteen islands. This is the last intact, fully-functioning barrier island ecosystem on the unglaciated coast.

In 1989, TNC initiated a broad approach to ecosystem conservation based in part on the area's United Nations Biosphere Reserve designation. The general mission of the project is to preserve biodiversity and the habitats that the species need to survive. This is accomplished through the following six initiatives: 1) protect the core natural area of the barrier islands; 2) ensure appropriate uses within the buffer area; 3) monitor the ecosystem with an emphasis on scientific research; 4) educate at all levels; 5) form partnerships with every sector; and 6) enhance the local economy through protection of the natural system.

Partnerships, highlighted throughout all the initiatives, have been forged in many areas. The Northampton Economic Forum was formed in 1989 to design an economic implementation plan that emphasizes sustainability of natural resources. The forum includes representatives from TNC, the National Association for the Advancement of Colored People (NAACP), local governments, and business people. In 1987, the University of Virginia was awarded a $2 million grant to begin a long-term ecological research project on the reserve. Scientists on this effort are studying natural processes with an emphasis on developing monitoring strategies for future management decisions. Monitoring of the reserve includes long-term scientific and problem-specific issues as well as sociological and economic factors.

STATUS AND OUTLOOK

Participants in the project believe strongly in developing a sustainable development model in the buffer area since this is the only way to ensure long-term protection of the core reserve. Currently, TNC is working on ways to support other groups in their sustainable development efforts and to demonstrate sustainable agriculture in pilot projects.

Factors Facilitating Progress

The community has strong ties tothe natural resources of the area and has a long history of living close to the land. This ethic has greatly facilitated the momentum of the project. The community realized early on that conservation and economics are complimentary in many ways and they generally support innovative and sustainable projects. In addition, the barrier islands that TNC purchased were in relatively pristine condition due to limited human impact.

Obstacles to Progress

TNC and the community have been through years of planning and implementing, and yet the ecological results are slow to appear. Measuring and quantifying ecological results are also difficult.

Contact information:

Ms. Terry Thompson
The Nature Conservancy
Virginia Coast Reserve
PO Box 158
Nassanadox, VA 23413
(804) 442-3049
Fax: (804) 442-5418
E-mail:
   TATHOMPSON@AOL.COM

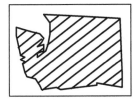

WILD STOCK INITIATIVE

Location:
Washington

Project size:
42 million acres

Initiators:
*Washington
Department of Fish
and Wildlife,
Northwest Treaty
Indian tribes*

PROJECT AREA DESCRIPTION

This project focuses on streams and
their watersheds throughout the state of
Washington in which salmon and steelhead trout populations have been identified. Healthy salmonid populations
need streams with cool, clean water, a
gravelly bottom, stable flows, adequate
summer flows, and migration access to
saltwater. The watersheds of such
streams vary from the forested Cascade
Mountains in west-central Washington,
to the agricultural Columbia basin and
the urbanized Puget Sound basin.
Several federal and state listed threatened or endangered species occur in this
state. They include the Snake River fall
chinook, Snake River spring-summer
chinook complex, and the Snake River
sockeye. In addition, petitions have
been submitted for the federal listing of
numerous salmonid stocks under the
Endangered Species Act (ESA).

Public lands include national forests,
national parks, grazing lands, state
forests, and state parks. In addition, a
significant portion of forest land is
owned by timber companies. Agriculture, including grazing and cultivation
of grain crops (oats, barley, etc.), is
especially important east of the Cascade
Mountains. In the Puget Sound basin,
urban development is growing.

ECOSYSTEM STRESSES

Hydrologic alterations have led to salmonid habitat loss, resulting in reduced
productivity and high mortality rates.
Such alterations are due to the fact that
in eastern Washington, water diversion
and over-allocation of water for
agricultural purposes have resulted in
inadequate stream flow. Major
hydroelectric development on the

Columbia River has resulted in high
upstream and downstream passage
mortalities through dams. Furthermore,
riparian buffer has been lost due to
grazing, resulting in high water
temperatures and reduced stream flow.
In western Washington, both roads and
timber management have affected the
periodicity and amplitude of floods,
and have caused severe erosion and
sedimentation. Urban development has
also led to habitat loss and degradation.
Finally, overfishing and hatchery
programs have affected salmonid
population levels.

PROJECT DESCRIPTION

The Wild Stock Initiative was launched
in 1992 by the Washington Department
of Fish and Wildlife (WDFW; formerly
the Departments of Fisheries and Wildlife) and the Northwest Treaty Indian
tribes in response to the growing realization that many wild salmonid stocks
are seriously depleted, and that if no
remedial action is taken, fisheries will
be closed and many more wild salmonids will be listed under the ESA. The
goal of the initiative is to protect and
increase the long-term productivity,
abundance, and diversity of wild
salmonids and their ecosystems to
sustain 1) ceremonial, subsistence,
recreational, and commercial fisheries;
2) non-consumptive fish benefits; and
3) related cultural and ecological
values.

The Wild Stock Initiative consists of
three integral components: 1) a Salmon
and Steelhead Stock Inventory (SASSI);
2) development of a wild salmonid
policy; and 3) protection and restoration
of stocks and their habitat. Specific
restoration actions may include

habitat restoration, modification of hatchery practices, captive broodstock projects, and new harvest management strategies. Also included in the initiative is monitoring. Although part of the original initiative, the development of a wild salmonid policy was spurred on by legislative mandates in 1994 to create a statewide rather than agency or tribal policy. Subsequently, the governor required all other state agencies with authority over salmonid habitat to become involved in the development of this policy.

STATUS AND OUTLOOK

By 1994, SASSI had classified 435 wild salmon and steelhead stocks. In addition, an inventory of salmonid habitat and critical physical habitat components is ongoing. Policy development has included the completion of an extensive policy discussion paper

in May 1994; meetings with the Oregon Department of Fish and Wildlife, tribes, and forty key constituents (representing fishing, environmental, and land and water use interests) in May and June 1994; and the completion of a working draft environmental impact statement in July 1995. The policy development process has led to increased cooperation and coordination among state agencies and with constituents. Recovery planning activities are currently underway for most stocks identified as having critically low abundance according to SASSI.

Factors Facilitating Progress

It has been fortunate that no one has challenged the existence of serious resource problems needing resolution. A specific legislative mandate for policy development has also been helpful. In addition, the legislature has allo-

cated funds to WDFW for the Wild Stock Initiative, resulting in focused resources for the project, such as staff and funding.

Obstacles to Progress

Development of the wild salmonid policy and its associated recovery plans has been much slower than anticipated. Dealing with a complex array of major resource management issues at a broad, conceptual level has proven challenging. It may also be challenging to reach agreement with all tribes and agencies on the policy while building public consensus.

Contact information:

Mr. Rich Lincoln
Washington Department of Fish
and Wildlife
600 Capitol Way North
Olympia, WA 98501-1091
(206) 902-2750

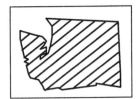

WILDLIFE AREA PLANNING

Location:
Washington

Project size:
840,000 acres

Initiator:
*Washington
Department of Fish
and Wildlife*

PROJECT AREA DESCRIPTION

All lands under the jurisdiction of the Washington Department of Fish and Wildlife (WDFW) are included in the project area. Within their 840,000-acre jurisdiction, individual WDFW holdings range from 400 acres in western Washington to 100,000 acres in the eastern part of the state. Many different ecosystems are included.

Much of the WDFW's forested lands contain a mix of coniferous and deciduous tree species. On the west side of the Cascade range, Douglas-fir, western red cedar, and hemlock are the dominant species. On the east side, Douglas-fir and ponderosa pine in the higher moisture habitats transition to several hundred thousand acres of grass and shrub steppe under WDFW's management authority. Big sage, stiff sage, and bitter brush are a few shrubs that grow on the steppe.

The lands are primarily managed for wildlife protection and habitat preservation. Ten to twenty federally listed threatened and endangered species, such as the bald eagle, peregrine falcon, northern spotted owl, and salmon species, reside on WDFW wildlife areas. Recreation is the second management focus, including hunting, fishing, horseback riding, rock climbing, and bird watching, among others. Motorcycling and snowmobiling are permitted by WDFW on some existing roads.

ECOSYSTEM STRESSES

Given the diversity of WDFW's lands and the geographic distance between them, stresses unevenly affect wildlife areas. Localized areas are very strongly impacted by a specific stress which may not register as very significant on a statewide scale. For instance, non-point pollution sources, eutrophication, timber management, and drought are severe stresses on specific areas, yet none of these stresses have a widespread impact on several areas.

The presence of exotic species, however, is a major stress on many WDFW lands, especially on the east side of the Cascade range. Russian olive, originally planted by WDFW to create riparian habitat cover, is now choking out native plants and decreasing biological diversity in some wildlife areas. Another exotic plant, purple loosestrife, is destroying some wetlands; WDFW is trying to address this problem through a control program.

PROJECT DESCRIPTION

WDFW began purchasing lands in the 1940s for recreational hunting and fishing. There are currently twenty-five primary wildlife management areas in the state. The Department's adoption of ecosystem-based approaches was an evolutionary policy to stabilize species populations in light of mounting demands on the wildlife areas by the public. In Spring 1992, the program manager of WDFW's Wildlife Area Division directed that four-year management plans be developed for all wildlife areas so that WDFW would have a defensible framework by which to approve or reject activities on the lands. Planning also would ensure that consistent, statewide, biologically-based activities were being proposed. Each management plan is based on agency goals and objectives, public input, and WDFW's legal responsibilities. Plans will be reviewed and updated every four years to determine if social or ecological conditions that influence management have changed.

The plan's management activities focus on specific components of the ecosystem, with the overall goal of improving habitat and maintaining species populations. Typical ecosystem-based management activities on wildlife areas include, among others, 1) sedimentation control through road maintenance; 2) fencing to protect riparian areas and uplands from grazing; 3) routing human use outside of sensitive areas; 4) riparian and upland restoration including shrub; 5) tree and herbaceous plantings to provide soil stabilization; 6) food; 7) cover for wildlife; and 8) fire control.

STATUS AND OUTLOOK

WDFW has continued to purchase lands in the last five years, often with mitigation funds that come from the dams operated by Tacoma Power and Light, Bonneville Power Administration, and other utilities. Resources have been mapped for each of the wildlife areas using geographic information systems (GIS) technology.

Factors Facilitating Progress

Cooperation within the agency and public involvement are credited with allowing the planning processes to progress. The establishment of citizens advisory groups has allowed WDFW to receive support from the public, including citizens who previously did not understand or support management activities on WDFW wildlife areas.

Obstacles to Progress

Convincing internal staff and citizens that the project planning process was going to be useful and successful was challenging. At times, members of the public have not agreed with WDFW's proposal for management activities on a specific wildlife area. Applying additional, wide-ranging ecosystem-based approaches will be a future challenge.

Contact information:

Mr. Paul Dahmer
Wildlife Area Inventory and
  Planning Coordinator
Washington Department of Fish
  and Wildlife
600 Capitol Way North
Olympia, WA 98501-1091
(360) 664-0705
Fax: (360) 902-2946
E-mail: dahmepad@dfw.wa.gov

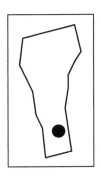

WILDLIFE HABITAT IMPROVEMENT GROUP

Location:
Southern Vermont

Project size:
5,600 acres

Initiators:
Private landowners

PROJECT AREA DESCRIPTION

The project area is defined by the property boundaries of more than forty contiguous landowners in the towns of Newfane, Wardsboro, and Townshend, Vermont. The overall ecosystem that corresponds with the project area is described as a system of high wooded wetlands. The project area has been further divided into ten ecotypes, which include mixed cover, older forest, and wetlands. The vegetative cover is a mix of northern hardwoods and conifers and is secondary growth resulting from the decline of agriculture in the region. At an altitude of 1,600 feet, the area provides the headwaters for several streams. The 5,600-acre project area is predominantly privately-owned land that has passed down from generation to generation. The land use is mainly rural settlement, except for 200 acres owned by International Paper.

ECOSYSTEM STRESSES

Although this area has not been rated by the state as critical in terms of acid rain, the potential impacts of acidification on water quality and on growth of conifers in the area are still a major concern. A second threat is the population growth and increased demand for recreational land use. This poses a threat to wildlife habitat and is magnified by the recent trend toward development. A third threat to habitat is the improper management of woodlands and timber cuts, which can alter edge habitat for wildlife and cause loss of critical stands.

PROJECT DESCRIPTION

The project revolves around a voluntary consortium of private landowners with the goals of stewardship, preservation,

and improvement of wildlife habitat. The force behind the project's start-up was a single individual motivated by a three-day intensive training course offered by Vermont COVERTS, Inc. (Woodlands for Wildlife) financed by the Ruffed Grouse Society. The course promotes the management of woodlands to enhance wildlife habitat and emphasizes communication between neighboring landowners. In 1985, the Wildlife Habitat Improvement Group (WHIG) was formed from three core landowners owning parcels comprising 650 acres. Over the next several years, over 3,800 acres were added to the project through an active yet informal outreach campaign by the founding individual.

In 1989, geographic information systems (GIS) mapping was completed for the entire project area (4,600 acres at that point) with funding through a U.S.D.A. Forest Service Stewardship Grant and the Windham Foundation. In 1990, a more formal educational outreach program was initiated with five local elementary schools, centered on the relation of silviculture and habitat. The project also has been opened up for informative nature walks to promote to the public the ethic of land stewardship and the knowledge of ecosystem connectivity.

STATUS AND OUTLOOK

In December 1995, the state of Vermont purchased a 245-acre woodlot which connected the Townshend State Forest (the oldest in Vermont) and added another 1,000 acres to the project area, bringing the total project size to 5,600 acres. The state fish and wildlife agency will conduct a comparable survey of critical wildlife habitat on its 1,000 acre parcel, to be added to the GIS map. A celebration involving all landowners, government officials, politicians, and the public will be held on WHIG lands in August 1996.

There is increasing interest in the project from abutting property owners, and it is likely that the project area will continue to expand in all directions. Several thousand acres may become available for easement protection, some of which are at the core of the WHIG project area and to the west of the Green Mountain National Forest. A bill was introduced in the state's 1996 legislative session which would make it easier for landowners to grant conservation easements by reducing assessments when easements are granted.

The current focus is on long-range planning for the 5,600 acres. Although the forty landowners are not held together by a contractual agreement or an association, a legal agreement

may be on the horizon. WHIG also may serve as a model for similar landowner projects in the future. Finally, the Vermont Land Trust is making presentations to several landowners in the WHIG area with the goal of creating permanent easements.

Factors Facilitating Progress

A major facilitating force for the project has been the stability in landownership over many generations. People are therefore willing to spend time and resources on the project. Also, due to the rural character of the area, landowners already have a tradition of communicating with their neighbors. A source of technical expertise and advice has come from the U.S. Fish and Wildlife Service and the state forester, both having been supportive during numerous meetings and having helped with nature walks.

The project enjoys the support of numerous officials and politicians, such as the governor, other elected officials, the commissioner of Forests, Parks and

Recreation, and the commissioner of Fish and Wildlife. A WHIG founder and member, State Representative David Clarkson, has been very active in this effort.

Obstacles to Progress

Some landowners did not want to participate in the project. Others who have participated have remained passive because they are only seasonal residents. Still others want to become involved at a time when the project may not be able to expand due to resource constraints. A future threat to the project is the uncertainty that is introduced when land is turned over to new generations or to new owners.

Contact information:

Mr. David Clarkson
RR1 Box 2426
Newfane, VT 05345
(802) 365-4243
E-mail: DCLARKS@
LEG.STATE.VT.US

PART THREE

APPENDIXES

STATE-BY-STATE LISTING OF
619 ECOSYSTEM MANAGEMENT PROJECTS

(*Note:* Projects in multiple states are listed under each state. Projects featured in the "Catalog" section are denoted by "♦"; their ID numbers begin with "P".)

ID	PROJECT	LOCATION
	ALABAMA	
L106	ACF/ACT Comprehensive Study	AL, GA, FL
L163	Cahaba River Basin Project	AL
L181	Central Gulf Ecosystem	AL, MS
L267	Flint Creek	AL
P032 ♦	Georgia Mountain Ecosystem Management Project	AL
P033 ♦	Grand Bay Savanna	AL, MS
P038 ♦	Gulf of Mexico Program	FL, AL, MS, LA, TX
P043 ♦	Interior Low Plateau	KY, TN, AL
L369	Lower Mississippi Valley Joint Venture	LA, MS, AR, TN, KY, MO, IL, AL
L372	Lower Tennessee River - Cumberland River Ecosystem	TN, KY, AL
L402	Mobile Bay Restoration Demonstrations	AL
L532	Southern Appalachian Man and the Biosphere Program (SAMAB)	TN, NC, SC, GA, AL, WV
L536	Southern Forested Wetlands	MS, AL, AR, LA, SC, GA
	ALASKA	
L113	Andreafsky Weir Coho Salmon Escapement Monitoring	AK
L116	Arctic Tundra Long-Term Ecological Research Site	AK
L129	Bar-tailed Godwit/Large Shorebird Aerial Fall Surveys	AK
L152	Bonanza Creek Experimental Forest	AK
L214	Contaminants Monitoring in Salvaged Waterfowl Carcasses	AK
L215	Cooper Landing	AK
L228	Denali National Park and Preserve	AK
L283	Glacier Bay Ecosystem Partnership	AK
P046 ♦	Kenai River Watershed Project	AK
L341	Kwethluk Village Fisheries Monitoring Plan	AK
P066 ♦	Northeast Chichagof Island	AK
P080 ♦	Prince William Sound - Copper River Ecosystem Initiative	AK
L546	Squirrel River Integrated Activity Plan	AK
L575	Tuluksak River Fish Harvest Study	AK
	ARIZONA	
L117	Arizona FWS Partners for Wildlife Project	AZ
L201	Colorado Plateau Ecosystem Partnership Project	CO, UT, AZ, NM
L202	Colorado River Basin Salinity Control Program	CO, UT, AZ, WY, NV, CA, NM
L237	East Clear Creek	AZ
L321	Interior Basin Ecoregion	NV, ID, WY, UT, AZ
L323	International Sonoran Desert Alliance	AZ, Mexico
L361	Lone Mountain/San Rafael Ecosystem Project	AZ
P052 ♦	Malpai Borderlands Initiative	NM, AZ
L383	Maverick Project	AZ
L421	Navajo Mountain Natural Area	AZ, UT
L471	Pocket-Baker Ecosystem Analysis	AZ

ID	PROJECT	LOCATION
L497	Rio Grande Basin Landscape-Scale Assessment	TX, NM, AZ
L500	Riparian Recovery Plan Initiative	AZ, NM
P086 ♦	San Pedro River	AZ
L514	San Rafael Valley Association Planning Efforts	AZ
L515	San Simon River Ecosystem Project	AZ, NM
P101 ♦	Verde River Greenway	AZ
L603	West Clear Creek Ecosystem Management	AZ
L617	Yavapai Ecosystem	AZ
	ARKANSAS	
L118	Ark - Red River Ecosystem Team	CO, KS, OK, TX, AR, NM, MO
L136	Big Woods of Arkansas	AR
P012 ♦	Cache/Lower White Rivers Ecosystem Management Plan	AR
L218	Cossatot River State Park - Natural Area	AR
L233	Dorcheat Bayou Cooperative Management Project	AR
L368	Lower Mississippi Alluvial Valley Wetland Conservation Plan	AR, IL, KY, LA, MO, MS, TN
L369	Lower Mississippi Valley Joint Venture	LA, MS, AR, TN, KY, MO, IL, AL
L395	Mississippi River Alluvial Plain Bioreserve Project	AR, LA, MS, TN, KY
P072 ♦	Ouachita National Forest	AR, OK
L536	Southern Forested Wetlands	MS, AL, AR, LA, SC, GA
	CALIFORNIA	
L138	Biodiversity Research Consortium--species distribution in 8 states	OR, WA, CA, PA, MD, WV, VA, DE
L139	Bioregional Planning in California	CA
L144	Blacks Mountain Interdisciplinary Research Project	CA
P010 ♦	Butte Valley Basin	CA
L165	California Desert Ecosystem Management Plan	CA
L166	California Gnatcatcher - Coastal Sage Scrub NCCP - NBS Research	CA
L167	California Watershed Projects Inventory	CA
L184	Channel Islands Biosphere Reserve	CA
L199	Coles Levee/Arco Ecological Preserve	CA
L202	Colorado River Basin Salinity Control Program	CO, UT, AZ, WY, NV, CA, NM
L213	Consumnes River Watershed	CA
L216	Coordinated Resource Management and Planning Council	CA
L227	Delta Levee Protection Program	CA
L230	Desert Tortoise Technique Comparison Study	CA
P024 ♦	Dos Palmas Oasis	CA
L238	East Lassen Management Plan	CA, NV
L256	Elkhorn Slough	CA
L264	Fish Slough	CA
L275	Fort Ord	CA
L281	Giant Garter Snake - Multi-species Habitat Conservation Effort	CA
P036 ♦	Guadalupe-Nipomo Dunes Preserve	CA

ID	PROJECT	LOCATION
L304	Hayfork Adaptive Management Project	CA
L322	Intermountain West Ecosystem	WA, OR, CA, NV, UT
L330	Kern County Habitat Conservation Plan	CA
L333	Kings River Ecosystems Research Project	CA
L337	Klamath Basin Assessment	OR, CA
L338	Klamath River Basin Ecosystem Restoration Project	OR, CA
L388	Miami Basin	CA
L403	Mojave Desert Ecosystem Initiative	CA
L420	Natural Community Conservation Planning (NCCP)	CA
L440	Northwest Forest Ecosystem Plan, Research Support	WA, OR, CA
L454	PACFISH	OR, WA, CA
L492	Redding Resource Management Plan	CA
L505	San Francisco Bay Plan	CA
L506	San Francisco Bay/Sacramento-San Joaquin Delta Estuary	CA
L507	San Joaquin River Management Program	CA
L508	San Joaquin Valley Regional Ecosystem Protection Planning Group	CA
L509	San Joaquin Valley: Strat. for Balancing Biodiversity and Economy	CA
L512	San Luis Rey River Corridor Management Plan	CA
P087 ♦	Santa Catalina Island Ecological Restoration Program	CA
P088 ♦	Santa Margarita River	CA
L517	Santa Monica Bay National Estuary Program	CA
L524	Sierra Nevada Ecosystem Project	CA
L526	Silverspot Butterfly Recovery Efforts	OR, WA, CA
L535	Southern California Ecoregion - U.S. Fish and Wildlife Service	CA
L572	Trinity River Restoration Project	CA
L577	USFS Participation in Local Agency Planning	CA
L585	Upper Feather River Watershed Projects	CA
L590	Upper Sacramento River Riparian Habitat Management	CA
L619	Yreka River - Siskiyou Forest Management Roundtable	CA
L533	Yuba Watershed Institute	CA

COLORADO

ID	PROJECT	LOCATION
L109	Air Force Academy	CO
L114	Animas River Basin Watershed Project	CO
L118	Ark - Red River Ecosystem Team	CO, KS, OK, TX, AR, NM, MO
L119	Arkansas River Water Needs Assessment	CO
L123	Axial Basin Coordinated Resource Management Plan	CO
L126	Badger Creek Watershed Management Area	CO
L128	Bald Mountain Basin Coordinated Resource Management Plan	CO
L137	Biodiversity Assessment, South-central CO & North-central NM	CO, NM
L153	Book Cliffs Conservation Initiative	UT, CO
L182	Central Plains Experimental Range	CO
L188	Cherry Creek Landscape Analysis	CO
L194	Clear Creek Watershed Forum	CO
L200	Colorado Front Range Ecosystem Management Res. and Dem. Project	CO
L201	Colorado Plateau Ecosystem Partnership Project	CO, UT, AZ, NM
L202	Colorado River Basin Salinity Control Program	CO, UT, AZ, WY, NV, CA, NM
L203	Colorado River Endangered Fish Recovery Program	CO
L204	Colorado Rockies Regional Cooperative	CO

ID	PROJECT	LOCATION
P021 ♦	Colorado State Forest Ecosystem Planning Project	CO
L234	Douglas Basin Ecosystem Management Area	CO
L235	Dry Creek Basin Coordinated Resource Management Plan	CO
L236	Eagle River Multi-Objective Management Plan	CO
L255	Elevenmile Ecosystem Management Project	CO
L277	Four Mile/Divide Creek Analysis	CO
L284	Glade Landscape Analysis	CO
L294	Greater Gunnison Gorge Ecosystem Management Plan	CO
L297	Gunnison Basin Ecological Classification and Inventory	CO
L299	Habitat Partnership Program	CO
P056 ♦	Mesa Creek Coordinated Resource Management Plan	CO
L405	Montezuma County Federal Lands Program	CO
L407	Muddy Creek Landscape Analysis	CO
L418	National Hierarchy of Ecological Units	CO
L430	Niwot Ridge Biosphere Reserve	CO
L439	Northwest Colorado Riparian Task Force	CO
P073 ♦	Owl Mountain Partnership	CO
L466	Pines Project	CO
L468	Pinos Ecosystem Analysis	CO
L470	Playa Lakes Joint Venture	TX, OK, KS, CO, NM
L476	Powderhorn Wilderness Management Plan	CO
P084 ♦	Ruby Canyon and Black Ridge Ecosystem Management Plan	CO
L502	Sage Grouse Habitat Improvement Initiative	CO
P085 ♦	San Luis Valley Comprehensive Ecosystem Management Plan	CO
L513	San Miguel River Multi-Objective Plan	CO
L540	Southwest Colorado Interagency LANDSAT Vegetation Classification Project	CO
L549	State Involvement in National Forest Plan Revisions throughout Colorado	CO
L570	Trapper Creek Aquatic and Riparian Restoration Project	CO
P098 ♦	Trout Mountain Roadless Area	CO
L578	USFS/BLM Ecosystem Management Team	CO
L580	Uncompahgre Riverway	CO
L583	Upper Arkansas Watershed Initiative and Forum	CO
P100 ♦	Upper Huerfano Ecosystem	CO
L592	Upper/Middle Rio Grande Ecosystem	NM, CO
L604	West Elk Wilderness HRM/AMP	CO
L616	Yampa Valley Alliance	CO

CONNECTICUT

ID	PROJECT	LOCATION
L162	COVERTS	VT, CT
L231	Disturbance of eastern forest ecosystems by stressor/host/pathogen interactions	CT
L272	Forest Insect Biology and Biocontrol	CT
L367	Lower Connecticut River Special Area Management Plan	CT, VT, NH, MA
L408	Mudge Pond	CT
L424	New England - New York ECOMAP	ME, NH, VT, MA, CT, RI, NY
L425	New England Resource Protection Project	NH, CT, RI
L534	Southern Berkshires Bioreserve	MA, CT
P096 ♦	Tidelands of the Connecticut River	CT
P099 ♦	Upper Farmington River Management Plan	CT, MA

DELAWARE

ID	PROJECT	LOCATION
L138	Biodiversity Research Consortium--species distribution in 8 states	OR, WA, CA, PA, MD, WV, VA, DE

ID	PROJECT	LOCATION
P017 ♦	Chesapeake Bay Program	MD, VA, PA, DC, DE, NY, WV
L190	Chesapeake Bay/Mid-Atlantic Highlands/Mid-Atlantic Landscape Assessment	NY, NJ, PA, WV, MD, VA, NC, DE
L198	Coastal Plain Ponds	MA, NJ, DE, MD, RI
L222	Delaware Bayshores Bioreserve	NJ, DE
L223	Delaware Estuary Program	NJ, DE
L224	Delaware Inland Bays Estuary Program	DE
L414	Nanticoke/Blackwater Rivers Bioreserve	MD, DE
P067 ♦	Northern Delaware Wetlands Rehabilitation Program	DE

DISTRICT OF COLUMBIA

ID	PROJECT	LOCATION
L112	Anacostia River	MD, DC
P017 ♦	Chesapeake Bay Program	MD, VA, PA, DC, DE, NY, WV
L417	National Capital Region Cons. Data Center/DC Natural Heritage Prog.	MD, DC, VA, WV, PA

FLORIDA

ID	PROJECT	LOCATION
L106	ACF/ACT Comprehensive Study	AL, GA, FL
L115	Apalachicola River and Bay Ecosystem Management Program	FL
L180	Central Florida Native Grassland Management Project	FL
L196	Closing the Gap in Florida's Wildlife Habitat Conservation System	FL
L197	Coastal Barrier Island Ecosystem Effort	FL
L249	Ecosystem Management Initiative	FL
P031 ♦	Florida Bay Ecosystem Management Area	FL
L268	Florida Keys National Marine Sanctuary	FL
L269	Florida Keys Project	FL
P038 ♦	Gulf of Mexico Program	FL, AL, MS, LA, TX
L306	Hillsborough River Ecosystem Management Area	FL
L317	Indian River Lagoon National Estuary Program	FL
L363	Long Leaf Pine-Eglin Air Force Base	FL
P051 ♦	Lower St. Johns River Ecosystem Management Area	FL
L374	Loxahatchee River Basin Wetland Planning Project	FL
L379	Mangrove Rehabilitation Program	FL
L462	Pensacola Bay Watershed Ecological Evaluation	FL
L516	Sand Pine-Scrub Oak	FL
L518	Sarasota Bay National Estuary Program	FL
P091 ♦	South Florida/Everglades Ecosystem Restoration Initiative	FL
L538	Southern Phosphate District	FL
L558	Suwannee River Ecosystem Management Area	FL
L561	Tampa Bay National Estuary Program	FL
L589	Upper Peace River	FL
L600	Wekiva River Basin	FL
L614	Xeric Oak Scrub Ecological Survey	FL

GEORGIA

ID	PROJECT	LOCATION
L106	ACF/ACT Comprehensive Study	AL, GA, FL
L111	Altamaha River Bioreserve	GA
P015 ♦	Chattooga River Project	GA, SC, NC
L207	Conasauga River	GA, TN
L491	Red-cockaded woodpecker	GA
L519	Savannah River Basin	NC, SC, GA
L520	Savannah River Basin Watershed Project	GA, SC
L532	Southern Appalachian Man and the Biosphere Program (SAMAB)	TN, NC, SC, GA, AL, WV

ID	PROJECT	LOCATION
L536	Southern Forested Wetlands	MS, AL, AR, LA, SC, GA
L539	Southlands Experimental Forest	GA
L609	Whole Farm/Ranch Planning	GA, ID, MN, NE, NY, PA

HAWAII

ID	PROJECT	LOCATION
L239	East Maui Watershed Partnership	HI
L302	Hawaiian Forest Challenge	HI
L327	Kapunakea Preserve	HI
L331	Kilauea Forest - Puu Maka'ala Fence Construction	HI
L332	Kilauea-Olaa Working Group	HI
L352	Land Use District Boundary Review	HI
P060 ♦	Molokai Preserves	HI
L419	Natural Areas Reserve System	HI
P061 ♦	Natural Resource Roundtable	HI
L442	Nu'upia Ponds	HI
L455	Pacific Air Force Command	HI
P081 ♦	Pu'u Kukui Watershed Management Area	HI
L554	Stream Protection and Management (SPAM) Program	HI
L576	U.S. Fish & Wildlife Service--Pacific Islands Ecoregion	HI
L593	Waikamoi Preserve	HI
L594	Waimea Canyon, Kokee & Polihale, and Na Pali Coast State Parks	HI

IDAHO

ID	PROJECT	LOCATION
L131	Bear River Watershed	WY, UT, ID
L148	Boise Cascade Ecosystem Management Demonstration Project - Idaho	ID
L151	Boise River Wildfire Recovery	ID
L211	Conservation Agreement for Bonneville Cutthroat Trout	ID
L221	Deadwood Landscape Analysis	ID
L280	Garden Creek/Craig Mountain	ID
P034 ♦	Greater Yellowstone Ecosystem	WY, MT, ID
L305	Henry's Fork Watershed Council	ID, WY
L311	Idaho Ecosystem Management Project	ID
L312	Idaho Panhandle National Forest Aquatic Ecosystem Strategy	ID
L321	Interior Basin Ecoregion	NV, ID, WY, UT, AZ
P042 ♦	Interior Columbia Basin Ecosystem Management Project	WA, OR, ID, MT, WY, NV
L340	Kootenay River Network	MT, ID, BC
P055 ♦	McPherson Ecosystem Enhancement Project	ID
L389	Mica Creek Watershed Study	ID
L456	Pacific Northwest Watershed Project	OR, WA, ID
L495	Revision of the Forest Plan for the Targhee National Forest	ID, WY
L525	Silver Creek	ID
L545	Spruce Creek and/or Logging Gulch - NEPA documents	ID
L562	Teanaway Ecosystem Demonstration Project	ID
L565	Thousand Springs Preserve	ID
P097 ♦	Trail Creek Ecosystem Analysis	ID
L609	Whole Farm/Ranch Planning	GA, ID, MN, NE, NY, PA

ILLINOIS

ID	PROJECT	LOCATION
P011 ♦	Cache River Wetlands	IL
P019 ♦	Chicago Region Biodiversity Council	IL
L210	Conservation 2000 - Ecosystem-Based Management	IL
P027 ♦	Ecosystem Charter for the Great Lakes-St. Lawrence Basin	MI, MN, WI, IN, IL, OH, PA, NY, Ont, Que

ID	PROJECT	LOCATION
L291	Great Lakes Program / EPA Great Lakes National Program Office	MI, MN, WI, IN, IL, NY, OH, PA
L328	Kaskaskia Private Lands Initiative	IL
L345	Lake Michigan Lakewide Management Plan	IL, IN, MI, WI
L368	Lower Mississippi Alluvial Valley Wetland Conservation Plan	AR, IL, KY, LA, MO, MS, TN
L369	Lower Mississippi Valley Joint Venture	LA, MS, AR, TN, KY, MO, IL, AL
L371	Lower Missouri River - Data Collection	MO, MN, WI, IA, IL
L376	MacKinaw River Project	IL
L444	Oak-Savanna Ecosystem Project	IL, IN, MN, MI, OH, WI, IA, MO
P070 ♦	Ohio River Valley Ecosystem	IL, IN, OH, PA, NY, WV, KY, TN, VA, MD
L457	Panther-Cox Creek Watershed Management Plan	IL
L553	Strategic Plan for the Illinois Department of Conservation	IL
L586	Upper Mississippi River System Environmental Management Program	MN, IA, WI, IL, MO
L587	Upper Mississippi River/Tallgrass Prairie Ecosystem	MN, IA, WI, IL, MO

INDIANA

ID	PROJECT	LOCATION
L146	Blue River Corridor	IN
L177	Cedar Creek Watershed Habitat Restoration	IN
P027 ♦	Ecosystem Charter for the Great Lakes-St. Lawrence Basin	MI, MN, WI, IN, IL, OH, PA, NY, Ont, Que
P030 ♦	Fish Creek Watershed Project	IN, OH
L290	Great Lakes Basin Ecosystem Team	MI, MN, WI, IN, other
L291	Great Lakes Program / EPA Great Lakes National Program Office	MI, MN, WI, IN, IL, NY, OH, PA
L310	ICEM Oak Savannah Project	MI, IN, MN, MO, WI
L318	Indiana Coastal Coordination Program	IN
L319	Indiana Coordinated Resource Management	IN
P040 ♦	Indiana Grand Kankakee Marsh Restoration Project	IN
L345	Lake Michigan Lakewide Management Plan	IL, IN, MI, WI
L373	Lower Wabash Habitat Restoration	IN
L441	Northwest Indiana Environmental Initiative	IN
L444	Oak-Savanna Ecosystem Project	IL, IN, MN, MI, OH, WI, IA, MO
P070 ♦	Ohio River Valley Ecosystem	IL, IN, OH, PA, NY, WV, KY, TN, VA, MD
L458	Partners for Wildlife	IN
L591	Upper Wabash Habitat Restoration Project	IN

IOWA

ID	PROJECT	LOCATION
L133	Beeds Lake Water Quality Project	IA
L135	Big Spring Basin	IA
L157	Broken Kettle Grassland	IA
L178	Centerville City Reservoir Water Quality Protection Project	IA
L621	Clear Lake Enhancement and Restoration (C.L.E.A.R.)	IA
L292	Great Plains Partnership	MN, MT, ND, WY, SD, IA, NE
P035 ♦	Green Valley State Park Ecosystem Management Plan	IA
P044 ♦	Iowa River Corridor Project	IA
L370	Lower Missouri River	KS, NE, IA, MO

ID	PROJECT	LOCATION
L371	Lower Missouri River - Data Collection	MO, MN, WI, IA, IL
P059 ♦	Missouri River Mitigation Project	KS, NE, IA, MO
L400	Missouri River Natural Resource Group	MT, ND, SD, MO, IA, NE, KA
L444	Oak-Savanna Ecosystem Project	IL, IN, MN, MI, OH, WI, IA, MO
L464	Pine Creek Water Quality Project	IA
P079 ♦	Prairie Pothole Joint Venture	ND, SD, MN, IA, MT
L551	Stone State Park Ecosystem Management Plan	IA
L552	Storm Lake Water Quality Protection Project	IA
L584	Upper Big Mill Creek	IA
L586	Upper Mississippi River System Environmental Management Program	MN, IA, WI, IL, MO
L587	Upper Mississippi River/Tallgrass Prairie Ecosystem	MN, IA, WI, IL, MO
L595	Walnut Creek National Wildlife Refuge - Prairie Learning Center	IA

KANSAS

ID	PROJECT	LOCATION
L118	Ark - Red River Ecosystem Team	CO, KS, OK, TX, AR, NM, MO
L187	Cheney Lake - N. F. Ninnescah Watershed Water Quality Project	KS
P018 ♦	Cheyenne Bottoms Wildlife Area	KS
L307	Hillsdale Water Quality Project	KS
L326	Kansas - FWS Partners for Wildlife	KS
P047 ♦	Konza Prairie Research Natural Area	KS
L370	Lower Missouri River	KS, NE, IA, MO
P059 ♦	Missouri River Mitigation Project	KS, NE, IA, MO
L470	Playa Lakes Joint Venture	TX, OK, KS, CO, NM
L544	Spring Straight and Cedar Creek Ecosystem Based Planning Projects	KS

KENTUCKY

ID	PROJECT	LOCATION
P043 ♦	Interior Low Plateau	KY, TN, AL
L368	Lower Mississippi Alluvial Valley Wetland Conservation Plan	AR, IL, KY, LA, MO, MS, TN
L369	Lower Mississippi Valley Joint Venture	LA, MS, AR, TN, KY, MO, IL, AL
L372	Lower Tennessee River - Cumberland River Ecosystem	TN, KY, AL
L378	Mammoth Cave Area Biosphere Reserve	KY
L395	Mississippi River Alluvial Plain Bioreserve Project	AR, LA, MS, TN, KY
P070 ♦	Ohio River Valley Ecosystem	IL, IN, OH, PA, NY, WV, KY, TN, VA, MD

LOUISIANA

ID	PROJECT	LOCATION
P005 ♦	Barataria-Terrebonne National Estuary Program	LA
L164	Calcasieu-Sabine Cooperative River Basin Study	LA
L295	Gulf Coast Bird Observatory Network	TX, LA
P038 ♦	Gulf of Mexico Program	FL, AL, MS, LA, TX
L347	Lake Ponchartrain Basin Restoration	LA
L366	Louisiana Coastal Wetlands Planning, Protection and Restoration Act	LA
L368	Lower Mississippi Alluvial Valley Wetland Conservation Plan	AR, IL, KY, LA, MO, MS, TN
L369	Lower Mississippi Valley Joint Venture	LA, MS, AR, TN, KY, MO, IL, AL
L395	Mississippi River Alluvial Plain Bioreserve Project	AR, LA, MS, TN, KY
L467	Pineywoods Conservation Initiative	LA, TX

ID	PROJECT	LOCATION
L536	Southern Forested Wetlands	MS, AL, AR, LA, SC, GA
P095 ♦	Tensas River Basin Initiative	LA
L608	Wetlands Productivity Study	LA

MAINE

ID	PROJECT	LOCATION
L175	Casco Bay Estuary Project	ME
L244	Ecology and Management of Northern Conifer and Associated Ecosystems	ME
L296	Gulf of Maine Council	ME, NH, MA
P037 ♦	Gulf of Maine Rivers Ecosystem Plan	ME, NH, MA
L329	Kennebunk Plains	ME
L377	Maine Forest Biodiversity Project	ME
L424	New England - New York ECOMAP	ME, NH, VT, MA, CT, RI, NY
P068 ♦	Northern Forest Lands Council	VT, NY, NH, ME
L504	Salmon Habitat and River Enhancement (SHARE)	ME
L547	St. Croix International Waterway Commission	ME
L598	Waterboro Barrens Preserve	ME

MARYLAND

ID	PROJECT	LOCATION
L112	Anacostia River	MD, DC
L138	Biodiversity Research Consortium--species distribution in 8 states	OR, WA, CA, PA, MD, WV, VA, DE
P017 ♦	Chesapeake Bay Program	MD, VA, PA, DC, DE, NY, WV
L190	Chesapeake Bay/Mid-Atlantic Highlands/Mid-Atlantic Landscape Assessment	NY, NJ, PA, WV, MD, VA, NC, DE
L191	Chesapeake Rivers	MD
L198	Coastal Plain Ponds	MA, NJ, DE, MD, RI
L206	Comprehensive Plan for Maryland's Wildlife Management Areas	MD
L413	Nanjemoy Creek Ecosystem Initiative	MD
L414	Nanticoke/Blackwater Rivers Bioreserve	MD, DE
L416	Nassawango Creek Ecosystem Initiative	MD
L417	National Capital Region Cons. Data Center/DC Natural Heritage Prog.	MD, DC, VA, WV, PA
P070 ♦	Ohio River Valley Ecosystem	IL, IN, OH, PA, NY, WV, KY, TN, VA, MD
P089 ♦	Sideling Hill Creek Bioreserve	MD, PA
P093 ♦	State Lines Serpentine Barrens	PA, MD
L564	The Nature Conservancy Bioreserve Protection Program	MD
L569	Town Creek Ecosystem Stewardship Project	MD

MASSACHUSETTS

ID	PROJECT	LOCATION
L122	Atlantic Salmon Restoration Ecology and Management Research	MA
L161	Buzzards Bay Program	MA
L172	Cape Cod, Martha's Vineyard, Nantucket	MA
L198	Coastal Plain Ponds	MA, NJ, DE, MD, RI
L209	Connecticut River Corridor	MA
L296	Gulf of Maine Council	ME, NH, MA
P037 ♦	Gulf of Maine Rivers Ecosystem Plan	ME, NH, MA
L301	Harvard Forest LTER Site	MA
L309	Hyannis Ponds	MA
L367	Lower Connecticut River Special Area Management Plan	CT, VT, NH, MA
L381	Massachusetts Bays Program	MA
L387	Merrimack River	NH, MA
L415	Narragansett Bay Project	MA, RI
L422	Neponset River Watershed Project	MA

ID	PROJECT	LOCATION
L424	New England - New York ECOMAP	ME, NH, VT, MA, CT, RI, NY
P078 ♦	Plainfield Project	MA
L527	Silvio Conte Refuge Environmental Impact Statement	MA and others
L534	Southern Berkshires Bioreserve	MA, CT
P099 ♦	Upper Farmington River Management Plan	CT, MA
L596	Waquoit Bay National Estuarine Reserve	MA
L610	Wildlife Communities and Habitat Relationships in New England	MA

MICHIGAN

ID	PROJECT	LOCATION
P003 ♦	Allegan State Game Area	MI
L195	Clinton River Area of Concern	MI
P026 ♦	Eastern Upper Peninsula Ecosystem Management Consortium	MI
P027 ♦	Ecosystem Charter for the Great Lakes-St. Lawrence Basin	MI, MN, WI, IN, IL, OH, PA, NY, Ont, Que
P029 ♦	Escanaba River State Forest	MI
L286	Grand Traverse Bay Watershed Pilot Project--Whole Farm/Ranch Planning	MI
L290	Great Lakes Basin Ecosystem Team	MI, MN, WI, IN, other
L291	Great Lakes Program / EPA Great Lakes National Program Office	MI, MN, WI, IN, IL, NY, OH, PA
L310	ICEM Oak Savannah Project	MI, IN, MN, MO, WI
L324	Isle Royale Biosphere Reserve	MI
L336	Kirtland's Warbler Recovery Plan	MI
L345	Lake Michigan Lakewide Management Plan	IL, IN, MI, WI
L348	Lake Superior Basin Biosphere Proposed Biosphere Reserve	MI, MN, WI, Canada
L349	Lake Superior Binational Program Habitat Projects	MI, MN, WI, Canada
L350	Lake Superior EMAP - Great Lakes Assessment	MN, MI
L409	Mulligan Creek Project	MI
L436	Northern Grey Wolf	MN, WI, MI
L437	Northern Lake Huron Bioreserve	MI
P069 ♦	Northern Lower Michigan Ecosystem Management Project	MI
L444	Oak-Savanna Ecosystem Project	IL, IN, MN, MI, OH, WI, IA, MO
L503	Saginaw Bay Area of Concern	MI
L531	Southeast Michigan Initiative	MI
L537	Southern Lake Michigan Initiative	MI
P092 ♦	St. Marys River Remedial Action Plan	MI, Ont
L568	Total Ecosystem Management Strategies	MI

MINNESOTA

ID	PROJECT	LOCATION
L147	Blufflands Initiative	MN
L150	Boise Cascade Ecosystem Management Project - Minnesota	MN
L170	Cannon River Watershed Partnership	MN
L171	Cannon Valley Big Woods Ecosystem Conservation Initiative	MN
L176	Cedar Creek Natural History Area Long-Term Ecological Research Site	MN
L220	DNR Regional Planning: Region V Prototype	MN
L242	Ecological Classification & Inventory Demonstration Area	MN
P027 ♦	Ecosystem Charter for the Great Lakes-St. Lawrence Basin	MI, MN, WI, IN, IL, OH, PA, NY, Ont, Que
L270	Forest Bird Diversity Initiative	MN
L282	Glacial Lake Agassiz Interbeach Area Stewardship Project	MN
L290	Great Lakes Basin Ecosystem Team	MI, MN, WI, IN, other

ID	PROJECT	LOCATION
L291	Great Lakes Program / EPA Great Lakes National Program Office	MI, MN, WI, IN, IL, NY, OH, PA
L292	Great Plains Partnership	MN, MT, ND, WY, SD, IA, NE
L310	ICEM Oak Savannah Project	MI, IN, MN, MO, WI
L348	Lake Superior Basin Biosphere Proposed Biosphere Reserve	MI, MN, WI, Canada
L349	Lake Superior Binational Program Habitat Projects	MI, MN, WI, Canada
L350	Lake Superior EMAP - Great Lakes Assessment	MN, MI
L360	Lk Superior Highlands/Nemadji River Basin Project	MN
L371	Lower Missouri River - Data Collection	MO, MN, WI, IA, IL
L392	Minnesota County Biological Survey	MN
L393	Minnesota Environmental Indicators Initiative	MN
P057 ♦	Minnesota Peatlands	MN
L436	Northern Grey Wolf	MN, WI, MI
L444	Oak-Savanna Ecosystem Project	IL, IN, MN, MI, OH, WI, IA, MO
P076 ♦	Phalen Chain of Lakes Watershed Project	MN
L465	Pine Flats Ecosystem Management Project	MN
P079 ♦	Prairie Pothole Joint Venture	ND, SD, MN, IA, MT
L489	Red River Watershed	ND, MN
L571	Tri-County Leech Lake Watershed Project	MN
L586	Upper Mississippi River System Environmental Management Program	MN, IA, WI, IL, MO
L587	Upper Mississippi River/Tallgrass Prairie Ecosystem	MN, IA, WI, IL, MO
L601	Wells Creek Watershed Partnership	MN
L609	Whole Farm/Ranch Planning	GA, ID, MN, NE, NY, PA

MISSISSIPPI

ID	PROJECT	LOCATION
L124	Back Bay Biloxi Ecosystem Assessment	MS
L181	Central Gulf Ecosystem	AL, MS
P033 ♦	Grand Bay Savanna	AL, MS
P038 ♦	Gulf of Mexico Program	FL, AL, MS, LA, TX
L368	Lower Mississippi Alluvial Valley Wetland Conservation Plan	AR, IL, KY, LA, MO, MS, TN
L369	Lower Mississippi Valley Joint Venture	LA, MS, AR, TN, KY, MO, IL, AL
L395	Mississippi River Alluvial Plain Bioreserve Project	AR, LA, MS, TN, KY
L536	Southern Forested Wetlands	MS, AL, AR, LA, SC, GA
L618	Yazoo Basin	MS

MISSOURI

ID	PROJECT	LOCATION
L118	Ark - Red River Ecosystem Team	CO, KS, OK, TX, AR, NM, MO
L158	Brush Creek EARTH Project	MO
L262	Farm of the Future	MO
L298	Ha Ha Tonka State Park	MO
L310	ICEM Oak Savannah Project	MI, IN, MN, MO, WI
L368	Lower Mississippi Alluvial Valley Wetland Conservation Plan	AR, IL, KY, LA, MO, MS, TN
L369	Lower Mississippi Valley Joint Venture	LA, MS, AR, TN, KY, MO, IL, AL
L370	Lower Missouri River	KS, NE, IA, MO
L371	Lower Missouri River - Data Collection	MO, MN, WI, IA, IL
L380	Mark Twain Watershed Project	MO

ID	PROJECT	LOCATION
L386	Meramec River	MO
P058 ♦	Missouri Coordinated Resource Management	MO
L396	Missouri Masterpieces	MO
L397	Missouri Ozark Forest Ecosystem Project (MOFEP)	MO
L398	Missouri Resource Assessment Partnership	MO
P059 ♦	Missouri River Mitigation Project	KS, NE, IA, MO
L400	Missouri River Natural Resource Group	MT, ND, SD, MO, IA, NE, KA
L401	Missouri River Post-Flood Evaluation (MRPE)	MO
L444	Oak-Savanna Ecosystem Project	IL, IN, MN, MI, OH, WI, IA, MO
P074 ♦	Partners for Prairie Wildlife	MO
L477	Prairie State Park Research Program	MO
L499	Riparian Ecosystem Assessment and Management (REAM)	MO
L542	Special Ecological Stewardship	MO
P094 ♦	Stegall Mountain Natural Area	MO
L581	Union Ridge Conservation Area	MO
L586	Upper Mississippi River System Environmental Management Program	MN, IA, WI, IL, MO
L587	Upper Mississippi River/Tallgrass Prairie Ecosystem	MN, IA, WI, IL, MO
L588	Upper Niangua River Hydrologic Unit Area	MO
L613	Wilson Creek National Battlefield	MO

MONTANA

ID	PROJECT	LOCATION
L132	Beartree Challenge	MT
P007 ♦	Bitterroot Ecosystem Management Research Project	MT
L143	Blackfoot Challenge	MT
L189	Cherry Creek Watershed Coop. Management Plan & Water Quality Special Project	MT
L257	Elkhorns Mountains Cooperative Management Area	MT
L266	Flathead County Master Plan	MT
L288	Grassland Ecosystem Comparison Project	ND, SD, MT, WY
L292	Great Plains Partnership	MN, MT, ND, WY, SD, IA, NE
P034 ♦	Greater Yellowstone Ecosystem	WY, MT, ID
P042 ♦	Interior Columbia Basin Ecosystem Management Project	WA, OR, ID, MT, WY, NV
L340	Kootenay River Network	MT, ID, BC
L400	Missouri River Natural Resource Group	MT, ND, SD, MO, IA, NE, KA
L410	Multi-Agency Approach to Planning and Evaluation (MAAPE)	ND, SD, MT
L433	North Dakota-Montana Paddlefish Management Plan	ND, MT
P079 ♦	Prairie Pothole Joint Venture	ND, SD, MN, IA, MT

NEBRASKA

ID	PROJECT	LOCATION
L241	Eastern Nebraska Saline Wetlands	NE
L258	Elm Creek Watershed Section 319 Nonpoint Source Project	NE
L292	Great Plains Partnership	MN, MT, ND, WY, SD, IA, NE
L370	Lower Missouri River	KS, NE, IA, MO
L399	Missouri River Division - U.S. Army Corps of Engineers	NE, others
P059 ♦	Missouri River Mitigation Project	KS, NE, IA, MO
L400	Missouri River Natural Resource Group	MT, ND, SD, MO, IA, NE, KA
P062 ♦	Nebraska Sandhills Ecosystem	NE
L447	Omaha Stretch of the Missouri River	NE

ID	PROJECT	LOCATION
L469	Platte River	NE
P082 ◆	Rainwater Basin Joint Venture	NE
L620	Salt Valley Lakes Project—Wildwood Lake	NE
L609	Whole Farm/Ranch Planning	GA, ID, MN, NE, NY, PA

New Hampshire

ID	PROJECT	LOCATION
L208	Concord Pine Barrens	NH
L245	Ecology and Management of Northern Hardwoods	NH
L248	Ecosystem Dynamics in Mature and Harvested Forests of New England	NH
L296	Gulf of Maine Council	ME, NH, MA
P037 ◆	Gulf of Maine Rivers Ecosystem Plan	ME, NH, MA
L308	Hubbard Brook Experimental Forest	NH
L315	Implementing Ecosystem Based Forest Mgmt. - "Exemplary Forestry Init."	NH
L367	Lower Connecticut River Special Area Management Plan	CT, VT, NH, MA
L385	Measuremt, anal., & modeling of forest ecosystems in a changing env.	NH
L387	Merrimack River	NH, MA
L424	New England - New York ECOMAP	ME, NH, VT, MA, CT, RI, NY
L425	New England Resource Protection Project	NH, CT, RI
P064 ◆	New Hampshire Forest Resources Plan	NH
P068 ◆	Northern Forest Lands Council	VT, NY, NH, ME
L557	Supersanctuary (Harris Center for Conservation Education)	NH

New Jersey

ID	PROJECT	LOCATION
L190	Chesapeake Bay/Mid-Atlantic Highlands/Mid-Atlantic Landscape Assessment	NY, NJ, PA, WV, MD, VA, NC, DE
L198	Coastal Plain Ponds	MA, NJ, DE, MD, RI
L222	Delaware Bayshores Bioreserve	NJ, DE
L223	Delaware Estuary Program	NJ, DE
L225	Delaware River/Delmarva Coastal Watershed	NJ
L300	Hackensack Meadowlands District	NJ
P039 ◆	Hudson River/New York Bight Ecosystem	NY, NJ
L354	Landscape Project	NJ
P065 ◆	New Jersey Pinelands	NJ

New Mexico

ID	PROJECT	LOCATION
L118	Ark - Red River Ecosystem Team	CO, KS, OK, TX, AR, NM, MO
L137	Biodiversity Assessment, South-central CO & North-central NM	CO, NM
L155	Bosque Riparian Restoration	NM
L169	Canadian River Commission	OK, TX, NM
L173	Carson National Forest Planning Project	NM
L192	Chijuillita Watershed	NM
L201	Colorado Plateau Ecosystem Partnership Project	CO, UT, AZ, NM
L202	Colorado River Basin Salinity Control Program	CO, UT, AZ, WY, NV, CA, NM
L252	Effects of PAH on Colorado Squawfish	NM
L271	Forest Ecosystem Management Plan	NM
L276	Fort Stanton Special Management Area	NM
L335	Kiowa Grasslands Integrated Resource Management Program	NM
L355	Largo Canyon Watershed Management and Erosion Control Plan	NM
L356	Largo-Aqua Fria Watershed Project	NM
P052 ◆	Malpai Borderlands Initiative	NM, AZ
L384	McGregor Coordinated Resource Management Plan	NM

ID	PROJECT	LOCATION
P063 ◆	Negrito Project	NM
L470	Playa Lakes Joint Venture	TX, OK, KS, CO, NM
L497	Rio Grande Basin Landscape-Scale Assessment	TX, NM, AZ
L498	Rio Puerco Watershed Stabilization Initiative	NM
L500	Riparian Recovery Plan Initiative	AZ, NM
L510	San Juan Basin Unlined Pit Closure and Remediation	NM
L515	San Simon River Ecosystem Project	AZ, NM
L521	Sevilleta National Wildlife Refuge Long-Term Ecological Research Site	NM
L592	Upper/Middle Rio Grande Ecosystem	NM, CO

New York

ID	PROJECT	LOCATION
L108	Adirondack Park/Northwest Flow	NY
P002 ◆	Albany Pine Bush	NY
L160	Buffalo River Area of Concern	NY
P017 ◆	Chesapeake Bay Program	MD, VA, PA, DC, DE, NY, WV
L190	Chesapeake Bay/Mid-Atlantic Highlands/Mid-Atlantic Landscape Assessment	NY, NJ, PA, WV, MD, VA, NC, DE
L240	Eastern Lake Ontario Conservation Initiative	NY
P027 ◆	Ecosystem Charter for the Great Lakes-St. Lawrence Basin	MI, MN, WI, IN, IL, OH, PA, NY, Ont, Que
L253	Eighteenmile Creek Area of Concern	NY
L278	French Creek Bioreserve	NY, PA
L291	Great Lakes Program / EPA Great Lakes National Program Office	MI, MN, WI, IN, IL, NY, OH, PA
L293	Great Swamp Ecosystem Initiative	NY
P039 ◆	Hudson River/New York Bight Ecosystem	NY, NJ
L344	Lake Champlain Wetlands	VT, NY
L362	Long Island Sound	NY
L423	Neversink River Ecosystem Initiative	NY
L424	New England - New York ECOMAP	ME, NH, VT, MA, CT, RI, NY
L427	Niagara River Area of Concern	NY
P068 ◆	Northern Forest Lands Council	VT, NY, NH, ME
P070 ◆	Ohio River Valley Ecosystem	IL, IN, OH, PA, NY, WV, KY, TN, VA, MD
L448	Onondaga Lake	NY
L451	Oswego River Harbor Area of Concern	NY
L459	Peconic Bay	NY
L460	Peconic Bioreserve	NY
L475	Poultney River Conservation Program	NY, VT
L501	Rochester Embayment Area of Concern	NY
L522	Shawangunk Ridge Biodiversity Partnership	NY
L548	St. Lawrence River Area of Concern	NY
L555	Structure and Function of Urban Forests	NY
L609	Whole Farm/Ranch Planning	GA, ID, MN, NE, NY, PA

Nevada

ID	PROJECT	LOCATION
L142	Black Rock/High Rock Interdistrict Management Area	NV
L202	Colorado River Basin Salinity Control Program	CO, UT, AZ, WY, NV, CA, NM
L238	East Lassen Management Plan	CA, NV
L263	Fish Creek Restoration Project	NV
L321	Interior Basin Ecoregion	NV, ID, WY, UT, AZ
P042 ◆	Interior Columbia Basin Ecosystem Management Project	WA, OR, ID, MT, WY, NV
L322	Intermountain West Ecosystem	WA, OR, CA, NV, UT

ID	PROJECT	LOCATION
P054 ♦	Marys River Riparian/Aquatic Restoration Project	NV
L450	Oregon High Desert Bioreserve	OR, NV
P077 ♦	Piute/El Dorado Desert Wildlife Management Area	NV
L483	Pyramid Lake/Stillwater Marsh Project	NV
L486	Quinn River Riparian Improvement and Demonstration Project	NV
L487	Railroad Valley Wetlands Enhancement	NV
L602	Wells Resource Management Plan, Elk Amendment	NV

NORTH CAROLINA

ID	PROJECT	LOCATION
L110	Albemarle-Pamlico Sound National Estuary Program	NC, VA
L121	Atlantic Cedar Restoration Program	NC
P015 ♦	Chattooga River Project	GA, SC, NC
L190	Chesapeake Bay/Mid-Atlantic Highlands/Mid-Atlantic Landscape Assessment	NY, NJ, PA, WV, MD, VA, NC, DE
L219	Coweeta Hydrologic Laboratory	NC
L273	Fort Bragg Integrated Natural Resources Planning	NC
L359	Little Tennessee River Group	NC
P050 ♦	Lower Roanoke River Bioreserve	NC
L406	Mt. Roan Balds Management	NC, TN
L426	New Hope Creek Corridor Project	NC
L490	Red Wolf Recovery Program	NC
L496	Richland Creek Corridor	NC
L519	Savannah River Basin	NC, SC, GA
L532	Southern Appalachian Man and the Biosphere Program (SAMAB)	TN, NC, SC, GA, AL, WV

NORTH DAKOTA

ID	PROJECT	LOCATION
L247	Ecoregions of North and South Dakota	ND
L288	Grassland Ecosystem Comparison Project	ND, SD, MT, WY
L292	Great Plains Partnership	MN, MT, ND, WY, SD, IA, NE
L400	Missouri River Natural Resource Group	MT, ND, SD, MO, IA, NE, KA
L410	Multi-Agency Approach to Planning and Evaluation (MAAPE)	ND, SD, MT
L432	North Dakota Conservation Reserve Program	ND
L433	North Dakota-Montana Paddlefish Management Plan	ND, MT
P079 ♦	Prairie Pothole Joint Venture	ND, SD, MN, IA, MT
L478	Prairie, Wetland, and Missouri River Mainstem Ecosystem	ND, SD
L489	Red River Watershed	ND, MN

OHIO

ID	PROJECT	LOCATION
L120	Ashtabula River Area of Concern	OH
P006 ♦	Big Darby Creek Partnership	OH
P027 ♦	Ecosystem Charter for the Great Lakes-St. Lawrence Basin	MI, MN, WI, IN, IL, OH, PA, NY, Ont, Que
P030 ♦	Fish Creek Watershed Project	IN, OH
L285	Grand River Partners	OH
L291	Great Lakes Program / EPA Great Lakes National Program Office	MI, MN, WI, IN, IL, NY, OH, PA
L314	Impact of Atmospheric Deposition & Global Change on Forest Health & Productivity	OH
L316	Indian Lake Hydrologic Unit Project	OH
L382	Maumee River Area of Concern	OH
L444	Oak-Savanna Ecosystem Project	IL, IN, MN, MI, OH, WI, IA, MO

ID	PROJECT	LOCATION
P070 ♦	Ohio River Valley Ecosystem	IL, IN, OH, PA, NY, WV, KY, TN, VA, MD
L452	Ottawa River Watershed Study	OH
L485	Quantitative Methods for Modeling Forest Ecosystems	OH
L494	Restoration of Ohio Oak Forests with Prescribed Fire	OH
L550	Stillwater Creek	OH

OKLAHOMA

ID	PROJECT	LOCATION
L118	Ark - Red River Ecosystem Team	CO, KS, OK, TX, AR, NM, MO
L169	Canadian River Commission	OK, TX, NM
L313	Illinois River - Battle Branch	OK
L445	Oklahoma Biodiversity Initiative	OK
L446	Oklahoma Natural Areas Registry Program	OK
P071 ♦	Oklahoma Tallgrass Prairie Preserve	OK
P072 ♦	Ouachita National Forest	AR, OK
L470	Playa Lakes Joint Venture	TX, OK, KS, CO, NM

OREGON

ID	PROJECT	LOCATION
P004 ♦	Applegate Partnership	OR
L138	Biodiversity Research Consortium--species distribution in 8 states	OR, WA, CA, PA, MD, WV, VA, DE
L145	Blue Mountains Elk Initiative	OR, WA
P009 ♦	Blue Mountains Natural Resources Institute	OR, WA
L174	Cascade Center for Ecosystem Management	OR
L179	Central Cascades Adaptive Management Area	OR
L205	Columbia River Gorge National Scenic Area	OR, WA
L217	Coos Bay/Coquille River Basins	OR
P028 ♦	Elliott State Forest Management Plan	OR
L287	Grande Ronde Model Watershed Program	OR, WA
P042 ♦	Interior Columbia Basin Ecosystem Management Project	WA, OR, ID, MT, WY, NV
L322	Intermountain West Ecosystem	WA, OR, CA, NV, UT
L337	Klamath Basin Assessment	OR, CA
L338	Klamath River Basin Ecosystem Restoration Project	OR, CA
L339	Klamath-Lake Partnership	OR
L440	Northwest Forest Ecosystem Plan, Research Support	WA, OR, CA
L449	Oregon Biodiversity Project	OR
L450	Oregon High Desert Bioreserve	OR, NV
L454	PACFISH	OR, WA, CA
L456	Pacific Northwest Watershed Project	OR, WA, ID
L479	Proposed Coquille Forest of Coquille Indian Tribe	OR
L526	Silverspot Butterfly Recovery Efforts	OR, WA, CA
L566	Tillamook Bay National Estuary Program	OR
L574	Trout Creek Mountain Working Group	OR
L579	Umpqua Basin Fisheries Restoration Initiative	OR
L605	West Eugene Wetlands Project	OR
L611	Willamette River Basin	OR

PENNSYLVANIA

ID	PROJECT	LOCATION
L138	Biodiversity Research Consortium--species distribution in 8 states	OR, WA, CA, PA, MD, WV, VA, DE
P017 ♦	Chesapeake Bay Program	MD, VA, PA, DC, DE, NY, WV
L190	Chesapeake Bay/Mid-Atlantic Highlands/Mid-Atlantic Landscape Assessment	NY, NJ, PA, WV, MD, VA, NC, DE
L212	Conserving Biodiversity in Pennsylvania	PA

ID	PROJECT	LOCATION
L243	Ecology and Management of Allegheny Hardwood Forests	PA
P027 ♦	Ecosystem Charter for the Great Lakes-St. Lawrence Basin	MI, MN, WI, IN, IL, OH, PA, NY, Ont, Que
L265	Fishing Creek	PA
L278	French Creek Bioreserve	NY, PA
L291	Great Lakes Program / EPA Great Lakes National Program Office	MI, MN, WI, IN, IL, NY, OH, PA
L303	Hawk Mountain Sanctuary	PA
L357	Laurels Reserve	PA
L364	Long Pond Barrens	PA
L417	National Capital Region Cons. Data Center/DC Natural Heritage Prog.	MD, DC, VA, WV, PA
L435	Northern Forest Health Monitoring Program	PA
L438	Northern Stations Global Change Research Program	PA
P070 ♦	Ohio River Valley Ecosystem	IL, IN, OH, PA, NY, WV, KY, TN, VA, MD
L461	Penn's Woods Strategic Plan	PA
L472	Pocono Habitat Demonstration Project	PA
L473	Pocono Mountains	PA
L493	Resource Characterization Study	PA
P083 ♦	Robbie Run Study Area	PA
P089 ♦	Sideling Hill Creek Bioreserve	MD, PA
L543	Spring Creek Corridor Study	PA
P093 ♦	State Lines Serpentine Barrens	PA, MD
L609	Whole Farm/Ranch Planning	GA, ID, MN, NE, NY, PA

PUERTO RICO

ID	PROJECT	LOCATION
L250	Ecosystem Plan for the Caribbean Watershed	PR, VI
P048 ♦	Lajas Valley Lagoon System	PR
L375	Luquillo Experimental Forest	PR
L481	Puerto Rico Forest Stewardship Program	PR
L511	San Juan Bay Estuary Program	PR

RHODE ISLAND

ID	PROJECT	LOCATION
P008 ♦	Block Island Refuge	RI
L198	Coastal Plain Ponds	MA, NJ, DE, MD, RI
L415	Narragansett Bay Project	MA, RI
L424	New England - New York ECOMAP	ME, NH, VT, MA, CT, RI, NY
L425	New England Resource Protection Project	NH, CT, RI

SOUTH CAROLINA

ID	PROJECT	LOCATION
P001 ♦	ACE Basin	SC
L185	Charleston Harbor Project	SC
P015 ♦	Chattooga River Project	GA, SC, NC
P022 ♦	Congaree River Corridor Water Quality Planning Assessment	SC
L474	Pocotaligo River and Swamp Restoration	SC
L519	Savannah River Basin	NC, SC, GA
L520	Savannah River Basin Watershed Project	GA, SC
L532	Southern Appalachian Man and the Biosphere Program (SAMAB)	TN, NC, SC, GA, AL, WV
L536	Southern Forested Wetlands	MS, AL, AR, LA, SC, GA
L556	Sumter and Francis Marion National Forests	SC
L607	Westvaco Corp., Timberlands Division, Southern Region/Southern Woodlands	SC

SOUTH DAKOTA

ID	PROJECT	LOCATION
L125	Bad River	SD
L141	Black Hills	SD, WY
L154	Bootstraps	SD

ID	PROJECT	LOCATION
L288	Grassland Ecosystem Comparison Project	ND, SD, MT, WY
L292	Great Plains Partnership	MN, MT, ND, WY, SD, IA, NE
L400	Missouri River Natural Resource Group	MT, ND, SD, MO, IA, NE, KA
L410	Multi-Agency Approach to Planning and Evaluation (MAAPE)	ND, SD, MT
L411	Multi-Objective Fllod Mitigation Plan Vermillion River Basin	SD
P079 ♦	Prairie Pothole Joint Venture	ND, SD, MN, IA, MT
L478	Prairie, Wetland, and Missouri River Mainstem Ecosystem	ND, SD

TENNESSEE

ID	PROJECT	LOCATION
P020 ♦	Clinch Valley Bioreserve	VA, TN
L207	Conasauga River	GA, TN
P043 ♦	Interior Low Plateau	KY, TN, AL
L368	Lower Mississippi Alluvial Valley Wetland Conservation Plan	AR, IL, KY, LA, MO, MS, TN
L369	Lower Mississippi Valley Joint Venture	LA, MS, AR, TN, KY, MO, IL, AL
L372	Lower Tennessee River - Cumberland River Ecosystem	TN, KY, AL
L395	Mississippi River Alluvial Plain Bioreserve Project	AR, LA, MS, TN, KY
L406	Mt. Roan Balds Management	NC, TN
P070 ♦	Ohio River Valley Ecosystem	IL, IN, OH, PA, NY, WV, KY, TN, VA, MD
L523	Shoreline Management Initiative	TN
L532	Southern Appalachian Man and the Biosphere Program (SAMAB)	TN, NC, SC, GA, AL, WV
L560	TVA Regional Natural Heritage Project	TN
L573	Tripartite Agreement for Fish and Wildlife Resources	TN
L118	Ark - Red River Ecosystem Team	CO, KS, OK, TX, AR, NM, MO

TEXAS

ID	PROJECT	LOCATION
L127	Balcones Canyon	TX
L134	Big Bend National Park Biosphere Reserve	TX
L169	Canadian River Commission	OK, TX, NM
P023 ♦	Corpus Christi Bay National Estuary Program	TX
L274	Fort Hood	TX
L279	Galveston Bay National Estuary Program	TX
L295	Gulf Coast Bird Observatory Network	TX, LA
P038 ♦	Gulf of Mexico Program	FL, AL, MS, LA, TX
P049 ♦	Lower Rio Grande Ecosystem Plan	TX
L467	Pineywoods Conservation Initiative	LA, TX
L470	Playa Lakes Joint Venture	TX, OK, KS, CO, NM
L488	Rattlesnake Island Marsh Project	TX
L497	Rio Grande Basin Landscape-Scale Assessment	TX, NM, AZ
L563	Texas Hill Country	TX
L582	Upland Wildlife Ecology Program	TX

UTAH

ID	PROJECT	LOCATION
L131	Bear River Watershed	WY, UT, ID
L153	Book Cliffs Conservation Initiative	UT, CO
P014 ♦	Canyon Country Partnership	UT
L183	Chalk Creek Coordinated Resource Management Plan	UT
L201	Colorado Plateau Ecosystem Partnership Project	CO, UT, AZ, NM

ID	PROJECT	LOCATION
L202	Colorado River Basin Salinity Control Program	CO, UT, AZ, WY, NV, CA, NM
L229	Desert Experimental Range	UT
L261	Escalante/Kanab Resource Management Plan	UT
L289	Great Basin Ecosystem Initiative	UT
L321	Interior Basin Ecoregion	NV, ID, WY, UT, AZ
L322	Intermountain West Ecosystem	WA, OR, CA, NV, UT
L358	Little Bear River Watershed Project	UT
L390	Mill Creek Canyon Management Partnership	UT
L404	Monroe Mountain Livestock/Big Game Demonstration Area	UT
L421	Navajo Mountain Natural Area	AZ, UT
L453	Otter Creek Watershed Restoration Project	UT

VERMONT

ID	PROJECT	LOCATION
L162	COVERTS	VT, CT
P013 ♦	Camp Johnson Sandplain Restoration	VT
L320	Integrating the ecological and social dimensions of forest ecosystem management	VT
L343	Lake Champlain Basin Program	VT
L344	Lake Champlain Wetlands	VT, NY
L367	Lower Connecticut River Special Area Management Plan	CT, VT, NH, MA
L424	New England - New York ECOMAP	ME, NH, VT, MA, CT, RI, NY
P068 ♦	Northern Forest Lands Council	VT, NY, NH, ME
L443	Nulhegan Deer Wintering Area Agreement	VT
L475	Poultney River Conservation Program	NY, VT
P105 ♦	Wildlife Habitat Improvement Group	VT

VIRGIN ISLANDS

ID	PROJECT	LOCATION
L250	Ecosystem Plan for the Caribbean Watershed	PR, VI

VIRGINIA

ID	PROJECT	LOCATION
L110	Albemarle-Pamlico Sound National Estuary Program	NC, VA
L138	Biodiversity Research Consortium--species distribution in 8 states	OR, WA, CA, PA, MD, WV, VA, DE
L140	Black Creek Watershed	VA
P017 ♦	Chesapeake Bay Program	MD, VA, PA, DC, DE, NY, WV
L190	Chesapeake Bay/Mid-Atlantic Highlands/Mid-Atlantic Landscape Assessment	NY, NJ, PA, WV, MD, VA, NC, DE
P020 ♦	Clinch Valley Bioreserve	VA, TN
L260	Ely Creek Watershed	VA
L417	National Capital Region Cons. Data Center/DC Natural Heritage Prog.	MD, DC, VA, WV, PA
P070 ♦	Ohio River Valley Ecosystem	IL, IN, OH, PA, NY, WV, KY, TN, VA, MD
P102 ♦	Virginia Coast Reserve	VA

WASHINGTON

ID	PROJECT	LOCATION
L138	Biodiversity Research Consortium--species distribution in 8 states	OR, WA, CA, PA, MD, WV, VA, DE
L145	Blue Mountains Elk Initiative	OR, WA
P009 ♦	Blue Mountains Natural Resources Institute	OR, WA
L149	Boise Cascade Ecosystem Management Project - Central Washington	WA
L186	Chehalis River Basin	WA
L193	Clallam River Landscape Plan	WA
L205	Columbia River Gorge National Scenic Area	OR, WA
L254	Elbe Hills	WA

ID	PROJECT	LOCATION
L259	Elochoman	WA
L287	Grande Ronde Model Watershed Program	OR, WA
P041 ♦	Integrated Landscape Management for Fish and Wildlife	WA
P042 ♦	Interior Columbia Basin Ecosystem Management Project	WA, OR, ID, MT, WY, NV
L322	Intermountain West Ecosystem	WA, OR, CA, NV, UT
L351	Lake Whatcom	WA
L353	Landowner Forum	WA
L365	Loomis State Forest	WA
L412	Multi-species Forest Management Program	WA
L429	Nisqually River Management Plan	WA
L431	Nooksack River Watershed Initiatives	WA, BC
L440	Northwest Forest Ecosystem Plan, Research Support	WA, OR, CA
L454	PACFISH	OR, WA, CA
L456	Pacific Northwest Watershed Project	OR, WA, ID
L463	Personal Use Firewood	WA
L480	Protection of Forest Health and Productivity Research	WA
L482	Puget Sound Estuary	WA
L484	Pysht River Cooperative	WA
L526	Silverspot Butterfly Recovery Efforts	OR, WA, CA
L528	Siouxon	WA
L530	Soleduck Watershed Analysis	WA
L567	Tonasket Citizen's Council	WA
L597	Washington State Ecosystems Conservation Project--Riparian & Wetland Program	WA
L251	Washington State Ecosystems Conservation Project--Upland Wildlife Restoration	WA
L599	Watershed Restoration Partnership Program (WRPP)	WA
P103 ♦	Wild Stock Initiative	WA
P104 ♦	Wildlife Area Planning	WA
L612	Willapa Alliance Natural Resource Program	WA
L615	Yakima River Watershed Council	WA

WEST VIRGINIA

ID	PROJECT	LOCATION
L138	Biodiversity Research Consortium--species distribution in 8 states	OR, WA, CA, PA, MD, WV, VA, DE
L168	Canaan Valley	WV
P017 ♦	Chesapeake Bay Program	MD, VA, PA, DC, DE, NY, WV
L190	Chesapeake Bay/Mid-Atlantic Highlands/Mid-Atlantic Landscape Assessment	NY, NJ, PA, WV, MD, VA, NC, DE
L246	Ecology and Management of Timber and Water Resources in the Central Appalachians	WV
L417	National Capital Region Cons. Data Center/DC Natural Heritage Prog.	MD, DC, VA, WV, PA
P070 ♦	Ohio River Valley Ecosystem	IL, IN, OH, PA, NY, WV, KY, TN, VA, MD
L529	Smoke Hole/North Fork Mountain Project	WV
L532	Southern Appalachian Man and the Biosphere Program (SAMAB)	TN, NC, SC, GA, AL, WV
L559	Systems to Integrate Harvesting with Other Resource Mgmt Objectives	WV

WISCONSIN

ID	PROJECT	LOCATION
L130	Baraboo Hills Bioreserve	WI
P016 ♦	Chequamegon National Forest Landscape Analysis and Design	WI
L232	Door Peninsula Conservation Initiative	WI
P027 ♦	Ecosystem Charter for the Great Lakes-St. Lawrence Basin	MI, MN, WI, IN, IL, OH, PA, NY, Ont, Que

ID	PROJECT	LOCATION
L290	Great Lakes Basin Ecosystem Team	MI, MN, WI, IN, other
L291	Great Lakes Program / EPA Great Lakes National Program Office	MI, MN, WI, IN, IL, NY, OH, PA
L310	ICEM Oak Savannah Project	MI, IN, MN, MO, WI
L325	Kakagon Sloughs	WI
P045 ♦	Karner Blue Butterfly Habitat Conservation Plan	WI
L334	Kinnickinnic River Watershed	WI
L345	Lake Michigan Lakewide Management Plan	IL, IN, MI, WI
L348	Lake Superior Basin Biosphere Proposed Biosphere Reserve	MI, MN, WI, Canada
L349	Lake Superior Binational Program Habitat Projects	MI, MN, WI, Canada
L371	Lower Missouri River - Data Collection	MO, MN, WI, IA, IL
P053 ♦	Marathon County Forests	WI
L391	Milwaukee Estuary Area of Concern	WI
L428	Nicolet National Forest	WI
L434	North Temperate Lakes Long-Term Ecological Research Site	WI
L436	Northern Grey Wolf	MN, WI, MI
L444	Oak-Savanna Ecosystem Project	IL, IN, MN, MI, OH, WI, IA, MO
P075 ♦	Patrick Marsh Wetland Mitigation Bank Site	WI
L586	Upper Mississippi River System Environmental Management Program	MN, IA, WI, IL, MO
L587	Upper Mississippi River/Tallgrass Prairie Ecosystem	MN, IA, WI, IL, MO

ID	PROJECT	LOCATION
	WYOMING	
L107	Absaroka Front in Northwestern Wyoming: A Multiple-Use Challenge	WY
L131	Bear River Watershed	WY, UT, ID
L141	Black Hills	SD, WY
L156	Bridger-Teton Forest Plan	WY
L159	Buffalo Resource Area Ecosystem Management Planning Prototype	WY
L202	Colorado River Basin Salinity Control Program	CO, UT, AZ, WY, NV, CA, NM
L226	Delineations of Landtype Associations	WY
P025 ♦	East Fork Management Plan	WY
L288	Grassland Ecosystem Comparison Project	ND, SD, MT, WY
L292	Great Plains Partnership	MN, MT, ND, WY, SD, IA, NE
P034 ♦	Greater Yellowstone Ecosystem	WY, MT, ID
L305	Henry's Fork Watershed Council	ID, WY
L321	Interior Basin Ecoregion	NV, ID, WY, UT, AZ
P042 ♦	Interior Columbia Basin Ecosystem Management Project	WA, OR, ID, MT, WY, NV
L342	LaBarge Watershed Cooperative Management	WY
L495	Revision of the Forest Plan for the Targhee National Forest	ID, WY
P090 ♦	Snake River Corridor Project	WY
L541	Southwest Wyoming Resource Evaluation	WY
L606	West Fork Bear River Ecosystem Management Project	WY

CONTACT INFORMATION FOR
619 ECOSYSTEM MANAGEMENT PROJECTS

(*Note:* Projects are in alphabetical order. ID numbers of projects featured in the "Catalog" section start with "P".)

P001
ACE Basin
State(s): SC
Contact: Mr. Mike Prevost
　ACE Basin Bioreserve Project Director
　The Nature Consevancy
　PO Box 848
　8675 Willtown Road
　Hollywood, SC 29449
　Phone: (803)889-2427
　Fax: (803)889-3282

L106
ACF/ACT Comprehensive Study
State(s): AL, GA, FL
Contact: Mr. Mike Eubanks
　U.S. Army Corps of Enginners
　Mobile COE
　Inland Environment
　Mobile, AL

L107
Absaroka Front in Northwestern Wyoming:
　A Multiple-Use Challenge
State(s): WY
Contact: Ms. Marian Atkins
　Wildlife Biologist
　Bureau of Land Management
　Grass Creek Resource Area
　PO Box 119
　Worland, WY 82401
　Phone: (307)347-9871
　Fax: (307)347-0195

L108
Adirondack Park/Northwest Flow
State(s): NY
Contact: Mr. Tim Barnett
　Executive Director
　The Nature Conservancy
　New York - Adirondack Field Office
　PO Box 65
　Keenevalley, NY 12943
　Phone: (518)576-2082
　Fax: (518)576-4203

L109
Air Force Academy
State(s): CO
Contact:
　U.S. Air Force
　Washington, DC

P002
Albany Pine Bush
State(s): NY
Contact: Ms. Stephanie Gebauer
　The Nature Conservancy
　Albany Pine Bush
　1653 Central Avenue
　Albany, NY 12205
　Phone: (518)464-6496
　Fax: (518)464-6761

L110
Albemarle-Pamlico Sound National Estuary
　Program
State(s): NC, VA
Contact: Mr. Guy Stefanski
　North Carolina Department of Environment,
　　Health & Natural Resources
　Albemarle-Pamlico Estuary Study
　PO Box 27687
　Raleigh, NC 27611
　Phone: (919)715-4084
　Fax: (919)715-1616

P003
Allegan State Game Area
State(s): MI
Contact: Mr. John Lerg
　Wildlife Biologist
　Michigan Department of Natural Resources
　Allegan State Game Area
　4590 118th Avenue
　Allegan, MI 49010
　Phone: (616)673-2430

L111
Altamaha River Bioreserve
State(s): GA
Contact: Ms. Christi Lambert
　Project Director
　The Nature Conservancy
　Altamaha River Bioreserve
　PO Box 484, 202 Broad Street
　Darien, GA 31305
　Phone: (912)437-2161
　Fax: (912)437-2161
　E-mail: clambert@tnc.org

L112
Anacostia River
State(s): MD, DC
Contact: Mr. Jon Capacasa
　U.S. Environmental Protection Agency
　Region III (3DA00)
　841 Chestnut Building
　Philadelphia, PA 19107
　Phone: (215)597-6529
　Fax: (215)597-8255

L113
Andreafsky Weir Coho Salmon Escapement
　Monitoring
State(s): AK
Contact:
　U.S. Fish & Wildlife Service
　Kenai, AK

L114
Animas River Basin Watershed Project
State(s): CO
Contact: Mr. Bill Simon
　Animas Basin Coordinator
　PO Box 401
　Silverton, CO 81433

L115
Apalachicola River and Bay Ecosystem
　Management Program
State(s): FL
Contact: Mr. John Abendroth
　Environmental Specialist
　Florida Department of Environmental
　　Protection
　3900 Commonwealth Blvd.
　Mail Station 46
　Tallahassee, FL 32399
　Phone: (904)488-0784
　Fax: (904)922-5380

P004
Applegate Partnership
State(s): OR
Contact: Mr. Jack Shipley
　Board Member
　Applegate Partnership
　1340 Missouri Flat Road
　Grants Pass, OR 97527
　Phone: (541)846-6917

L116
Arctic Tundra Long-Term Ecological
　Research Site
State(s): AK
Contact: Dr. John Hobbie
　PI, Director
　The Ecosystem Center
　Marine Biological Laboratory
　Woods Hole, MA 02543
　Phone: (508)548-3705 ext. 473
　Fax: (508)457-1548

L117
Arizona FWS Partners for Wildlife Project
State(s): AZ
Contact: Ms. Marie Sullivan
　U.S. Fish & Wildlife Service
　2321 W. Royal Palm Road, Suite 103
　Phoenix, AZ 85021-4951
　Phone: (602)640-2720
　Fax: (602)640-2730
　E-mail: R2FWE_PH@mail.fws.gov

L118
Ark - Red River Ecosystem Team
State(s): CO, KS, OK, TX, AR, NM, MO
Contact:
　U.S. Fish & Wildlife Service
　Tulsa, OK

L119
Arkansas River Water Needs Assessment
State(s): CO
Contact: Mr. Dan Muller
　Bureau of Land Management
　SC-212A
　Denver Federal Center
　PO Box 25047
　Denver, CO 80225-0047
　Phone: (303)236-7198
　Fax: (303)236-3508

L120
Ashtabula River Area of Concern
State(s): OH
Contact: Ms. Amy Pelka
U.S. Environmental Protection Agency
Region V
77 W. Jackson Boulevard
Chicago, IL 60603-3507
Phone: (312)886-0135
Fax: (312)886-7804

L121
Atlantic Cedar Restoration Program
State(s): NC
Contact:
U.S. Air Force
Washington, DC

L122
Atlantic Salmon Restoration Ecology and
Management Research
State(s): MA
Contact: Dr. Richard DeGraaf
Acting Project Leader
USDA Forest Service
Holdsworth Hall
University of Massachusetts
Amherst, PA 01003
Phone: (413)545-0357

L123
Axial Basin Coordinated Resource
Management Plan
State(s): CO
Contact: Mr. Ozzie Kerste
Range Conservationist
Bureau of Land Management
Little Snake Resource Area
1280 Industrial Ave.
Craig, CO 81625
Phone: (303)824-4441
Fax: (303)824-8881

L124
Back Bay Biloxi Ecosystem Assessment
State(s): MS
Contact:
Acting Director, GED/NHEERL
U.S. Environmental Protection Agency
1 Sabine Island Drive
Gulf Breeze, FL 32561-5299
Phone: (904)934-9382
Fax: (904)943-2403

L125
Bad River
State(s): SD
Contact:
USDA Natural Resources Conservation
Service
Pierre, SD

L126
Badger Creek Watershed Management Area
State(s): CO
Contact: Mr. John Carochi
Bureau of Land Management
Canon City District Office
PO Box 2200
Canon City, CO 81215-2200
Phone: (719)275-0631

L127
Balcones Canyon
State(s): TX
Contact:
Texas Parks & Wildlife Department
Austin, TX

L128
Bald Mountain Basin Coordinated Resource
Management Plan
State(s): CO
Contact: Mr. Ozzie Kerste
Range Conservationist
Bureau of Land Management
Little Snake Resource Area
1280 Industrial Ave.
Craig, CO 81625
Phone: (303)824-4441

L129
Bar-tailed Godwit/Large Shorebird Aerial
Fall Surveys
State(s): AK
Contact: Mr. Brian J. McCaffery
U.S. Fish & Wildlife Service
Yukon Delta National Wildlife Refuge
PO Box 346
Bethel, AK 99559
Phone: (907)543-3151
Fax: (907)543-4413

L130
Baraboo Hills Bioreserve
State(s): WI
Contact:
The Nature Conservancy
Baraboo, WI

P005
Barataria-Terrebonne National Estuary
Program
State(s): LA
Contact: Mr. Richard DeMay
Science and Technology Coordinator
Barataria Terrebonne National Estuary
Program
PO Box 2663
Thibodaux, LA 70310
Phone: (800)259-0869; (504)447-0868
Fax: (504)447-0870
E-mail: BTEP-RD@nich-nsunet.nich.edu

L131
Bear River Watershed
State(s): WY, UT, ID
Contact: Ms. Barbara Russell
Bear River RC&D Council
1260 N. 200 East, Suite 4
Logan, UT 84321
Phone: (801)753-3871
Fax: (801)753-4037

L132
Beartree Challenge
State(s): MT
Contact:
USDA Forest Service

L133
Beeds Lake Water Quality Project
State(s): IA
Contact: Mr. Randall L. Cooney
Franklin SWCD
115 2nd Avenue, N.W.
Hampton, IA 50441
Phone: (515)456-3530
Fax: (515)456-3762

L134
Big Bend National Park Biosphere Reserve
State(s): TX
Contact: W. Philip Koepp
Chief, Science and Resources Management
Division
National Park Service
Big Bend National Park
PO Box 129
Big Bend National Park, TX 79834-0129
Phone: (915)477-2251 ext. 141
Fax: (915)477-2251 ext. 153
E-mail: BIBE_RESOURCE_MANAGEMENT
@NPS.GOV

P006
Big Darby Creek Partnership
State(s): OH
Contact: Mr. Stuart Lewis
Assistant Chief
Ohio Department of Natural Resources
1889 Fountain Square
Columbus, OH 43224
Phone: (614)265-6453

L135
Big Spring Basin
State(s): IA
Contact: Dr. George Hallberg
University of Iowa Hygienic Laboratory
102 Oakdale Campus #H101 OH
Iowa City, IA 52242
Phone: (319)335-4500
Fax: (319)335-4555

L136
Big Woods of Arkansas
State(s): AR
Contact: Ms. Leslee Ditzig Spraggins
Project Manager
The Nature Conservancy
Arkansas Field Office
601 N. University
Little Rock, AR 72205
Phone: (501)663-6699
Fax: (501)663-8332

L137
Biodiversity Assessment, South-central CO
& North-central NM
State(s): CO, NM
Contact: Mr. Larry D. Mullen
USDA Forest Service
Rocky Mountain Region
PO Box 25127
Lakewood, CO 80225
Phone: (303)275-5006
Fax: (303)275-5075

L138
Biodiversity Research Consortium--species
   distribution in 8 states
State(s): OR, WA, CA, PA, MD, WV, VA, DE
Contact: Dr. Larry Masters
   Chief Zoologist
   The Nature Conservancy
   201 Devonshire Street
   Boston, MA 02110
   Phone: (617)542-1908
   Fax: (617)482-5866
   E-mail: lmaster@tnc.org

L139
Bioregional Planning in California
State(s): CA
Contact:
   California Department of Forestry
   Sacramento, CA

P007
Bitterroot Ecosystem Management Research
   Project
State(s): MT
Contact: Dr. Clint Carlson
   Team Leader
   USDA Forest Service
   Forestry Science Laboratory
   PO Box 8089
   Missoula, MT 59807
   Phone: (406)329-3485
   Fax: (406)543-2663

L140
Black Creek Watershed
State(s): VA
Contact:
   Virginia Department of Mines, Minerals and
      Energy
   Big Stone Gap, VA

L141
Black Hills
State(s): SD, WY
Contact:
   The Nature Conservancy
   Bismarck, ND

L142
Black Rock/High Rock Interdistrict
   Management Area
State(s): NV
Contact: Mr. Bud Cribley
   Area Manager
   Bureau of Land Management
   Sonoma/Gerlach Resource Area
   705 East 4th Street
   Winnemucca, NV 89445
   Phone: (702)623-1500
   Fax: (702)623-1503

L143
Blackfoot Challenge
State(s): MT
Contact: Mr. George Hirschenberger
   Bureau of Land Management
   Garnet Resource Area
   3255 Fort Missoula Road
   Missoula, MT 59801-7293
   Phone: (406)329-3908
   Fax: (406)549-1562

L144
Blacks Mountain Interdisciplinary Research
   Project
State(s): CA
Contact: Ms. Kathy Harcksen
   Project Coordinator
   USDA Forest Service
   Pacific Southwest Research Station
   2400 Washington Avenue
   Redding, CA 96001
   Phone: (916)246-5455

P008
Block Island Refuge
State(s): RI
Contact: Mr. Chris Littlefield
   Bioreserve Manager
   The Nature Conservancy
   Block Island Refuge
   PO Box 1287
   Block Island, RI 02807
   Phone: (401)466-2129

L145
Blue Mountains Elk Initiative
State(s): OR, WA
Contact:
   Blue Mountains Elk Initiative
   LaGrande, OR

P009
Blue Mountains Natural Resources Institute
State(s): OR, WA
Contact: Ms. Lynn Starr
   Blue Mountains Natural Resources Institute
   1401 Gekeler Lane
   La Grande, OR 97850
   Phone: (503)962-6529
   Fax: (503)962-6504

L146
Blue River Corridor
State(s): IN
Contact: Mr. Allen Pursell
   Project Manager
   The Nature Conservancy
   PO Box 5
   Corydon, IN 47112
   Phone: (812)738-2087

L147
Blufflands Initiative
State(s): MN
Contact:
   Minnesota Department of Natural Resources
   Rochester, MN

L148
Boise Cascade Ecosystem Management
   Demonstration Project - Idaho
State(s): ID
Contact: Mr. Jonathan Haufler
   Corporate Contact
   Boise Cascade Corporation
   PO Box 50
   Boise, ID 83728-0001
   Phone: (208)384-6013
   Fax: (208)384-7699

L149
Boise Cascade Ecosystem Management
   Project - Central Washington
State(s): WA
Contact: Mr. Jonathan Haufler
   Corporate Contact
   Boise Cascade Corporation
   PO Box 50
   Boise, ID 83728-0001
   Phone: (208)384-6013
   Fax: (208)384-7699

L150
Boise Cascade Ecosystem Management
   Project - Minnesota
State(s): MN
Contact: Mr. Jonathan Haufler
   Corporate Contact
   Boise Cascade Corporation
   PO Box 50
   Boise, ID 83728-0001
   Phone: (208)384-6013
   Fax: (208)384-7699

L151
Boise River Wildfire Recovery
State(s): ID
Contact: Ms. Lyn Morelan
   Ecosystem Implementation Coordinator
   USDA Forest Service
   1750 Front Street
   Boise, ID 83702
   Phone: (208)364-4170

L152
Bonanza Creek Experimental Forest
State(s): AK
Contact:
   University of Alaska
   Fairbanks, AK

L153
Book Cliffs Conservation Initiative
State(s): UT, CO
Contact: Mr. Paul Andrews
   Book Cliffs Area Manager
   170 South East
   Vernal, UT 84078
   Phone: (801)789-1362
   Fax: (801)781-4410

L154
Bootstraps
State(s): SD
Contact:
   USDA Natural Resources Conservation
      Service
   White River, SD

L155
Bosque Riparian Restoration
State(s): NM
Contact: Mr. Phill Norton
   Refuge Manager
   U.S. Fish & Wildlife Service
   Bosque del Apache National Wildlife Refuge
   PO Box 1246
   Socorro, NM 87801
   Phone: (505)835-1828
   Fax: (505)835-0314

L156
Bridger-Teton Forest Plan
State(s): WY
Contact:
   USDA Forest Service, WY

L157
Broken Kettle Grassland
State(s): IA
Contact: Mr. Jerry Selby
　Director of Science and Stewardship
　The Nature Conservancy
　Iowa Field Office
　431 E. Locust, Suite 200
　Des Moines, IA 50309
　Phone: (515)244-5044
　Fax: (515)244-8890
　E-mail: IAFD@netins.net

L158
Brush Creek EARTH Project
State(s): MO
Contact: Mr. Ron Dent
　Fisheries Man. District Supervisor
　Missouri Department of Conservation
　1014 Thompson Blvd.
　Sedalia, MO 65301-2243
　Phone: (816)530-5500
　Fax: (816)530-5504

L159
Buffalo Resource Area Ecosystem
　Management Planning Prototype
State(s): WY
Contact: Mr. Bruce Daughton
　Project Team Leader
　Bureau of Land Management
　Buffalo Resource Area
　189 North Cedar
　Buffalo, WY 82834
　Phone: (307)684-5586

L160
Buffalo River Area of Concern
State(s): NY
Contact: Ms. Ellen Heath
　U.S. Environmental Protection Agency
　Region II
　26 Federal Plaza
　New York, NY 10278
　Phone: (212)264-5352
　Fax: (212)264-2194

P010
Butte Valley Basin
State(s): CA
Contact: Mr. Jim Stout
　Resource Officer
　USDA Forest Service
　Klamath National Forest
　Goosenest Ranger District
　37805 Highway 97
　Macdoel, CA 96058
　Phone: (916)398-4391
　Fax: (916)398-4599

L161
Buzzards Bay Program
State(s): MA
Contact: Mr. Joseph E. Costa
　Buzzards Bay Project
　2 Spring Street
　Marion, MA 02738
　Phone: (508)748-3600
　Fax: (508)748-3962

L162
COVERTS
State(s): VT, CT

P011
Cache River Wetlands
State(s): IL
Contact: Mr. John Penberthy
　Project Manager
　The Nature Conservancy
　Cache River Office
　Route 1, Box 53E
　Ullin, IL 62992
　Phone: (618)634-2524
　Fax: (618)634-9656
　E-mail: cachebio@aol.com

P012
Cache/Lower White Rivers Ecosystem
　Management Plan
State(s): AR
Contact: Dr. Scott Yaich
　Wildlife Management Biologist
　U.S. Fish & Wildlife Service
　Wildlife & Habitat Management Office
　PO Box 396
　St. Charles, AR 72140
　Phone: (501)282-3213
　Fax: (501)282-3391

L163
Cahaba River Basin Project
State(s): AL
Contact: Ms. Mary Kay Lynch
　U.S. Environmental Protection Agency
　Region IV
　345 Courtland Street, NE
　Atlanta, GA 30365
　Phone: (404)347-3555 ext. 6607

L164
Calcasieu-Sabine Cooperative River Basin
　Study
State(s): LA
Contact: Mr. Donald W. Gohmert
　State Conservationist
　USDA Natural Resources Conservation
　　Service
　3737 Government St.
　Alexandria, LA 71302-3727
　Phone: (318)473-7751
　Fax: (318)473-7682

L165
California Desert Ecosystem Management
　Plan
State(s): CA
Contact:
　Bureau of Land Management, CA

L166
California Gnatcatcher - Coastal Sage Scrub
　NCCP - NBS Research
State(s): CA
Contact: Mr. Stephen Veirs
　Unit Leader
　National Biological Service
　University of California
　CPSU-DES
　Davis, CA 95616-8576
　Phone: (916)757-7119
　Fax: (916)752-3350
　E-mail: SDVEIRS@UCDAVIS.EDU

L167
California Watershed Projects Inventory
State(s): CA
Contact:
　University of California-Davis
　Davis, CA

P013
Camp Johnson Sandplain Restoration
State(s): VT
Contact: Mr. Robert Popp
　Nongame and Natural Heritage Program
　103 S. Main Street
　Waterbury, VT 05671-0501
　Phone: (802)241-3718

L168
Canaan Valley
State(s): WV
Contact: Mr. John Forren
　U.S. Environmental Protection Agency
　Region III (3ES42)
　Philadelphia, PA 19107
　Phone: (215)597-3361
　Fax: (215)597-7906

L169
Canadian River Commission
State(s): OK, TX, NM
Contact: Mr. Leland D. Tillman
　Chairman and U.S. Commissioner
　Canadian River Commission
　Eastern Plains Council of Governments
　104 West 2nd
　Clovis, NM 88101
　Phone: (505)762-7714
　Fax: (505)762-7715

L170
Cannon River Watershed Partnership
State(s): MN
Contact: Ms. Allene Moesler
　Executive Director
　Cannon River Watershed Partnership
　PO Box 501
　Faribault, MN 55021
　Phone: (507)332-0488
　Fax: (507)332-0513

L171
Cannon Valley Big Woods Ecosystem
　Conservation Initiative
State(s): MN
Contact: Ms. Nancy Falkum
　SE Area Coordinator
　The Nature Conservancy
　Cannon Valley Office
　328 Central Avenue
　Faribault, MN 55021
　Phone: (507)332-0525
　Fax: (507)334-4448

P014
Canyon Country Partnership
State(s): UT
Contact: Mr. Joel Tuhy
　Colorado Plateau Public Lands Director
　The Nature Conservancy
　PO Box 1329
　Moab, UT 84532
　Phone: (801)259-4629; (801)259-2551
　Fax: (801)259-2677

L172
Cape Cod, Martha's Vineyard, Nantucket
State(s): MA
Contact:
　The Nature Conservancy
　Boston, MA

L173
Carson National Forest Planning Project
State(s): NM
Contact:
   USDA Forest Service
   Taos, NM

L174
Cascade Center for Ecosystem Management
State(s): OR
Contact: Mr. Fred Swanson
   Research Geologist
   Forestry Sciences Lab
   3200 Jefferson Way
   Corvallis, OR 97331
   Phone: (503)750-7355
   Fax: (503)750-7329
   E-mail: swanson@fsl.orst.edu

L175
Casco Bay Estuary Project
State(s): ME
Contact: Ms. Patricia Harrington
   Casco Bay Estuary Project
   312 Canco Road
   Portland, ME 04103
   Phone: (207)828-1043
   Fax: (207)828-4001

L176
Cedar Creek Natural History Area Long-Term Ecological Research Site
State(s): MN
Contact: Dr. Johannes Knops
   Site Research Director
   University of Minnesota
   Department of Ecology, Evolution and Behavior
   100 Ecology Building
   1987 Upper Buford Circle
   St. Paul, MN 55108-6097
   Phone: (612)625-5700
   Fax: (612)624-6777
   E-mail: KNOPS@LTER.UMN.EDU

L177
Cedar Creek Watershed Habitat Restoration
State(s): IN
Contact: Mr. Scott Fetters
   U.S. Fish & Wildlife Service
   Northern Indiana Suboffice
   120 South Lake Street
   Suite 230
   Warsaw, IN 46580
   Phone: (219)267-6620
   Fax: (219)269-7432

L178
Centerville City Reservoir Water Quality Protection Project
State(s): IA
Contact: T. Sue Snyder
   Iowa Department of Natural Resources
   12 & Washington
   Ag Bldg.
   Centerville, IA 51544
   Phone: (515)856-3893
   Fax: (515)586-6048

L179
Central Cascades Adaptive Management Area
State(s): OR

L180
Central Florida Native Grassland Management Project
State(s): FL
Contact: Mr. John Walewski
   U.S. Air Force
   AF/ECVP
   1260 Air Force Pentagon
   Washington, DC 20330-1260

L181
Central Gulf Ecosystem
State(s): AL, MS
Contact: Mr. Bob Strader
   U.S. Fish & Wildlife Service
   6578 Dogwood View Parkway, Suite B
   Jackson, MS 39213
   Phone: (601)965-4903
   Fax: (601)965-4010

L182
Central Plains Experimental Range
State(s): CO
Contact: Mr. Gary Frasier
   USDA - Agricultural Research Service
   Rangeland Resources Research Unit
   1701 Center Avenue
   Fort Collins, CO 80526
   Phone: (970)498-4230
   Fax: (970)482-2909

L183
Chalk Creek Coordinated Resource Management Plan
State(s): UT
Contact: Mr. Roy Gunnell
   Utah Department of Environmental Quality
   Division of Water Quality
   PO Box 144870
   Salt Lake City, UT 84114-4870
   Phone: (801)538-6146
   Fax: (801)538-6016

L184
Channel Islands Biosphere Reserve
State(s): CA
Contact:
   National Park Service
   Santa Barbara, CA

L185
Charleston Harbor Project
State(s): SC
Contact: Ms. Shirley D. Conner
   Environmental Planner
   Charleston Harbor Project
   Ocean and Coastal Resource Management
   4130 Faber Place, Suite 302
   Charleston, SC 29405
   Phone: (803)747-4323
   Fax: (803)747-8234

P015
Chattooga River Project
State(s): GA, SC, NC
Contact: Mr. David Meriwether
   USDA Forest Service
   R-8
   1720 Peachtree Rd. NW
   Atlanta, GA 30367
   Phone: (404)347-4663

L186
Chehalis River Basin
State(s): WA
Contact: Mr. Dave Palmer
   Chairman
   Chehalis River Council
   PO Box 586
   Ockville, WA 98568
   Phone: (206)273-8117

L187
Cheney Lake - N. F. Ninnescah Watershed Water Quality Project
State(s): KS
Contact: Mr. Lyle D. Frees
   USDA Natural Resources Conservation Service
   314 North Poplar
   South Hutchinson, KS 67505-1297
   Phone: (316)665-0231
   Fax: (316)669-5496

P016
Chequamegon National Forest Landscape Analysis and Design
State(s): WI
Contact: Ms. Linda Parker
   Ecologist
   USDA Forest Service
   Chequamegon National Forest
   1170 4th Avenue
   Park Falls, WI 54552
   Phone: (715)762-5169

L188
Cherry Creek Landscape Analysis
State(s): CO
Contact: Mr. Phil Kemp
   USDA Forest Service
   San Juan National Forest
   Dolores Ranger District
   PO Box 210
   Dolores, CO 81323
   Phone: (970)882-7296
   Fax: (970)882-7582

L189
Cherry Creek Watershed Coop. Management Plan & Water Quality Special Project
State(s): MT
Contact:
   Prairie County Conservation District
   410 E. Spring
   PO Box 622
   Terry, MT 59349
   Phone: (406)637-5381

P017
Chesapeake Bay Program
State(s): MD, VA, PA, DC, DE, NY, WV
Contact:
   Chesapeake Bay Foundation
   162 Prince George Street
   Annapolis, MD 21401
   Phone: (410)268-8816

L190

Chesapeake Bay/Mid-Atlantic Highlands/Mid-Atlantic Landscape Assessment

State(s): NY, NJ, PA, WV, MD, VA, NC, DE
Contact: Mr. K. Bruce Jones
U.S. Environmental Protection Agency
EMSL-LV/MSD
PO Box 93478
Las Vegas, NV 89193-3478
Phone: (702)798-2671
Fax: (702)798-2208
E-mail: msdkbj@vegas1.las.epa.gov

L191

Chesapeake Rivers

State(s): MD
Contact:
The Nature Conservancy
Chevy Chase, MD

P018

Cheyenne Bottoms Wildlife Area

State(s): KS
Contact: Mr. Karl Grover
Area Manager
Kansas Department of Wildlife and Parks
Cheyenne Bottoms Wildlife Area
Rt. 3
Great Bend, KS 67530
Phone: (316)793-3066

P019

Chicago Wilderness

State(s): IL
Contact: Ms. Laurel M. Ross
Bioreserve Program Director
The Nature Conservancy
Illinois Field Office
8 S. Michigan Ave., Suite 900
Chicago, IL 60603
Phone: (312)346-8166 ext. 14
Fax: (312)346-5606
E-mail: lross@mcs.com

L192

Chijuillita Watershed

State(s): NM
Contact: Mr. Brett O'Haver
Bureau of Land Management
Cuba Field Station
435 Montano Road, NE
Albuquerque, NM 87107
Phone: (505)289-3748
Fax: (505)761-8911

L193

Clallam River Landscape Plan

State(s): WA
Contact: Mr. Mark Johnsen
Ozette District Manager
Washington Department of Natural Resources
Olympia Region
411 Tillicum Lane
Forks, WA 98331
Phone: (360)374-6131
Fax: (360)374-5202

L194

Clear Creek Watershed Forum

State(s): CO
Contact: Mr. Carl Norbeck
Clear Creek Watershed Coordinator
Colorado Water Quality Control Division
4200 Cherry Creek Drive South
Denver, CO 80222-1530
Phone: (303)692-3513
Fax: (303)782-0390

L621

Clear Lake Enhancement & Restoration (C.L.E.A.R.) Water Quality Project

State(s): IA
Contact: Mr. Ric Zarwell
Project Coordinator
Iowa Department of Natural Resources
1415 South Monroe, Suite E
Mason City, IA 50401-5615

P020

Clinch Valley Bioreserve

State(s): VA, TN
Contact: Mr. Bill Kittrell
The Nature Conservancy
102 South Court Street
Abingdon, VA 24210
Phone: (540)676-2209
Fax: (540)676-3819

L195

Clinton River Area of Concern

State(s): MI
Contact: Ms. Callie Bolattino
U.S. Environmental Protection Agency
Region V (GLNPO)
77 W. Jackson Boulevard
Chicago, IL 60604-3507
Phone: (312)353-3490
Fax: (312)353-2018

L196

Closing the Gap in Florida's Wildlife Habitat Conservation System

State(s): FL
Contact:
Florida Game and Fresh Water Fish Commission
Tallahassee, FL

L197

Coastal Barrier Island Ecosystem Effort

State(s): FL
Contact: Mr. John Walewski
U.S. Air Force
AF/ECVP
1260 Air Force Pentagon
Washington, DC 20330-1260

L198

Coastal Plain Ponds

State(s): MA, NJ, DE, MD, RI
Contact: Mr. Tim Simmons
Director of Science and Stewardship
The Nature Conservancy
79 Milk Street
Boston, MA 02109
Phone: (617)423-2545
Fax: (617)423-8690

L199

Coles Levee/Arco Ecological Preserve

State(s): CA
Contact:
California Department of Water Resources
Fresno, CA

L200

Colorado Front Range Ecosystem Management Research & Demonstration Project

State(s): CO
Contact: Mr. Carl Edminster
Rocky Mountain Station
240 W. Prospect
Ft. Collins, CO 80526
Phone: (303)498-1264

L201

Colorado Plateau Ecosystem Partnership Project

State(s): CO, UT, AZ, NM
Contact: Mr. Doug Johnson
U.S. Environmental Protection Agency
Region VIII (8PM-SI)
999 18th Street
Denver, CO 80202
Phone: (303)293-1469
Fax: (303)293-1647

L202

Colorado River Basin Salinity Control Program

State(s): CO, UT, AZ, WY, NV, CA, NM
Contact: Mr. Jack Barnett
Colorado River Basin Salinity Control Forum
106 W. 500 South, Suite 101
Bountiful, UT 84010
Phone: (801)292-4663
Fax: (801)524-6320

L203

Colorado River Endangered Fish Recovery Program

State(s): CO
Contact: Mr. John Hamill
PO Box 25486
Denver Federal Center
Denver, CO 80225
Phone: (303)236-8154

L204

Colorado Rockies Regional Cooperative

State(s): CO
Contact: Mr. Howard Alden
Partnership Coordinator
800 E. Co. Rd. 58
Ft. Collins, CO 80524
Phone: (303)482-0983

P021

Colorado State Forest Ecosystem Planning Project

State(s): CO
Contact: Mr. Jeff Jones
Special Program Coordinator
Colorado State Forest Service
203 Forestry Building
Colorado State University
Ft. Collins, CO 80523
Phone: (303)491-7287
Fax: (303)491-7736

L205
Columbia River Gorge National Scenic Area
State(s): OR, WA
Contact: Mr. Allen Bell
Columbia River Gorge Commission
PO Box 730
White Salmon, WA 98672
Phone: (509)493-3323
Fax: (509)493-2229

L206
Comprehensive Plan for Maryland's
Wildlife Management Areas
State(s): MD
Contact: Mr. Jim Mullan
Chief of Land Management
Maryland Department of Natural Resources
Wildlife Division
3 Pershing Street, Rm. 110
Cumberland, MD 21502
Phone: (301)777-2136

L207
Conasauga River
State(s): GA, TN
Contact: Mr. Gregory Huber
USDA Natural Resources Conservation
Service
1401 Dean Street, Suite I
Rome, GA 30161
Phone: (706)291-5652
Fax: (706)291-5658

L208
Concord Pine Barrens
State(s): NH
Contact: Mr. David Van Luven
Concord Pine Barrens Ecologist
The Nature Conservancy
2 1/2 Beacon Street, Ste #6
Concord, NH 03301
Phone: (603)224-5853
Fax: (603)228-2459

P022
Congaree River Corridor Water Quality
Planning Assessment
State(s): SC
Contact: Mr. Richard A. Clark
Resource Management Specialist
National Park Service
Congaree Swamp National Monument
200 Caroline Sims Road
Hopkins, SC 29061
Phone: (803)776-4396 ext. 307
Fax: (803)783-4241

L209
Connecticut River Corridor
State(s): MA
Contact: Mr. Terry Blunt
Director
Connecticut Department of Environmental
Management
Connecticut Valley Action Program
136 Damon Rd.
Northampton, MA 01060
Phone: (413)586-8706
Fax: (413)784-1663

L210
Conservation 2000 - Ecosystem-Based
Management
State(s): IL
Contact: Mr. Carl Becker
Illinois Division of Natural Heritage
524 South Second Street
Springfield, IL 62701-1787
Phone: (217)785-8774
Fax: (217)785-8277

L211
Conservation Agreement for Bonneville
Cutthroat Trout
State(s): ID
Contact: Mr. Mark Booth
Supervisory Rangeland Management
Specialist
USDA Forest Service
431 Clay St.
Montpelier, ID 83254
Phone: (208)847-0375

L212
Conserving Biodiversity in Pennsylvania
State(s): PA
Contact: Dr. Kim Steiner
Pennsylvania State University
School of Forest Resources
213 Ferguson Building
University Park, PA 16802-4300
Phone: (814)865-9351
Fax: (814)865-3725

L213
Consumnes River Watershed
State(s): CA
Contact: Mr. Rich Reiner
13501 Franklin Blvd.
Galt, CA 95632

L214
Contaminants Monitoring in Salvaged
Waterfowl Carcasses
State(s): AK
Contact: Kim Trust
Ecological Services Anchorage
605 W. Fourth Avenue, Rm. G-62
Anchorage, AK 99501
Phone: (907)271-2888

L215
Cooper Landing
State(s): AK
Contact: Mr. Duane Harp
District Ranger
USDA Forest Service
Chugach National Forest
Seward Ranger District
PO Box 390, 334 Fourth Avenue
Seward, AK 99664
Phone: (907)224-3374
Fax: (907)224-3268
E-mail: USFS: R10F04D03A

L216
Coordinated Resource Management and
Planning Council
State(s): CA
Contact: Ms. Lisa Taricco
CRMP Council Coordinator
CARCD
801 K Street, Suite 318
Sacramento, CA 95814
Phone: (916)447-7237
Fax: (916)447-2532

L217
Coos Bay/Coquille River Basins
State(s): OR
Contact: Mr. Mike Rylko
U.S. Environmental Protection Agency
1200 Sixth Avenue
Seattle, WA 98101
Phone: (206)553-4014
Fax: (206)553-1775

P023
Corpus Christi Bay National Estuary
Program
State(s): TX
Contact: Dr. Hudson DeYoe
Corpus Christi Bay National Estuary Program
Texas AMU - Corpus Christi Campus
Campus Box 290
6300 Ocean Blvd.
Corpus Christi, TX 78412
Phone: (512)985-6767 ext. 6301
E-mail: Deyoe@tamucc.edu

L218
Cossatot River State Park - Natural Area
State(s): AR
Contact: Mr. Randy Roberson
Resource Management Specialist
Arkansas State Parks
One Capitol Mall
Little Rock, AR 72201
Phone: (501)682-6938
Fax: (501)682-1364

L219
Coweeta Hydrologic Laboratory
State(s): NC
Contact:
Coweeta Hydrologic Laboratory
Otto, NC

L220
DNR Regional Planning: Region V
Prototype
State(s): MN
Contact:
Ms. Peggy Thomas
Minnesota Department of Natural Resources
2300 Silver Creek Road NE
Rochester, MN 55906
Phone: (507)285-7428; (507)297-3518

L221
Deadwood Landscape Analysis
State(s): ID
Contact:
USDA Forest Service
McCall, ID

L222
Delaware Bayshores Bioreserve
State(s): NJ, DE
Contact:
The Nature Conservancy
Port Norris, NJ

L223
Delaware Estuary Program
State(s): NJ, DE
Contact: Mr. Robert Tudor
    Program Coordinator
    U.S. Environmental Protection Agency
    Region III
    841 Chestnut Building
    Philadelphia, PA 19107
    Phone: (215)597-9977
    Fax: (215)597-7906

L224
Delaware Inland Bays Estuary Program
State(s): DE
Contact: Mr. John Schneider
    Delaware Inland Bays Estuary Program
    Delaware Department of Natural Resources
        and Environmental Control
    PO Box 1401, 89 Kings Highway
    Dover, DE 19903
    Phone: (302)739-4590
    Fax: (302)739-6140

L225
**Delaware River/Delmarva Coastal
    Watershed**
State(s): NJ
Contact:
    U.S. Fish & Wildlife Service
    Cape May Court House, NJ

L226
Delineations of Landtype Associations
State(s): WY
Contact: Mr. Bill Daniels
    Bureau of Land Management
    PO Box 1828
    Cheyenne, WY 82003
    Phone: (307)775-6105
    Fax: (307)775-6082

L227
Delta Levee Protection Program
State(s): CA
Contact:
    California Department of Water Resources
    Sacramento, CA

L228
Denali National Park and Preserve
State(s): AK
Contact: Mr. Gordon Olson
    National Park Service
    Denali National Park and Preserve
    PO Box 9
    Denali Park, AK 99755
    Phone: (907)683-2294
    Fax: (907)683-2720

L229
Desert Experimental Range
State(s): UT
Contact: Mr. Stanley G. Kitchen
    Manager - Desert Experimental Range
    USDA Forest Service
    Shrub Sciences Laboratory
    735 North 500 East
    Provo, UT 84606
    Phone: (801)377-5717

L230
**Desert Tortoise Technique Comparison
    Study**
State(s): CA
Contact:
    U.S. Air Force
    Washington, DC

L231
**Disturbance of Eastern Forest Ecosystems
    by Stressor/Host/Pathogen Interactions**
State(s): CT
Contact: Dr. Philip Wargo
    USDA Forest Service
    Northeastern Center for Forest Health
        Research
    51 Mill Pond Road
    Hamden, CT 06514
    Phone: (203)230-4312
    Fax: (203)230-4315
    E-mail:
    FSWA/S=P.WARGO/OU=24SL07A@MHS.A
    TTMAIL.COM

L232
Door Peninsula Conservation Initiative
State(s): WI
Contact: Mr. Mike Grimm
    The Nature Conservancy
    Door Peninsula Conservation Initiative
    653 County U
    Algoma, WI 54201
    Phone: (414)743-8695
    Fax: (414)743-8695

L233
**Dorcheat Bayou Cooperative Management
    Project**
State(s): AR
Contact: Mr. Thomas Foti
    Chief of Research
    Arkansas Natural Heritage Commission
    Suite 1500, Tower Building
    Little Rock, AR 72201
    Phone: (501)324-9761
    Fax: (501)324-9618

P024
Dos Palmas Oasis
State(s): CA
Contact: Mr. Cameron Barrows
    Southern California Area Manager
    The Nature Conservancy
    PO Box 188
    Thousand Palms, CA 92276
    Phone: (619)343-1234
    Fax: (619)343-0393

L234
Douglas Basin Ecosystem Management Area
State(s): CO
Contact: Mr. Bob Fowler
    White River Resource Area Manager
    Bureau of Land Management
    PO Box 928
    Meeker, CO 81641
    Phone: (303)878-3601
    Fax: (303)878-5717

L235
**Dry Creek Basin Coordinated Resource
    Management Plan**
State(s): CO
Contact: Mr. Clyde B. Johnson
    Range Conservationist
    Bureau of Land Management
    San Juan Resource Area
    Federal Building
    701 Camino Del Rio
    Durango, CO 81301
    Phone: (970)247-4082
    Fax: (970)385-4818

L236
**Eagle River Multi-Objective Management
    Plan**
State(s): CO
Contact: Ms. Kay Salazar
    National Park Service
    RMR-PPO
    PO Box 25287
    Denver, CO 80225
    Phone: (303)969-2857
    Fax: (303)987-6676

L237
East Clear Creek
State(s): AZ
Contact:
    USDA Forest Service
    Happy Jack, AZ

P025
East Fork Management Plan
State(s): WY
Contact: Mr. Chuck Clarke
    Habitat Management Coordinator
    Wyoming Game and Fish Department
    260 Buena Vista
    Lander, WY 82520
    Phone: (307)332-2688

L238
East Lassen Management Plan
State(s): CA, NV
Contact: Mr. Francis Berg
    Bureau of Land Management
    355 Hemsted Dr.
    Redding, CA 96002
    Phone: (916)224-2100
    Fax: (916)224-2172

L239
East Maui Watershed Partnership
State(s): HI
Contact: Mr. Mark White
    Maui Project Director
    The Nature Conservancy of Hawaii
    PO Box 1716
    Makawao, HI 96768
    Phone: (808)572-7849
    Fax: (808)572-5950

L240
**Eastern Lake Ontario Conservation
    Initiative**
State(s): NY
Contact: Ms. Sandra Bonanno
    Stewardship Ecologist
    The Nature Conservancy
    Central and Western New York Office
    315 Alexander Street - 2nd Floor
    Rochester, NY 14604
    Phone: (716)546-8030
    Fax: (716)546-7825

L241

Eastern Nebraska Saline Wetlands

State(s): NE

Contact: Mr. Thomas J. Taylor
U.S. Environmental Protection Agency
Region VII
Wetlands Protection Section
726 Minnesota Ave.
Kansas City, KS 66101
Phone: (913)551-7226
Fax: (913)551-7863

P026

**Eastern Upper Peninsula Ecosystem
Management Consortium**

State(s): MI

Contact: Mr. Les Homan
District Forest Planner
Michigan Department of Natural Resources
PO Box 77
Newberry, MI 49868
Phone: (906)293-5131
Fax: (906)293-8728

L242

**Ecological Classification & Inventory
Demonstration Area**

State(s): MN

Contact: Mr. John Almendinger
ECS Coordinator
USDA Forest Service
Deer River Ranger District
PO Box 308
Deer River, MN 56536
Phone: (218)246-2123

L243

**Ecology and Management of Allegheny
Hardwood Forests**

State(s): PA

Contact: Dr. Susan Stout
USDA Forest Service
PO Box 928
Warren, PA 16365
Phone: (814)563-1040

L244

**Ecology and Management of Northern
Conifer and Associated Ecosystems**

State(s): ME

Contact: Dr. John C. Brissette
USDA Forest Service
Northeastern Forest Experiment Station
5 Godfrey Drive
Orono, ME 04473
Phone: (207)866-7260
Fax: (207)866-7262

L245

**Ecology and Management of Northern
Hardwoods**

State(s): NH

Contact: Dr. Marie-Louise Smith
USDA Forest Service
PO Box 640
Durham, NH 03824
Phone: (603)868-7652
Fax: (603)868-7604

L246

**Ecology and Management of Timber and
Water Resources in the Central
Appalachians**

State(s): WV

Contact: Dr. Mary Beth Adams
USDA Forest Service
PO 404 - Nursery Bottom
Parsons, WV 26287
Phone: (304)478-2000

L247

Ecoregions of North and South Dakota

State(s): ND

Contact: Mr. James M. Omernik
Research Geographer
U.S. Environmental Protection Agency
Environmental Research Lab - Corvallis
200 SW 35th St.
Corvallis, OR 97333
Phone: (503)754-4458
Fax: (503)754-4716

P027

**Ecosystem Charter for the Great Lakes-St.
Lawrence Basin**

State(s): MI, MN, WI, IN, IL, OH, PA, NY, Ont,
Que

Contact: Ms. Victoria Pebbles
Program Specialist
Great Lakes Commission
Argus Building, 400 Fourth
Ann Arbor, MI 48103-4816
Phone: (313)665-9135
Fax: (313)665-4370
E-mail: pebbles@glc.org

L248

**Ecosystem Dynamics in Mature and
Harvested Forests of New England**

State(s): NH

Contact: Dr. Chris Eagar
USDA Forest Service
PO Box 640
Durham, NH 03824
Phone: (603)868-7636
Fax: (603)868-7604
E-mail: ceagur@asrr.arsusda.gov

L249

Ecosystem Management Initiative

State(s): FL

Contact: Ms. Pamela P. McVety
Executive Coordinator For Ecosystem
Management
Florida Department of Environmental
Protection
3900 Commonwealth Blvd.
Mail Station 45
Tallahassee, FL 32399-3000
Phone: (904)488-7454
Fax: (904)488-7093

L250

**Ecosystem Plan for the Caribbean
Watershed**

State(s): PR, VI

Contact: Mr. James Oland
Field Supervisor
U.S. Fish & Wildlife Service
Caribbean Field Office
PO Box 491
Boqueron, PR 00622
Phone: (809)851-7297
Fax: (809)851-7440

L252

Effects of PAH on Colorado Squawfish

State(s): NM

Contact: Ms. Stephanie Odell
Bureau of Land Management
Farmington District
1235 La Plata Highway
Farmington, NM 87401
Phone: (505)599-6314
Fax: (505)599-8998
E-mail: sodell@nm0151wp.fdo.nm.blm.gov

L253

Eighteenmile Creek Area of Concern

State(s): NY

Contact: Ms. Alice Yeh
U.S. Environmental Protection Agency
Region II
26 Federal Plaza
New York, NY 10278
Phone: (212)264-1865
Fax: (212)264-2194

L254

Elbe Hills

State(s): WA

Contact:
Washington Department of Natural Resources
Enumclaw, WA

L255

Elevenmile Ecosystem Management Project

State(s): CO

Contact: Ms. Sharon Kyhl
Project Manager
USDA Forest Service
Pike/San Isabel National Forest
South Park Ranger District
Box 219
Fairplay, CO 80440
Phone: (719)836-2031

L256

Elkhorn Slough

State(s): CA

Contact: Ms. Suzanne Marr
U.S. Environmental Protection Agency
Region IX (W-3-1)
75 Hawthorne Street
San Francisco, CA 94105-3901
Phone: (415)744-1974
Fax: (415)744-1078

L257

**Elkhorns Mountains Cooperative
Management Area**

State(s): MT

Contact: Mr. Merle Good
Area Manager
Headwaters Resource Area, MT
Phone: (406)494-5059

P028

Elliott State Forest Management Plan

State(s): OR

Contact: Ms. Jill Bowling
State Lands Program Director
Oregon Department of Forestry
2600 State Street
Salem, OR 97310
Phone: (503)945-7348

L258
Elm Creek Watershed Section 319 Nonpoint Source Project
State(s): NE
Contact: Mr. Dave Jensen
Nebraska Department of Environmental Quality
Suite 400, The Atrium
PO Box 98922
Lincoln, NE 68509-8922
Phone: (402)471-3196
Fax: (402)471-2909

L259
Elochoman
State(s): WA
Contact:
Washington Department of Natural Resources
Olympia, WA

L260
Ely Creek Watershed
State(s): VA
Contact:
Virginia Department of Mines, Minerals and Energy
Big Stone Gap, VA

L261
Escalante/Kanab Resource Management Plan
State(s): UT
Contact: Mr. Pete Wilkins
Bureau of Land Management
Cedar City District
176 East D. L. Sangent Dr.
Cedar City, UT 84720
Phone: (801)865-3034
Fax: (801)865-3058

P029
Escanaba River State Forest
State(s): MI
Contact: Mr. Lee Evison
Michigan Department of Natural Resources
Forest Management Division
6833 Highways 2, 41, and M35
Gladstone, MI 49837
Phone: (906)786-2351
Fax: (906)786-1300
E-mail: evisonl@dnr.state.mi.us

L262
Farm of the Future
State(s): MO
Contact: Mr. David Quarles
Greenley Resource Center
Route 1
Novelty, MO 63460
Phone: (816)739-4410

L263
Fish Creek Restoration Project
State(s): NV
Contact: Mr. Wayne King
Manager
Bureau of Land Management
Shoshone-Eureka Resource Area
Battle Mountain District
PO Box 1420, 50 Bastian Road
Battle Mountain, NV 89820
Phone: (702)635-4000
Fax: (702)635-4034

P030
Fish Creek Watershed Project
State(s): IN, OH
Contact: Mr. Larry Clemens
Project Manager
The Nature Conservancy
Fish Creek Watershed Project Office
Peachtree Plaza, Suite B2
1220 North 200 West
Angola, IN 46703
Phone: (219)665-9141
Fax: (219)665-9141

L264
Fish Slough
State(s): CA
Contact: Ms. Joy Fatooh
Wildlife Biologist
Bureau of Land Management
Bishop Resource Area
785 North Main Street, Suite E
Bishop, CA 93514
Phone: (619)872-4881
Fax: (619)872-2894

L265
Fishing Creek
State(s): PA
Contact:
North Central Pennsylvania Conservancy

L266
Flathead County Master Plan
State(s): MT
Contact:
Flathead County Planning Department, MT

L267
Flint Creek
State(s): AL
Contact: Mr. Charles Sweatt
U.S. Environmental Protection Agency
Region IV
345 Courtland Street, NE
Atlanta, GA 30365
Phone: (205)386-2614
Fax: (205)386-3331

P031
Florida Bay Ecosystem Management Area
State(s): FL
Contact:
Florida Department of Environmental Protection
3900 Commonwealth Boulevard
Tallahassee, FL 32399
Phone: (904)488-4892

L268
Florida Keys National Marine Sanctuary
State(s): FL
Contact: Mr. Fred McManus
U.S. Environmental Protection Agency
Region IV
345 Courtland Street, NE
Atlanta, GA 30365
Phone: (404)347-1740 ext. 4299
Fax: (404)347-1797

L269
Florida Keys Project
State(s): FL
Contact: Mr. Mark Robertson
The Nature Conservancy
Florida Keys Field Office
201 Front Street, Suite 222
Key West, FL 33040
Phone: (305)296-3880
Fax: (305)292-1763

L270
Forest Bird Diversity Initiative
State(s): MN
Contact: Mr. Lee A. Pfannmuller
Section of Ecological Services
Box 25, DNR Building
500 Lafayette Road
St. Paul, MN 55155-4025
Phone: (612)296-0783
Fax: (612)296-1811

L271
Forest Ecosystem Management Plan
State(s): NM
Contact:
U.S. Air Force
Washington, DC

L272
Forest Insect Biology and Biocontrol
State(s): CT
Contact: Dr. Michael Montgomery
USDA Forest Service
Northeastern Center for Forest Health Research
51 Mill Pond Road
Hamden, CT 06514
Phone: (203)230-4331
Fax: (203)230-4315
E-mail:
FSWA/S=M.Montgomery/ou=S24L07A@MHS.ATTMAIL.COM

L273
Fort Bragg Integrated Natural Resources Planning
State(s): NC
Contact: Mr. Alan Schultz
Wildlife Biologist
U.S. Army
DPWE, Wildlife Branch (Schultz)
Ft. Bragg, NC 28307-5000
Phone: (910))396-7022
Fax: (910)396-9474

L274
Fort Hood
State(s): TX
Contact: David Tazik & Tim Hayden
U.S. Army
Construction Engineer Research Lab
2902 Newmark Drive
Champaign, IL 61821
Phone: (217)352-6511
E-mail: d.tazik@cecer.army.mil;
t.hayden@cecer.army.mi

L275

Fort Ord

State(s): CA

Contact: Mr. Steve Addington
Fort Ord Project Manager
Bureau of Land Management
Hollister Resource Area
Hollister, CA 95023
Phone: (408)394-8314
Fax: (408)394-8346

L276

Fort Stanton Special Management Area

State(s): NM

Contact: Mr. Tim Kreager
Area manager
Bureau of Land Management
Roswell District Office
PO Drawer 1857
Roswell, NM 88202-1857
Phone: (505)624-1790

L277

Four Mile/Divide Creek Analysis

State(s): CO

Contact: Ms. Cindy Hockelberg
USDA Forest Service
Rifle Ranger District
094 County Road 244
Rifle, CO 81650
Phone: (303)625-2371

L278

French Creek Bioreserve

State(s): NY, PA

Contact: Dr. Susan McAlpine
Director
The Nature Conservancy
French Creek Project Office
413 North Main Street
Jamestown, NY 14701
Phone: (716)484-6442

L279

Galveston Bay National Estuary Program

State(s): TX

Contact: Dr. Frank Shipley
Program Director
Galveston Bay National Estuary Program
Bay Plaza One, Suite 210
711 West Bay Area Blvd.
Webster, TX 77598
Phone: (713)332-9937
Fax: (713)332-8590

L280

Garden Creek/Craig Mountain

State(s): ID

Contact: Ms. Janice Hill
2990 St. Highway 3
Deary, ID 83823
Phone: (208)877-1179
Fax: (208)877-1179

P032

Georgia Mountain Ecosystem Management Project

State(s): AL

Contact: Mr. J. Ralph Jordan
Senior Nat. Res. Management Specialist
Tennessee Valley Authority
Ridgeway Road
Norris, TN 37828
Phone: (423)632-1604
Fax: (423)632-1534

L281

Giant Garter Snake - Multi-Species Habitat Conservation Effort

State(s): CA

Contact:
Dixon Research Station
Dixon, CA

L282

Glacial Lake Agassiz Interbeach Area Stewardship Project

State(s): MN

Contact: Mr. Peter Buesseler
DNR Prairie Biologist
Minnesota Department of Natural Resources
1221 East Fir Avenue
Fergus Falls, MN 56537
Phone: (218)739-7497
E-mail: pbuessel@fergus.cfa.org

L283

Glacier Bay Ecosystem Partnership

State(s): AK

Contact: Dr. Joy Geiselman
Coordinator, Glacier Bay Ecosystem Initiative
National Biological Service
1011 East Tudor Road
Anchorage, AK 99503-6199
Phone: (907)786-3668
Fax: (907)786-3636

L284

Glade Landscape Analysis

State(s): CO

Contact: Mr. Phil Kemp
USDA Forest Service
San Juan National Forest
Dolores Ranger District
PO Box 210
Dolores, CO 81323
Phone: (970)882-7296
Fax: (970)882-7582

P033

Grand Bay Savanna

State(s): AL, MS

Contact: Mr. Pat Patterson
The Nature Conservancy
P.O. Box 1028
Jackson, MS 39215-1028
Phone: (601)355-5357

L285

Grand River Partners

State(s): OH

Contact:
Ohio Department of Natural Resources
Columbus, OH

L286

Grand Traverse Bay Watershed Pilot Project--Whole Farm/Ranch Planning

State(s): MI

Contact: Mr. LeRoy Hall
Acting State Resource Conservationist
USDA Natural Resources Conservation Service
Michigan State Office
Room 101
1405 South Harrison Road
East Lansing, MI 48823-5243
Phone: (517)337-6701 x1221

L287

Grande Ronde Model Watershed Program

State(s): OR, WA

Contact:
Grande Ronde Model Watershed Program
LaGrande, OR

L288

Grassland Ecosystem Comparison Project

State(s): ND, SD, MT, WY

Contact:
Bureau of Land Management

L289

Great Basin Ecosystem Initiative

State(s): UT

Contact:
U.S. Air Force
Washington, DC

L290

Great Lakes Basin Ecosystem Team

State(s): MI, MN, WI, IN, other

Contact: Mr. Dale P. Burkett
Deputy Great Lakes Coordinator
U.S. Fish & Wildlife Service
1405 S. Harrison Road, Rm. 308
East Lansing, MI 48823
Phone: (517)337-6807
Fax: (517)337-6812

L291

Great Lakes Program / EPA Great Lakes National Program Office

State(s): MI, MN, WI, IN, IL, NY, OH, PA

Contact: Ms. Karen Holland
Ecological Protection & Restoration Team Leader
U.S. Environmental Protection Agency
Great Lakes Program National Office (G-9J)
77 W. Jackson Boulevard
Chicago, IL 60604-3590
Phone: (312)353-2690
Fax: (312)353-2018
E-mail: holland.karen@epamail.epa.gov

L292

Great Plains Partnership

State(s): MN, MT, ND, WY, SD, IA, NE

Contact: Ms. Jo Clark
Western Governor's Association
600 17th Street
Suite 1705 South Tower
Denver, CO 80202-5452
Phone: (303)623-9378

L293

Great Swamp Ecosystem Initiative

State(s): NY

Contact: Mr. Dave Tobias
Director of Protection Programs
The Nature Conservancy
Lower Hudson Field Chapter
41 South Moger Avenue
Mt. Kisco, NY 10549
Phone: (914)244-3271
Fax: (914)244-3275

L294
Greater Gunnison Gorge Ecosystem
Management Plan
State(s): CO
Contact: Ms. Karen Tucker
  Recreation Planner
  Bureau of Land Management
  UBRA
  2505 South Townsend
  Montrose, CO 81401
  Phone: (303)249-6047;FTS(700)322-7317

P034
Greater Yellowstone Ecosystem
State(s): WY, MT, ID
Contact:
  Greater Yellowstone Coalition
  P.O. Box 1874
  Bozeman, MT 59715
  Phone: (406)586-1593

P035
Green Valley State Park Ecosystem
Management Plan
State(s): IA
Contact: Mr. Jim Scheffler
  Wetland Project Coordinator
  Iowa Department of Natural Resources
  Wallace State Office Building
  Des Moines, IA 50319
  Phone: (515)281-6157
  Fax: (515)281-6794

P036
Guadalupe-Nipomo Dunes Preserve
State(s): CA
Contact: Ms. Nancy Warner
  Field Representative
  The Nature Conservancy
  PO Box 15810
  San Luis Obispo, CA 93406
  Phone: (805)545-9925
  Fax: (805)545-8510

L295
Gulf Coast Bird Observatory Network
State(s): TX, LA
Contact: Mr. Ray Johnson
  Gulf Coast Bird Observatory
  1903 Port Royal Drive
  Nassau Bay, TX 77058
  Phone: (713)335-9040
  Fax: (713)335-9826

L296
Gulf of Maine Council
State(s): ME
Contact: Mr. Dave Kelley
  State Planning Office
  SHS #38
  Augusta, ME 04011
  Phone: (207)287-3261

P037
Gulf of Maine Rivers Ecosystem Plan
State(s): ME, NH, MA
Contact: Mr. Vic Segarich
  Ecosystem Team Leader
  U.S. Fish & Wildlife Service
  Nashua National Fish Hatchery
  151 Broad Street
  Nashua, NH 03063
  Phone: (603)886-7719

P038
Gulf of Mexico Program
State(s): FL, AL, MS, LA, TX
Contact: Dr. Douglas A. Lipka
  Director
  Gulf of Mexico Program
  Building 1103, Room 202
  Stennis Space Center, MS 39529
  Phone: (601)688-1172
  Fax: (601)688-2709

L297
Gunnison Basin Ecological Classification
and Inventory
State(s): CO
Contact: Ms. Sandy Hayes
  Ecologist
  Bureau of Land Management
  Gunnison Basin Resource Area
  216 N. Colorado
  Gunnison, CO 81230
  Phone: (303)641-0471;FTS(700)859-4447

L298
Ha Ha Tonka State Park
State(s): MO
Contact:
  Missouri Department of Natural Resources
  Jefferson City, MO

L299
Habitat Partnership Program
State(s): CO
Contact: Mr. David Bray
  Assistant Area Manager
  Bureau of Land Management
  Little Snake Resource Area
  1280 Industrial Ave.
  Craig, CO 81625
  Phone: (303)824-4441

L300
Hackensack Meadowlands District
State(s): NJ
Contact: Ms. Mary Anne Thiesing
  U.S. Environmental Protection Agency
  Region II
  Water Management Division
  New York, NY 10278
  Phone: (212)264-8793
  Fax: (212)264-4690

L301
Harvard Forest LTER Site
State(s): MA
Contact:
  Harvard University
  Petersham, MA

L302
Hawaiian Forest Challenge
State(s): HI
Contact:
  The Nature Conservancy of Hawaii
  Honolulu, HI

L303
Hawk Mountain Sanctuary
State(s): PA
Contact: Mr. Keith L. Bildstein
  Director of Research
  Hawk Mountain Sanctuary Association
  RR 2, Box 191
  Kempton, PA 19529-9449
  Phone: (610)756-6961
  Fax: (610)756-4468

L304
Hayfork Adaptive Management Project
State(s): CA
Contact:
  USDA Forest Service
  Albany, CA

L305
Henry's Fork Watershed Council
State(s): ID, WY
Contact:
  The Nature Conservancy
  Sun Valley, ID

L306
Hillsborough River Ecosystem Management
Area
State(s): FL
Contact: Pat Fricano
  Environmental Manager
  Florida Department of Environmental
    Protection
  3900 Commonwealth Blvd.
  Tallahassee, FL 32399
  Phone: (904)488-0784

L307
Hillsdale Water Quality Project
State(s): KS
Contact: Mr. Thomas Lorenz
  U.S. Environmental Protection Agency
  Region VII
  726 Minnesota Avenue
  Kansas City, KS 66101
  Phone: (913)551-7292
  Fax: (913)551-7765

L308
Hubbard Brook Experimental Forest
State(s): NH
Contact:
  Syracuse University
  Syracuse, NY

P039
Hudson River/New York Bight Ecosystem
State(s): NY, NJ
Contact: Ms. Elizabeth Herland
  Ecosystem Team Coordinator
  U.S. Fish & Wildlife Service
  Walkill River National Wildlife Refuge
  P.O. Box 3836
  Sussex, NJ 07461
  Phone: (201)702-7266
  Fax: (201)702-7286

L309
Hyannis Ponds
State(s): MA
Contact:
  The Nature Conservancy
  Boston, MA

L310
ICEM Oak Savannah Project
State(s): MI, IN, MN, MO, WI
Contact:
  Michigan Department of Natural Resources
  Lansing, MI

L311

Idaho Ecosystem Management Project

State(s): ID

Contact: Mr. Brad Holt
Boise Cascade
1111 West Hefferson Street
Boise, ID 83702
Phone: (208)793-2242
Fax: (208)384-7699

L312

Idaho Panhandle National Forest Aquatic
Ecosystem Strategy

State(s): ID

Contact:
USDA Forest Service
Silverton, ID

L313

Illinois River - Battle Branch

State(s): OK

Contact: Mr. Scott Smith
U. S. Environmental Protection Agency
Region VI (6W-QS)
1445 Ross Avenue
Dallas, TX 75202-2733

L314

Impact of Atmospheric Deposition & Global
Change on Forest Health & Productivity

State(s): OH

Contact: Dr. Robert Long
USDA Forest Service
359 Main Road
Delaware, OH 43015
Phone: (614)368-0050
Fax: (614)368-0152
E-mail: R.LONG@ASPR.ARSUSDA.GOV

L315

Implementing Ecosystem Based Forest
Management - "Exemplary Forestry
Initiative"

State(s): NH

Contact: Mr. Paul A. Doscher
Vice President
Society for the Protection of New Hampshire
Forests
54 Portsmouth St.
Concord, NH 03301
Phone: (603)224-9945

L316

Indian Lake Hydrologic Unit Project

State(s): OH

Contact: Mr. Greg Nageotte
Project Administrator
Natural Resources Conservation Service
324 Road 11
Bellefontaine, OH 43311
Phone: (513)593-2946
Fax: (513)592-3350

L317

Indian River Lagoon National Estuary
Program

State(s): FL

Contact: Mr. Drew Kendall
U.S. Environmental Protection Agency
Region IV
Coastal Programs
345 Courtland Street, NE
Atlanta, GA 30365
Phone: (404)347-1740 ext. 4301
Fax: (404)347-1797

L318

Indiana Coastal Coordination Program

State(s): IN

Contact:
Indiana Department of Natural Resources
Indianapolis, IN

L319

Indiana Coordinated Resource Management

State(s): IN

Contact:
USDA Natural Resources Conservation
Service
Indianapolis, IN

P040

Indiana Grand Kankakee Marsh Restoration
Project

State(s): IN

Contact: Mr. Jim Ruwaldt
Assistant Field Supervisor
U.S. Fish & Wildlife Service
620 S. Walker
Bloomington, IN 47403-2121
Phone: (812)334-4261 ext. 213
Fax: (812)334-4273
E-mail: James_ruwaldt@mail.fws.gov

P041

Integrated Landscape Management for Fish
and Wildlife

State(s): WA

Contact: Mr. Rollie Geppert
Washington Department of Fish and Wildlife
600 Capitol Way North
Olympia, WA 98501-1091
Phone: (360)902-2587

L320

Integrating the Ecological & Social
Dimensions of Forest Ecosystem
Management

State(s): VT

Contact: Dr. Mark Twery
Project Leader
USDA Forest Service
PO Box 968
Burlington, VT 05402
Phone: (802)951-6771
E-mail: MTWERY@ASRR.ARSUDA.GOV

L321

Interior Basin Ecoregion

State(s): NV, ID, WY, UT, AZ

Contact: Carlos Mendoza & Ronald Anglin
Ecoregion Team Co-Leaders
U.S. Fish & Wildlife Service
4600 Kietzke Lane
Building C, Rm 125
Reno, NV 89502
Phone: (702)784-5227
Fax: (702)784-5870

P042

Interior Columbia Basin Ecosystem
Management Project

State(s): WA, OR, ID, MT, WY, NV

Contact: Mr. Tom Quigley
USDA Forest Service
Pacific Northwest Research Station
P.O. Box 3890
Portland, OR 97208-2890
Phone: (503)326-5640

P043

Interior Low Plateau

State(s): KY, TN, AL

Contact: Mr. Bob Ford
Project Manager
Tennessee Conservation League
300 Orlando Avenue
Nashville, TN 37209-3200
Phone: (615)353-1133
Fax: (615)353-0083
E-mail: conserve.tcl@nashville.com

L322

Intermountain West Ecosystem

State(s): WA, OR, CA, NV, UT

Contact:
U.S. Fish & Wildlife Service

L323

International Sonoran Desert Alliance

State(s): AZ, Mexico

Contact:
Hia-Ced O-odham
Sells, AZ

P044

Iowa River Corridor Project

State(s): IA

Contact: Mr. James R. Munson
Iowa Private Lands Coordinator
U.S. Fish & Wildlife Service
PO Box 399
Prairie City, IA 50228
Phone: (515)994-2415
Fax: (515)994-2104

L324

Isle Royale Biosphere Reserve

State(s): MI

Contact:
National Park Service
Houghton, MI

L325

Kakagon Sloughs

State(s): WI

Contact: Mr. Matt Dallmon
Watershed Coordinator
The Nature Conservancy
Kakagon Sloughs Project Office
618 Main Street West - Suite B
Ashland, WI 54806
Phone: (715)682-5789
Fax: (715)682-5832

L326

Kansas - FWS Partners for Wildlife

State(s): KS

Contact: Mr. Jerre L. Gamble
U.S. Fish & Wildlife Service
PO Box 128
Hartford, KS 66854
Phone: (316)392-5553

L327

Kapunakea Preserve

State(s): HI

Contact: Mr. Mark White
Maui Project Director
The Nature Conservancy of Hawaii
PO Box 1716
Makawao, HI 96768
Phone: (808)572-7849
Fax: (808)572-5950

P045

Karner Blue Butterfly Habitat Conservation Plan

State(s): WI

Contact:
Wisconsin Department of Natural Resources
Box 7921
Madison, WI 53507

L328

Kaskaskia Private Lands Initiative

State(s): IL

Contact:
Okawville, IL

P046

Kenai River Watershed Project

State(s): AK

Contact: Mr. Randall H. Hagenstein
Associate State Director
The Nature Conservancy of Alaska
421 West 1st Avenue, Suite 200
Anchorage, AK 99501
Phone: (907)276-3133
Fax: (907)276-2584

L329

Kennebunk Plains

State(s): ME

Contact: Ms. Nancy Sferra
South Maine Preserves Manager
The Nature Conservancy
160 Main St.
Sanford, ME 04073
Phone: (207)490-4012
Fax: (207)490-4012

L330

Kern County Habitat Conservation Plan

State(s): CA

Contact:
California Department of Water Resources
Fresno, CA

L331

Kilauea Forest - Puu Maka'ala Fence Construction

State(s): HI

Contact:
Hawaii Department of Land & Natural Resources
Hilo, HI

L332

Kilauea-Olaa Working Group

State(s): HI

Contact:
Hawaii Division of Forestry and Wildlife
Hilo, HI

L333

Kings River Ecosystems Research Project

State(s): CA

Contact: Mr. Mark Smith
USDA Forest Service
Sierra National Forest
1600 Tollhouse Road
Clovis, CA 93611

L334

Kinnickinnic River Watershed

State(s): WI

Contact: Mr. Robert W. Chambers
Kinnickinnic River Land Trust
N8203 1130th St.
River Falls, WI 54022
Phone: (715)425-5738
Fax: (715)425-1746

L335

Kiowa Grasslands Integrated Resource Management Program

State(s): NM

Contact: Mr. Pam Brown
District Ranger
Kiowa National Grasslands
714 Main Street
Clayton, NM 88415
Phone: (505)374-9652

L336

Kirtland's Warbler Recovery Plan

State(s): MI

Contact:
USDA Forest Service
Cadillac, MI

L337

Klamath Basin Assessment

State(s): OR, CA

Contact:
USDA Forest Service
Yreka, CA

L338

Klamath River Basin Ecosystem Restoration Project

State(s): OR, CA

Contact: Mr. Steve Lewis
Project Supervisor
U.S. Fish & Wildlife Service
Klamath Basin Ecosystem Restoration Office
6600 Washburn Way
Klamath Falls, OR 97603-9365
Phone: (503)885-8481

L339

Klamath-Lake Partnership

State(s): OR

Contact:
Klamath Falls, OR

P047

Konza Prairie Research Natural Area

State(s): KS

Contact: Dr. David Hartnett
Kansas State University
Division of Biology
Ackert Hall
Manhattan, KS 66506
Phone: (913)532-5925
Fax: (913)532-6653
E-mail: dchart@ksuvm.ksu.edu

L340

Kootenay River Network

State(s): MT, ID, BC

Contact: Ms. Jill Davies
14 Old Bull River Road
Noxon, MT 59853
Phone: (406)847-2228

L341

Kwethluk Village Fisheries Monitoring Plan

State(s): AK

Contact:
U.S. Fish & Wildlife Service
Bethel, AK

L342

LaBarge Watershed Cooperative Management

State(s): WY

Contact: Mr. George Walker
Natural Resource Assistant
USDA Forest Service
Kemmerer Ranger District
PO Box 3
Kemmerer, WY 83101
Phone: (307)877-4415

P048

Lajas Valley Lagoon System

State(s): PR

Contact: Mr. James Oland
Field Supervisor
U.S. Fish & Wildlife Service
Caribbean Field Office
PO Box 491
Boqueron, PR 00622
Phone: (809)851-7297
Fax: (809)851-7440
E-mail: R4FWE_MAPR@MAIL.FWS.GOV

L343

Lake Champlain Basin Program

State(s): VT

Contact: Ms. Lisa Borre
Vermont Lake Champlain Coordinator
Lake Champlain Basin Program
Gordon-Center House
54 West Shore Road
Grand Isle, VT 05458
Phone: (802)372-3213
Fax: (802)372-6131

L344

Lake Champlain Wetlands

State(s): VT, NY

Contact: Mr. Jon Binhammer
The Nature Conservancy
Vermont Field Office
27 State Street
Montpelier, VT 05602
Phone: (802)229-4425
Fax: (802)229-1347

L345

Lake Michigan Lakewide Management Plan

State(s): IL, IN, MI, WI

Contact: Mr. Gary Kohlhepp
U.S. Environmental Protection Agency
Region V
77 W. Jackson Boulevard
Chicago, IL 60604
Phone: (312)886-4680
Fax: (312)886-7804

L347

Lake Ponchartrain Basin Restoration

State(s): LA

Contact: Mr. Carlton Dufrechou
Lake Ponchartrain Basin Foundation
PO Box 6965
Metairie, LA 70009
Phone: (504)836-2215
Fax: (504)836-7283

L348

Lake Superior Basin Biosphere Proposed Biosphere Reserve
State(s): MI, MN, WI, Canada
Contact: Mr. Robert Brander
Rt. 1, Box 146-2A
Washburn, WI 54891
Phone: (715)373-2988
Fax: (715)373-2938

L349

Lake Superior Binational Program Habitat Projects
State(s): MI, MN, WI, Canada
Contact:
National Wildlife Federation
Ann Arbor, MI

L350

Lake Superior EMAP - Great Lakes Assessment
State(s): MN, MI
Contact: Mr. Stephen Lozano
U.S. Environmental Protection Agency
Region V
Environmental Research Lab
6201 Congdon Blvd.
Deluth, MN 55804
Phone: (218)720-5594
Fax: (218)720-5539

L351

Lake Whatcom
State(s): WA
Contact: Mr. Tom Murphy
Washington Department of Natural Resources
919 North Township
Sedro Wooley, WA 98226
Phone: (360)856-3500
Fax: (360)856-2150

L352

Land Use District Boundary Review
State(s): HI
Contact: Ms. Mary Lou Kobayaski
Hawaii Office of State Planning
PO Box 3540
Honolulu, HI 96811-3540
Phone: (808)587-2808
Fax: (808)587-2824

L353

Landowner Forum
State(s): WA
Contact: Ms. Kaleen Cottingham
Supervisor
Washington Department of Natural Resources
PO Box 47014
Olympia, WA 98504-7014
Phone: (360)902-1360

L354

Landscape Project
State(s): NJ
Contact: Mr. Larry Niles
New Jersey Division of Fish, Game & Wildlife
Endangered & Nongame Species Program
CN 400
Trenton, NJ 08625-0400
Phone: (609)292-9400
Fax: (609)984-1414

L355

Largo Canyon Watershed Management and Erosion Control Plan
State(s): NM
Contact: Mr. Dale Wirth
Bureau of Land Management
Farmington District
1235 La Plata Highway
Farmington, NM 87401
Phone: (505)599-6320
Fax: (505)599-8998

L356

Largo-Aqua Fria Watershed Project
State(s): NM
Contact: Mr. Jim Stovall
Bureau of Land Management
198 Neel Avenue NW
Socorro, NM 87801
Phone: (505)835-0412

L357

Laurels Reserve
State(s): PA
Contact: Mr. Dan Hegarty
Assistant Land Manager
The Brandywine Conservancy
PO Box 141
Chadds Ford, PA 19317
Phone: (610)388-2700
Fax: (610)388-1575

L358

Little Bear River Watershed Project
State(s): UT
Contact: Mr. Roy Gunnell
Utah Department of Environmental Quality
Division of Water Quality
PO Box 144870
Salt Lake City, UT 84114-4870
Phone: (801)538-6146
Fax: (801)538-6016

L359

Little Tennessee River Group
State(s): NC
Contact:
Little Tennessee River Group
Franklin, NC

L360

Lk Superior Highlands/Nemadji River Basin Project
State(s): MN
Contact:
The Nature Conservancy
Minneapolis, MN

L361

Lone Mountain/San Rafael Ecosystem Project
State(s): AZ
Contact:
Lone Mountain Ranch
Burrus, AZ

L362

Long Island Sound
State(s): NY
Contact: Mr. Mark Tedesco
Long Island Sound Office
Stamford Government Center
Stamford, CT 06904
Phone: (203)977-1541
Fax: (203)977-1546

L363

Long Leaf Pine-Eglin Air Force Base
State(s): FL
Contact:
U.S. Air Force
Washington, DC

L364

Long Pond Barrens
State(s): PA
Contact: Dr. James F. Thorne
Director of Science and Stewardship
The Nature Conservancy
1211 Chestnut Street, 12th Floor
Philadelphia, PA 19107-4122
Phone: (215)963-1400
Fax: (215)963-1406
E-mail: jthorne@tnc.org

L365

Loomis State Forest
State(s): WA
Contact: Mr. Mark Mauren
Washington Department of Natural Resources
1111 Washington Street SE
PO Box 47014
Olympia, WA 98504-7014
Phone: (360)902-1747

L366

Louisiana Coastal Wetlands Planning, Protection and Restoration Act
State(s): LA
Contact:
U.S. Army Engineer District-New Orleans
New Orleans, LA

L367

Lower Connecticut River Special Area Management Plan
State(s): CT, VT, NH, MA
Contact: Mr. Ron Rozsa
Connecticut Department of Environmental Protection
Office of Long Island Sound Programs
79 Elm Street
Hartford, CT 06106-5127
Phone: (203)424-3034

L368

Lower Mississippi Alluvial Valley Wetland Conservation Plan
State(s): AR, IL, KY, LA, MO, MS, TN
Contact: Mr. Jay Gamble
U.S. Environmental Protection Agency
Region VI
1445 Ross Ave. (6E-FT)
Dallas, TX 75202-2733
Phone: (214)665-8339
Fax: (214)665-7446

L369

Lower Mississippi Valley Joint Venture
State(s): LA, MS, AR, TN, KY, MO, IL, AL
Contact: Mr. Charles Baxter
Coordinator
U.S. Fish & Wildlife Service
2424 South Frontage Road
Suite C
Vicksburg, MS 39810
Phone: (601)629-6600
Fax: (601)636-9541

L370
Lower Missouri River
State(s): KS, NE, IA, MO
Contact: Mr. J. C. Bryant
    U.S. Fish & Wildlife Service
    Big Muddy National Wildlife & Fish Refuge
    4200 New Haven Road
    Columbia, MO 65201-9634
    Phone: (314)876-1826
    Fax: (314)876-1839

L371
Lower Missouri River - Data Collection
State(s): MO, MN, WI, IA, IL
Contact: Mr. Bill Mauck
    Assistant Director
    Midwest Science Center
    4200 New Haven Road
    Columbia, MO 65201
    Phone: (314)875-5399
    Fax: (314)876-1896

P049
Lower Rio Grande Ecosystem Plan
State(s): TX
Contact: Mr. Art Coykendall
    Wildlife Biologist
    U.S. Fish & Wildlife Service
    320 N. Main St., Rm. 225
    McAllen, TX 78501
    Phone: (210)630-4636
    Fax: (210)630-1653

P050
Lower Roanoke River Bioreserve
State(s): NC
Contact: Ms. A. Este Stifel
    Director, Roanoke River Project
    The Nature Conservancy
    Suite 201
    4011 University Drive
    Durham, NC 27707
    Phone: (919)403-8558
    Fax: (919)403-0379
    E-mail: estifel@tnc.org

P051
Lower St. Johns River Ecosystem
    Management Area
State(s): FL
Contact: Ms. Jan Brewer
    Environmental Specialist
    Florida Department of Environmental
    Protection
    Ste 200B, 7825 Baymeadows Way
    Jacksonville, FL 32256-7577
    Phone: (904)448-4300
    Fax: (904)448-4366
    E-mail: Brewer_J@JAXI.DEP.STATE.FL.US

L372
Lower Tennessee River - Cumberland River
    Ecosystem
State(s): TN, KY, AL
Contact: Mr. John Taylor
    Manager
    U.S. Fish & Wildlife Service
    Tennessee National Wildlife Refuge
    PO Box 849
    Paris, TN 38242
    Phone: (901)642-2091

L373
Lower Wabash Habitat Restoration
State(s): IN
Contact: Mr. Jeff Kiefer
    U.S. Fish & Wildlife Service
    620 S. Walker
    Bloomington, IN 47403
    Phone: (812)334-4261 ext. 212

L374
Loxahatchee River Basin Wetland Planning
    Project
State(s): FL
Contact:
    Environmental Protection Agency
    Region IV
    Wetlands Planning Unit
    345 Courtland Street NE
    Atlanta, GA 30365
    Phone: (404)347-3871 ext. 6511
    Fax: (404)347-1798

L375
Luquillo Experimental Forest
State(s): PR
Contact: Dr. Ariel E. Lugo
    Director
    International Institute of Tropical Forestry
    P.O. Box 25000
    Rio Piedras, PR 00928-5000
    Phone: (809)766-5335
    Fax: (809)766-6263

L376
MacKinaw River Project
State(s): IL
Contact: Mr. James P. McMahon
    Mackinaw River Project Director
    The Nature Conservancy
    416 Main St., Suite 1600
    Peoria, IL 61602-1103
    Phone: (309)673-6689
    Fax: (305)673-8986

L377
Maine Forest Biodiversity Project
State(s): ME
Contact:
    Champion International Corp.
    Bucksport, ME

P052
Malpai Borderlands Initiative
State(s): NM, AZ
Contact: Dr. Ben Brown
    Program Director
    Animas Foundation
    HC 65, Box 179-B
    Animas, NM 88020
    Phone: (505)548-2622
    Fax: (505)548-2267
    E-mail: 6176022@mcimail.com

L378
Mammoth Cave Area Biosphere Reserve
State(s): KY
Contact: Mr. Jeff Bradybaugh
    Chief
    National Park Service
    Mammoth Cave National Park
    Division of Science & Research Management
    Mammoth Cave, KY 42259
    Phone: (502)749-2508
    Fax: (502)749-2916

L379
Mangrove Rehabilitation Program
State(s): FL
Contact:
    U.S. Air Force
    Washington, DC

P053
Marathon County Forests
State(s): WI
Contact: Mr. Mark Heyde
    Marathon County Forest Administrator
    Marathon County Forestry Department
    Courthouse
    500 Forest Street
    Wausau, WI 54403-5568
    Phone: (715)847-5267
    Fax: (715)848-9210

L380
Mark Twain Watershed Project
State(s): MO
Contact: Mr. Donald L. Schuster
    Project Manager
    USDA Natural Resources Conservation
    Service
    Mark Twain Water Quality Project Office
    28898 US Highway 63
    Macon, MO 63552-9587
    Phone: (816)385-6359
    Fax: (816)385-7269

P054
Marys River Riparian/Aquatic Restoration
    Project
State(s): NV
Contact: Mr. Bill Baker
    Bureau of Land Management
    Elko District
    PO Box 831
    3900 East Idaho Street
    Elko, NV 89802
    Phone: (702)753-0200

L381
Massachusetts Bays Program
State(s): MA
Contact: Dr. Diane Gould
    Massachusetts Bays Program
    100 Cambridge Street
    20th Floor
    Boston, MA 02202
    Phone: (617)727-9530 ext. 406
    Fax: (617)727-2754

L382
Maumee River Area of Concern
State(s): OH
Contact: Mr. Mark Messersmith
    U.S. Environmental Protection Agency
    Region V (WQB-16J)
    77 W. Jackson Boulevard
    Chicago, IL 60604-3507
    Phone: (312)353-2154
    Fax: (312)886-7804

L383
Maverick Project
State(s): AZ
Contact:
    USDA Forest Service
    Prescott, AZ

L384

**McGregor Coordinated Resource
Management Plan**

State(s): NM

Contact: Thersa M. Hanley
Archeologist
Bureau of Land Management
Caballo Resource Area
1800 Marquess
Las Cruces, NM 88005
Phone: (505)525-4342
Fax: (505)525-4412
E-mail: thanley@nm0857.lcdo.nm.blm.gov

P055

McPherson Ecosystem Enhancement Project

State(s): ID

Contact: Mr. Bruce Padian
USDA Forest Service
Caribou National Forest
Montpelier Ranger District
250 South 4th Avenue
Pocatello, ID 83201
Phone: (208)236-7500
Fax: (208)236-7503

L385

**Measurement, Analysis, & Modeling of
Forest Ecosystems in a Changing
Environment**

State(s): NH

Contact: Dr. Dale S. Solomon
USDA Forest Service
NE-4155
PO Box 640
Durham, NH 03824
Phone: (603)868-7666
Fax: (603)868-7604

L386

Meramec River

State(s): MO

Contact: Ms. Kathleen Mulder
U.S. Environmental Protection Agency
Region VII
726 Minnesota Avenue
Kansas City, KS 66101
Phone: (913)551-7542

L387

Merrimack River

State(s): NH, MA

Contact: Ms. Carolyn Jenkins
New England Interstate Water Pollution
Control Commission
255 Ballardvale St.
Wilmington, MA 01887
Phone: (508)658-0500
Fax: (508)658-5509

P056

**Mesa Creek Coordinated Resource
Management Plan**

State(s): CO

Contact: Mr. Jim Sazama
Range Conservationist
Bureau of Land Management
Uncompahgre Basin Resource Area
2505 S. Townsend
Montrose, CO 81401
Phone: (303)249-6047

L388

Miami Basin

State(s): CA

Contact: Mr. Tom Efird
District Ranger
USDA Forest Service
Sierra National Forest
Mariposa Ranger District
43060 Highway 41
Oakhurst, CA 93644
Phone: (209)683-4665
Fax: (209)683-7258

L389

Mica Creek Watershed Study

State(s): ID

Contact: Dr. Terry Cundy
Potlatch Corp.
805 Mill Rd.
PO Box 1016
Lewiston, ID 83501-1016
Phone: (208)799-4135
Fax: (208)799-1707

L390

**Mill Creek Canyon Management
Partnership**

State(s): UT

Contact:
USDA Forest Service, UT

L391

Milwaukee Estuary Area of Concern

State(s): WI

Contact: Ms. Marsha Jones
Wisconsin Department of Natural Resources
Southeast District
PO Box 12436
Milwaukee, WI 53212
Phone: (414)263-8708
Fax: (414)263-8483

L392

Minnesota County Biological Survey

State(s): MN

Contact:
Minnesota County Biological Survey
St. Paul, MN

L393

**Minnesota Environmental Indicators
Initiative**

State(s): MN

Contact:
Minnesota Department of Natural Resources
St. Paul, MN

P057

Minnesota Peatlands

State(s): MN

Contact: Mr. Bob Djupstrom
Scientific and Natural Area Supervisor
Minnesota Dept. of Natural Resources
Wildlife - SNA, Box 7
500 Lafayette Road
St. Paul, MN 55155
Phone: (612)297-2357
Fax: (612)297-4961
E-mail: bob.djupstrom@dnr.state.mn.us

L395

**Mississippi River Alluvial Plain Bioreserve
Project**

State(s): AR, LA, MS, TN, KY

Contact: Ms. Cindy Brown
Bioreserve Director
The Nature Conservancy
Louisiana Field Office
PO Box 4125
Baton Rouge, LA 70821
Phone: (504)338-1040
Fax: (504)338-1003

P058

**Missouri Coordinated Resource
Management**

State(s): MO

Contact: Ms. Sara Parker
Policy Analyst
Missouri Department of Conservation
PO Box 180
Jefferson City, MO 65102-0180
Phone: (573)751-4115 ext. 345
Fax: (573)526-4495

L396

Missouri Masterpieces

State(s): MO

Contact: Mr. Paul Nelson
Missouri Department of Natural Resources
Division of State Parks
PO Box 176
Jefferson City, MO 65102
Phone: (314)751-8360
Fax: (314)751-8656

L397

**Missouri Ozark Forest Ecosystem Project
(MOFEP)**

State(s): MO

Contact: Dr. Terry L. Robison
Missouri Department of Conservation
PO Box 180
Jefferson City, MO 65102-0180
Phone: (314)751-4115
Fax: (314)526-6670

L398

Missouri Resource Assessment Partnership

State(s): MO

Contact:
Missouri Department of Conservation
Jefferson City, MO

L399

**Missouri River Division - U.S. Army Corps
of Engineers**

State(s): NE, others

Contact:
U.S. Army Corps of Engineers
Omaha, NE

P059

Missouri River Mitigation Project

State(s): KS, NE, IA, MO

Contact: Mr. Steve Adams
Natural Resources Coordinator
Kansas Wildlife & Parks
900 SW Jackson, Suite 502
Topeka, KS 66612-1233
Phone: (913)296-2281
Fax: (913)296-6953

L400

Missouri River Natural Resource Group

State(s): MT, ND, SD, MO, IA, NE, KA

Contact: Mr. Greg Power
    North Dakota Game & Fish Department
    100 N. Bismarck Expressway
    Bismarck, ND 58501
    Phone: (701)328-6323

L401

**Missouri River Post-Flood Evaluation
    (MRPE)**

State(s): MO

Contact: Mr. John W. Smith
    Wildlife Research Supervisor
    Missouri Department of Conservation
    1110 S. College Ave.
    Columbia, MO 65201
    Phone: (314)882-9880
    Fax: (314)882-4517
    E-mail: smithj5@mail.conservation.state.mo.us

L402

Mobile Bay Restoration Demonstrations

State(s): AL

Contact: Dr. Douglas A. Lipka
    U.S. Environmental Protection Agency
    Gulf of Mexico Program
    Building 1103, Room 202
    Stennis Space Center, MS 30529
    Phone: (601)688-3726
    Fax: (601)688-2709

L403

Mojave Desert Ecosystem Initiative

State(s): CA

Contact: Mr. Steve Ahmann
    Natural and Cultural Resource Manager
    National Training Center
    PO Box 10026
    Attn: AFZJ-PW-EV
    Building 385
    Fort Irwin, CA 92310-5000
    Phone: (619)380-5291: (619)380-4760

P060

Molokai Preserves

State(s): HI

Contact: Mr. Ed Misaki
    Director of Programs
    The Nature Conservancy of Hawaii
    Molokai Preserves
    PO Box 220
    Kualapuu, HI 96757
    Phone: (808)553-5236

L404

**Monroe Mountain Livestock/Big Game
    Demonstration Area**

State(s): UT

Contact: Mr. Larry Greenwood
    Wildlife Biologist
    Bureau of Land Management
    150 East 900 North
    Richfield, UT 84701
    Phone: (801)896-8221
    Fax: (801)584-8268

L405

Montezuma County Federal Lands Program

State(s): CO

Contact:
    Fort Lewis College, CO

L406

Mt. Roan Balds Management

State(s): NC, TN

Contact:
    Project Leader -- District Ranger
    USDA Forest Service
    Pisgah National Forest
    Toecane Ranger District
    Box 128
    Burnsville, NC 28714
    Phone: (704)682-6146

L407

Muddy Creek Landscape Analysis

State(s): CO

Contact: Mr. David Van Norman
    USDA Forest Service
    Holy Cross Ranger District
    Box 190
    Minturn, CO 81645
    Phone: (303)827-5715

L408

Mudge Pond

State(s): CT

Contact:
    USDA Natural Resources Conservation
    Service

L409

Mulligan Creek Project

State(s): MI

Contact: Dr. Dean Premo
    White Water Associates, Inc.
    429 River Lane
    Amasa, MI 49903
    Phone: (906)822-7373
    Fax: (906)822-7977

L410

**Multi-Agency Approach to Planning and
    Evaluation (MAAPE)**

State(s): ND, SD, MT

Contact: Mr. Ron Reynolds
    U.S. Fish & Wildlife Service
    1500 E. Capitol Avenue
    Bismarck, ND 58501

L411

**Multi-Objective Fllod Mitigation Plan
    Vermillion River Basin**

State(s): SD

Contact: Mr. Duane Holmes
    National Park Service
    RMR-PPO
    P.O. Box 25287
    Denver, CO 80225
    Phone: (303)969-2855

L412

Multi-Species Forest Management Program

State(s): WA

Contact: Ms. Catherine L. Phillips
    Director External Affairs
    Weyerhaeuser Co.
    33405 Eighth Avenue South
    Federal Way, WA 98003
    Phone: (206)924-3172
    Fax: (206)924-3421
    E-mail: phillic@wdni.com

L413

Nanjemoy Creek Ecosystem Initiative

State(s): MD

Contact:
    The Nature Conservancy
    Chevy Chase, MD

L414

Nanticoke/Blackwater Rivers Bioreserve

State(s): MD, DE

Contact:
    The Nature Conservancy
    Salisbury, MD

L415

Narragansett Bay Project

State(s): MA, RI

Contact: Mr. Richard Ribb
    Rhode Island DEM
    Narragansett Bay Project
    291 Promenade Street
    Providence, RI 02908

L416

Nassawango Creek Ecosystem Initiative

State(s): MD

Contact:
    The Nature Conservancy
    Salisbury, MD

L417

**National Capital Region Cons. Data Center/
    DC Natural Heritage Prog.**

State(s): MD, DC, VA, WV, PA

Contact: Mr. Christopher Lea
    13025 Riley's Loch Road
    Poolesville, MD 20837

L418

National Hierarchy of Ecological Units

State(s): CO

Contact: Mr. Jeff Bruggink
    Forest Soil Scientist, Coordinator
    Phone: (710)545-8737

L419

Natural Areas Reserve System

State(s): HI

Contact: Mr. Peter T. Schuyler
    Natural Area Reserves Program Manager
    Hawaii Department of Land & Natural
        Resources
    Division of Forestry and Wildlife
    Kawaiahao Plaza, Suite 132
    567 South King Street
    Honolulu, HI 96813
    Phone: (808)587-0054
    Fax: (808)587-0064
    E-mail: pschuyl@pixi.com

L420

**Natural Community Conservation Planning
    (NCCP)**

State(s): CA

Contact:
    California Resources Agency
    1416 9th Street
    Sacramento, CA 95814

P061
Natural Resource Roundtable
State(s): HI
Contact: Mr. Scott A.K. Derrickson
    Hawaii Office of State Planning
    PO Box 3540
    Honolulu, HI 96811-3540
    Phone: (808)587-2805
    E-mail: sderric@pixi.com

L421
Navajo Mountain Natural Area
State(s): AZ, UT
Contact: Mr. Jack Meyer
    Program Manager
    Navajo Natural Heritage Program
    Navajo Fish & Wildlife Department
    PO Box 1480
    Window Rock, AZ 86515
    Phone: (602)871-7059; (602)871-6472
    Fax: (602)871-6177

P062
Nebraska Sandhills Ecosystem
State(s): NE
Contact: Mr. Gene Mack
    Sandhills Coordinator
    U.S. Fish & Wildlife Service
    Kearney Field Office
    PO Box 1686
    Kearney, NE 68848
    Phone: (308)236-5015
    Fax: (308)237-3899

P063
Negrito Project
State(s): NM
Contact: Mr. Don Weaver
    USDA Forest Service
    Gila National Forest
    Reserve Ranger Disitrict
    PO Box 170
    Reserve, NM 87830
    Phone: (505)533-6231

L422
Neponset River Watershed Project
State(s): MA
Contact:
    State of Massachusetts

L423
Neversink River Ecosystem Initiative
State(s): NY
Contact: Mr. Dave Tobias
    Director of Protection Programs
    The Nature Conservancy
    Lower Hudson Office
    41 South Moger Avenue
    Mt. Kisco, NY 10549
    Phone: (914)232-9431
    Fax: (914)232-1543

L424
New England - New York ECOMAP
State(s): ME, NH, VT, MA, CT, RI, NY
Contact: Ms. Marie-Louise Smith
    Research Ecologist
    Northeastern Forest Experimental Station
    PO Box 640
    Durham, NH 03824

L425
New England Resource Protection Project
State(s): NH, CT, RI
Contact: Ms. Rosemary Monahan
    U.S. Environmental Protection Agency
    New England
    J. F. Kennedy Building
    Boston, MA 02203
    Phone: (617)565-3518
    Fax: (617)565-4940

P064
New Hampshire Forest Resources Plan
State(s): NH
Contact: Ms. Susan Francher
    New Hampshire Division of Forests and
    Lands
    PO Box 1856
    Concord, NH 03302-1856
    Phone: (603)271-2214

L426
New Hope Creek Corridor Project
State(s): NC
Contact: Ms. Hildegard Ryals
    New Hope Creek Advisory Committee
    1620 University Drive
    Durham, NC 27707
    Phone: (919)489-5897

P065
New Jersey Pinelands
State(s): NJ
Contact: Mr. Don Kirchhoffer
    Project Manager
    Pinelands Preservation Alliance
    114 Hanover Street
    Pemberton, NJ 08068
    Phone: (609)894-8000
    Fax: (609)894-9455
    E-mail: dkirk100@aol.com

L427
Niagara River Area of Concern
State(s): NY
Contact: Ms. Ellen Heath
    U.S. Environmental Protection Agency
    Region II
    26 Federal Plaza
    New York, NY 10278
    Phone: (212)264-5352
    Fax: (212)264-2914

L428
Nicolet National Forest
State(s): WI
Contact: Ms. Linda Parker
    Ecologist
    USDA Forest Service
    1170 4th Ave.
    Park Falls, WI 54552
    Phone: (715)762-5169

L429
Nisqually River Management Plan
State(s): WA

L430
Niwot Ridge Biosphere Reserve
State(s): CO
Contact: Mr. William Bowman
    University of Colorado
    Mountain Research Station
    Campus Box 450
    Boulder, CO 80309-0450
    Phone: (303)429-8841
    Fax: (303)429-8699

L431
Nooksack River Watershed Initiatives
State(s): WA, BC
Contact:
    USDA Forest Service, WA

L432
**North Dakota Conservation Reserve
    Program**
State(s): ND
Contact: Mr. Randy Kreil
    North Dakota Game & Fish Department
    100 N. Bismarck Expressway
    Bismarck, ND 58501
    Phone: (701)328-6330
    Fax: (701)328-6352

L433
**North Dakota-Montana Paddlefish
    Management Plan**
State(s): ND, MT
Contact: Mr. Greg Power
    North Dakota Game & Fish Department
    100 N. Bismarck Expressway
    Bismarck, ND 58501
    Phone: (701)328-6323

L434
**North Temperate Lakes Long-Term
    Ecological Research Site**
State(s): WI
Contact:
    University of Wisconsin-Madison
    Madison, WI

P066
Northeast Chichagof Island
State(s): AK
Contact: Mr. Phil Mooney
    Habitat Biologist
    Alaska Department of Fish & Game
    Habitat and Restoration Division
    304 Lake Street
    Room 103
    Sitka, AK 99835-7563
    Phone: (907)747-5828
    Fax: (907)747-6239

P067
**Northern Delaware Wetlands Rehabilitation
    Program**
State(s): DE
Contact: Mr. Robert Hossler
    Delaware Division of Fish & Wildlife
    250 Bear/Christiance Road
    Bear, DE 19701-1041
    Phone: (302)323-4492
    Fax: (302)323-5314
    E-mail: rhossler@dnrec.state.de.us

L435
Northern Forest Health Monitoring Program
State(s): PA
Contact: Dr. Andrew Gillespie
USDA Forest Service
5 Radnor Corp Ctr
Suite 200
Radnor, PA 19087-4585
Phone: (610)975-4017
Fax: (610)975-4095
E-mail: gillesp@aol.com

P068
Northern Forest Lands Council
State(s): VT, NY, NH, ME
Contact: Mr. Charles Johnson
Vermont Department of Forests, Parks &
Recreation
103 S. Main St., 8 South
Waterbury, VT 05671-0601
Phone: (802)241-3652
Fax: (802)244-1481
E-mail: cjohnson@fr.anr.state.vt.us

L436
Northern Grey Wolf
State(s): MN, WI, MI
Contact: Mr. Richard Klukas
National Park Service
1709 Jackson St.
Omaha, NE 68102
Phone: (402)221-3603

L437
Northern Lake Huron Bioreserve
State(s): MI
Contact: Mr. Kent Gilges
The Nature Conservancy
Northern Lake Huron Project Office
PO Box 567
Cedarville, MI 49719
Phone: (906)484-9970
Fax: (906)484-9971

P069
**Northern Lower Michigan Ecosystem
Management Project**
State(s): MI
Contact: Mr. Michael T. Mang
Michigan Department of Natural Resources
PO Box 667
Gaylord, MI 49735
Phone: (517)732-3541 ext. 5042
Fax: (517)732-0794

L438
**Northern Stations Global Change Research
Program**
State(s): PA
Contact: Dr. Richard Birdsey
USDA Forest Service
5 Radnor Corp Ctr
Suite 200
Radnor, PA 19087-4585
Phone: (610)975-4092

L439
Northwest Colorado Riparian Task Force
State(s): CO
Contact: Mr. Dave Turcotte
Soil Scientist
Bureau of Land Management
Little Snake Resource Area office
1280 Industrial Ave.
Craig, CO 81625
Phone: (303)824-4441

L440
**Northwest Forest Ecosystem Plan, Research
Support**
State(s): WA, OR, CA
Contact: Mr. John Henshaw
USDA Forest Service
Pacific Northwest Research Station
PO Box 3890
Portland, OR 97208
Phone: (503)326-2081
Fax: (503)326-2455

L441
Northwest Indiana Environmental Initiative
State(s): IN
Contact:
Indiana Department of Environmental
Management
Gary, IN

L442
Nu'upia Ponds
State(s): HI
Contact:
U.S. Marine Corps Base Hawai'i
Kaneohe Bay, HI

L443
Nulhegan Deer Wintering Area Agreement
State(s): VT
Contact: Mr. Howard R. DeLano Jr.
Champion International Corp.
PO Box 70
West Stuartstown, NH 03597
Phone: (603)246-3331
Fax: (603)246-8885

L444
Oak-Savanna Ecosystem Project
State(s): IL, IN, MN, MI, OH, WI, IA, MO
Contact:
Great Lakes National Program Office
Chicago, IL

P070
Ohio River Valley Ecosystem
State(s): IL, IN, OH, PA, NY, WV, KY, TN, VA,
MD
Contact: Ms. Kari Duncan
Team Leader, Ohio River Ecosystem
U.S. Fish & Wildlife Service
White Sulphur Springs National Fish
Hatchery
400 East Main Street
White Sulphur Springs, WV 24986
Phone: (304)536-1361
Fax: (304)536-4634
E-mail:
RSFFA_WSSNFH_at_SHA~MAIN1@
mail.fws.gov

L445
Oklahoma Biodiversity Initiative
State(s): OK
Contact: Mr. Norman Murray
Biodiversity Coordinator
Oklahoma Department of Wildlife
Conservation
PO Box 53465
Oklahoma City, OK 73152
Phone: (405)521-4601

L446
Oklahoma Natural Areas Registry Program
State(s): OK
Contact: Ms. Melissa Nagel
Protection Specialist
The Nature Conservancy
Oklahoma City Office
1300 North Broadway Drive
Oklahoma City, OK 73103
Phone: (405)236-1044
Fax: (405)236-1045
E-mail: melnagel@aol.com

P071
Oklahoma Tallgrass Prairie Preserve
State(s): OK
Contact: Mr. Harvey Payne
Director, Tallgrass Prairie Preserve
The Nature Conservancy
PO Box 458
Pawhuska, OK 74056
Phone: (918)287-4803; (918)287-1290
Fax: (918)287-1296

L447
Omaha Stretch of the Missouri River
State(s): NE
Contact: Mr. Kerry B. Herndon
Great Plains Program Office
EPA Region VII
726 Minnesota Avenue
Kansas City, KS 66101
Phone: (913)551-7286
Fax: (913)551-7956
E-mail: herndon.kerry@epamail.epa.gov

L448
Onondaga Lake
State(s): NY
Contact: Mr. Christopher E. Dere
U.S. Environmental Protection Agency
Region II
26 Federal Plaza
New York, NY 10278
Phone: (212)264-5353
Fax: (212)264-2194

L449
Oregon Biodiversity Project
State(s): OR

L450
Oregon High Desert Bioreserve
State(s): OR, NV
Contact: Mr. Reid Schuller
Conservation Director
The Nature Conservancy
Bend Field Office
PO Box 1504
Bend, OR 97709
Phone: (505)317-1901

L451
Oswego River Harbor Area of Concern
State(s): NY
Contact: Ms. Alice Yeh
U.S. Environmental Protection Agency
Region II
26 Federal Plaza
New York, NY 10278
Phone: (212)264-1865
Fax: (212)264-2194

L452
Ottawa River Watershed Study
State(s): OH
Contact: Ms. Vicki Morrical
USDA Natural Resources Conservation
Service
200 N. High St. Rm. 522
Columbus, OH 43215

L453
Otter Creek Watershed Restoration Project
State(s): UT
Contact: Mr. Roy Gunnell
Utah Department of Environmental Quality
Division of Water Quality
PO Box 144870
Salt Lake City, UT 84114-4870
Phone: (801)538-6146
Fax: (801)538-6016

P072
Ouachita National Forest
State(s): AR, OK
Contact: Mr. Bill Pell
Ecosystem Management Coordinator
USDA Forest Service
Ouachita National Forest
PO Box 1270
Hot Springs, AR 71902
Phone: (501)321-5202
Fax: (501)321-5334
E-mail: S=W.PELL/OU1=R08F09A@MHS-
FSWA.ATTMAIL.COM

P073
Owl Mountain Partnership
State(s): CO
Contact: Mr. Jerry Jack
Project Manager
Bureau of Land Management
Kremmling Resource Area
1116 Park Avenue
PO Box 68
Kremmling, CO 80459
Phone: (970)724-3437
Fax: (970)724-9590

L454
PACFISH
State(s): OR, WA, CA
Contact:
Bureau of Land Management

L455
Pacific Air Force Command
State(s): HI
Contact:
U.S. Air Force
Washington, DC

L456
Pacific Northwest Watershed Project
State(s): OR, WA, ID
Contact: Ms. Domoni Glass
Manager, PNW Watershed Project
Boise Cascade Corporation
PO Box 50
Boise, ID 83728
Phone: (208)384-6670
Fax: (208)384-7699

L457
Panther-Cox Creek Watershed Management Plan
State(s): IL
Contact: Mr. James Reynolds
Landscape Architect
Illinois Department of Natural Resources
Divison of Planning
524 South Second St., Room 310
Springfield, IL 62704-1787
Phone: (217)782-3715
Fax: (217)782-9599

P074
Partners for Prairie Wildlife
State(s): MO
Contact: Mr. William D. McGuire
Private Land Coordinator
Missouri Department of Conservation
Wildlife Division
PO Box 180
Jefferson City, MO 65102-0180
Phone: (314)751-4115 ext. 148
Fax: (314)526-4663

L458
Partners for Wildlife
State(s): IN
Contact: Mr. James J. Ruwaldt
U.S. Fish & Wildlife Service
620 S. Walker
Bloomington, IN 47403
Phone: (812)334-4261 ext. 213

P075
Patrick Marsh Wetland Mitigation Bank Site
State(s): WI
Contact: Mr. Alan Crossley
Wildlife Biologist
Wisconsin Department of Natural Resources
3911 Fish Hatchery Road
Fitchburg, WI 53711
Phone: (608)275-3242
Fax: (608)275-3338

L459
Peconic Bay
State(s): NY
Contact: Mr. Vito Minei
Office of Ecology
Suffolk County
Department of County Center
Riverhead, NY 11401-3397
Phone: (516-852-2077
Fax: (516)852-2092

L460
Peconic Bioreserve
State(s): NY
Contact: Mr. Stuart Lowrie
Director of the Peconic Bioreserve
The Nature Conservancy
South Fork/Shelter Island Chapter
PO Box 5125
East Hampton, NY 11937
Phone: (516)329-7689
Fax: (516)329-0215
E-mail: SLOWRIE@aol.com

L461
Penn's Woods Strategic Plan
State(s): PA
Contact: Mr. Daniel Devlin Chief
Resources Planning
Pennsylvania Department of Conservation &
Natural Resources
Bureau of Forestry
PO Box 8552
Harrisburg, PA 17105-8552
Phone: (717)787-3444
Fax: (717)783-5109
E-mail: DEVLIN.DAN@a1.PADEA.GOV

L462
Pensacola Bay Watershed Ecological Evaluation
State(s): FL
Contact: Mr. Michael A. Lewis
U.S. Environmental Protection Agency
Environmental Research Laboratory
1 Sabine Island Drive
Gulf Breeze, FL 32561
Phone: (904)934-9382
Fax: (904)934-2403

L463
Personal Use Firewood
State(s): WA
Contact: Mr. James M. Pena
USDA Forest Service
Naches Ranger District
10061 Highway 12
Naches, WA 98937
Phone: (509)653-2205
Fax: (509)653-2638

P076
Phalen Chain of Lakes Watershed Project
State(s): MN
Contact: Ms. Sherri A. Buss
Phalen Watershed Project Coordinator
Ramsey-Washington Metro Watershed
District
1902 East County Road 13
Maplewood, MN 55109
Phone: (612)777-3665

L464
Pine Creek Water Quality Project
State(s): IA
Contact: Ms. Jennifer Welch
USDA Natural Resources Conservation
Service
1321 Edgington Avenue
Eldora, IA 50627
Phone: (515)858-5692
Fax: (515)858-3335

L465
Pine Flats Ecosystem Management Project
State(s): MN
Contact: Mr. Doug Haertzen
USDA Forest Service
Chippewa National Forest
Cass Lake Ranger District
Route 3, Box 219
Cass Lake, MN 56633
Phone: (218)335-2283
Fax: (218)335-6579

L466

Pines Project

State(s): CO

Contact:
Americorps
Mancos, CO

L467

Pineywoods Conservation Initiative

State(s): LA, TX

Contact: Mr. Ike McWhorter
Director
The Nature Conservancy
Sandylands Preserve
PO Box 909
Silsbee, TX 77656-0909
Phone: (409)385-0445
Fax: (409)385-4745

L468

Pinos Ecosystem Analysis

State(s): CO

Contact: Mr. Steve Hartvigsen
Facilitator
USDA Forest Service
Rio Grande National Forest
Del Norte Ranger District
Box 40
Del Norte, CO 81132
Phone: (719)657-3321

P077

Piute/El Dorado Desert Wildlife Management Area

State(s): NV

Contact: Mr. Jim Moore
Field Representative
The Nature Conservancy
1771 East Flamingo Rd, Ste 111B
Las Vegas, NV 89119
Phone: (702)737-8744
Fax: (702)737-5787

P078

Plainfield Project

State(s): MA

Contact: Ms. Susan Campbell
Stewardship Coordinator
Massachusetts Department of Environmental Management
463 West St.
Amherst, MA 01002
Phone: (413)256-1201

L469

Platte River

State(s): NE

Contact: Mr. David Bowman
Platte River Coordinator
U.S. Fish & Wildlife Service
PO Box 25486
Denver Federal Center
Denver, CO 80225
Phone: (303)236-8186

L470

Playa Lakes Joint Venture

State(s): TX, OK, KS, CO, NM

Contact: Mr. Chuck Mullins
U.S. Fish & Wildlife Service
2105 Osuna Road, NE
Albuquerque, NM 87113
Phone: (505)761-4525

L471

Pocket-Baker Ecosystem Analysis

State(s): AZ

Contact: Mr. John Gerritsma
USDA Forest Service
Coconino National Forest
Long Valley Ranger District
Long Valley Road, HC31 Box 68
Happy Jack, AZ 86024
Phone: (520)354-2216

L472

Pocono Habitat Demonstration Project

State(s): PA

Contact: Ms. Susan Dowell
U.S. Environmental Protection Agency
Region III (3ES43)
841 Chestnut Building
Philadelphia, PA 19107
Phone: (215)597-0355
Fax: (215)597-7906

L473

Pocono Mountains

State(s): PA

Contact: Mr. Ralph Cook
Vice President, Director
The Nature Conservancy
Poconos Mountains Office
PO Box 55
Long Pond Road
Long Pond, PA 18334
Phone: (717)643-7922
Fax: (717)643-7925

L474

Pocotaligo River and Swamp Restoration

State(s): SC

Contact: Mr. Robert H. Chappell
Study Manager, Engineering and Planning Division
U.S. Army Corps of Engineers
Charleston District
PO Box 919
Charleston, SC 29402-0919
Phone: (803)727-4594
Fax: (803)727-4260

L475

Poultney River Conservation Program

State(s): NY, VT

Contact: Mr. Michael S. Batcher
Director of Science and Stewardship
The Nature Conservancy
Eastern New York Chapter
251 River Street
Troy, NY 12180
Phone: (518)272-0195
Fax: (518)272-0298

L476

Powderhorn Wilderness Management Plan

State(s): CO

Contact: Mr. Bill Bottomly
Wilderness Plan Team Leader
Gunnison Basin Resource Area
2505 S. Townsend
Montrose, CO 81401
Phone: (303)249-6047;FTS(700)322-7327

P079

Prairie Pothole Joint Venture

State(s): ND, SD, MN, IA, MT

Contact: Mr. Mike McEnroe
Ecosystem Team Leader
U.S. Fish & Wildlife Service
Region 6
1500 E. Capitol Avenue
Bismarck, ND 58501
Phone: (701)250-4418
Fax: (701)250-4412

L477

Prairie State Park Research Program

State(s): MO

Contact:
Missouri Department of Conservation
Jefferson City, MO

L478

Prairie, Wetland, and Missouri River Mainstem Ecosystem

State(s): ND, SD

Contact: Mr. Mike McEnroe
Ecosystem Team Leader
U.S. Fish & Wildlife Service
1500 E. Capitol Avenue
Bismarck, ND 58501
Phone: (701)250-4418
Fax: (701)250-4412

P080

Prince William Sound - Copper River Ecosystem Initiative

State(s): AK

Contact: Ms. Lisa Thomas
Fish and Wildlife Biologist
National Biological Service
Alaska Science Center
1011 East Tudor
Anchorage, AK 99503
Phone: (907)786-3685
Fax: (907)786-3636

L479

Proposed Coquille Forest of Coquille Indian Tribe

State(s): OR

Contact: Mr. Ed Metcalf
Chair
Coquille Indian Tribe
P.O. Box 1435
Coos Bay, OR 97420
Phone: (800)622-5869

L480

Protection of Forest Health and Productivity Research

State(s): WA

Contact: Mr. Richard Everett
Science Team Leader
USDA Forest Service
Wenatchee Forest Service Laboratory
1133 N. Western Avenue
Wenatchee, WA 98801
Phone: (509)664-2742

P081
Pu'u Kukui Watershed Management Area
State(s): HI
Contact: Mr. Randal T. Bartlett
  Watershed Supervisor
  Maui Pineapple Company, Ltd.
  4900 Honoapi'ilani Highway
  Lahaina, HI 96761
  Phone: (808)669-5439
  Fax: (808)669-7089

L481
Puerto Rico Forest Stewardship Program
State(s): PR
Contact: Mr. Diego Jimenez
  State Forester
  Puerto Rico Department of Natural &
    Environmental Resources
  Forest Service Bureau
  P.O. Box 5887
  San Juan, PR 00906
  Phone: (809)724-3647; (809)724-3584
  Fax: (809)721-5984

L482
Puget Sound Estuary
State(s): WA
Contact: Mr. John Armstrong
  U.S. Environmental Protection Agency
  Region X, MS WD-139
  1200 6th Avenue
  Seattle, WA 98101
  Phone: (206)553-1368
  Fax: (206)553-0165

L483
Pyramid Lake/Stillwater Marsh Project
State(s): NV
Contact: Mr. Graham Chisholm
  The Nature Conservancy
  443 Marsh Avenue
  Reno, NV 89509
  Phone: (702)322-4990
  Fax: (702)322-5132

L484
Pysht River Cooperative
State(s): WA
Contact: Mr. Norm Schaaf
  General Manager
  Merrill & Ring Company
  PO Box 1058
  Port Angeles, WA 98362
  Phone: (360)452-2367
  Fax: (360)452-2015

L485
Quantitative Methods for Modeling Forest
  Ecosystems
State(s): OH
Contact: Dr. Charles Scott
  USDA Forest Service
  359 Main Road
  Delaware, OH 43015
  Phone: (614)368-0101
  Fax: (614)368-0152
  E-mail: scott@trees.neusfs4153.gov

L486
Quinn River Riparian Improvement and
  Demonstration Project
State(s): NV
Contact: Mr. Steve Williams
  Supervisory Range Conservationist
  USDA Forest Service
  Humboldt National Forest
  Santa Rosa Ranger District
  1200 East Winnemucca Boulevard
  Winnemucca, NV 89445
  Phone: (702)623-5025

L487
Railroad Valley Wetlands Enhancement
State(s): NV
Contact: Mr. Mark Biddlecomb
  Tonopah Resource Area Wildlife Biologist
  Bureau of Land Management
  Battle Mountain District
  Tonopah Resource Area
  PO Box 911
  Tonopah, NV 89049
  Phone: (702)482-7800
  Fax: (702)482-7810

P082
Rainwater Basin Joint Venture
State(s): NE
Contact: Mr. Steve Moran
  Coordinator
  Rainwater Basin Joint Venture
  1233 North Webb Rd., Suite 100
  Grand Island, NE 68803
  Phone: (308)385-6465
  Fax: (308)385-6469

L488
Rattlesnake Island Marsh Project
State(s): TX
Contact: Dr. M. Todd Meredino
  Texas Parks and Wildlife Department
  Matagorda County Courthouse, Rm. 101
  Bay City, TX 77414
  Phone: (409)244-7634
  Fax: (409)244-7628

L489
Red River Watershed
State(s): ND, MN
Contact: Mr. Paul Willman
  Red River RC&D Council
  1104 Hill Avenue
  Grafton, ND 58237
  Phone: (7010352-0127
  Fax: (701)352-3015

L490
Red Wolf Recovery Program
State(s): NC
Contact:
  U.S. Air Force
  Washington, DC

L491
Red-cockaded Woodpecker
State(s): GA
Contact:
  Georgia Pacific Corporation

L492
Redding Resource Management Plan
State(s): CA
Contact: Mr. Francis Berg
  Supervisor Resource Mgmt Specialist
  Bureau of Land Management
  Redding Resource Area
  355 Hemsted Drive
  Redding, CA 96002
  Phone: (916)224-2100
  Fax: (916)224-2172

L493
Resource Characterization Study
State(s): PA
Contact: Mr. John Arway
  Pennsylvania Fish and Boat Commission
  450 Robinson Lane
  Bellefonte, PA 16823
  Phone: (814)359-5140

L494
Restoration of Ohio Oak Forests with
  Prescribed Fire
State(s): OH
Contact: Dr. Elaine Kennedy Sutherland
  Research Ecologist
  USDA Forest Service
  Northeastern Forest Experiment Station
  359 Main Road
  Delaware, OH 43015
  Phone: (614)368-0090
  Fax: (614)368-0152
  E-mail: esutherland@asrr.arsusda.gov

L495
Revision of the Forest Plan for the Targhee
  National Forest
State(s): ID, WY
Contact: Mr. Dale Pekar
  Team Leader
  USDA Forest Service
  Targhee National Forest
  PO Box 208
  St. Anthony, ID 83445-0208
  Phone: (208)624-3151
  Fax: (208)624-7635

L496
Richland Creek Corridor
State(s): NC
Contact: Ms. Kate Dixon
  Executive Director
  Triangle Land Conservancy
  1100A Wake Forest Road
  Raleigh, NC 27604
  Phone: (919)833-3662
  Fax: (919)755-9356

L497
Rio Grande Basin Landscape-Scale
  Assessment
State(s): TX, NM, AZ
Contact: Mr. K. Bruce Jones
  U.S. Environmental Protection Agency
  EMSL-LV/MS
  PO Box 93478
  Las Vegas, NV 89193-3478
  Phone: (702)798-2671
  Fax: (702)798-2208
  E-mail: msdkbj@vegas1.las.epa.gov

L498

Rio Puerco Watershed Stabilization
   Initiative
State(s): NM
Contact: Mr. Hector A. Villalobos
   Area Manager, Rio Puerco Resource Area
   Bureau of Land Management
   Albuquerque District
   435 Montano NE
   Albuquerque, NM 87107
   Phone: (505)761-8797
   Fax: (505)761-8911

L499

Riparian Ecosystem Assessment and
   Management (REAM)
State(s): MO
Contact: Mr. Eric W. Kurzejeski
   Wildlife Research Biologist
   Missouri Department of Conservation
   1110 S. College Ave.
   Columbia, MO 65201
   Phone: (314)882-9880
   Fax: (314)882-4517
   E-mail: kurzee@mail.conservation.state.mo.us

L500

Riparian Recovery Plan Initiative
State(s): AZ, NM
Contact:
   U.S. Fish & Wildlife Service
   Albuquerque, NM

P083

Robbie Run Study Area
State(s): PA
Contact: Chris Nowak or Dave deCalesta
   USDA Forest Service
   Forestry Sciences Laboratory
   PO Box 928
   Warren, PA 16365
   Phone: (814)563-1040
   Fax: (814)563-1048

L501

Rochester Embayment Area of Concern
State(s): NY
Contact: Ms. Alice Yeh
   U.S. Environmental Protection Agency
   Region II
   26 Federal Plaza
   New York, NY 10278
   Phone: (212)264-1865
   Fax: (212)264-2194

P084

Ruby Canyon and Black Ridge Ecosystem
   Management Plan
State(s): CO
Contact: Mr. Harley Metz
   Grand Junction Resource Area
   2815 H Road
   Grand Junction, CO 81506
   Phone: (970)244-3076
   Fax: (970)244-3083

L502

Sage Grouse Habitat Improvement Initiative
State(s): CO
Contact: Mr. Joe Capodice
   Wildlife Biologist
   Bureau of Land Management
   Gunnison Basin Resource Area
   216 N. Colorado
   Gunnison, CO 81230
   Phone: (303)641-0471;FTS(700)859-4450

L503

Saginaw Bay Area of Concern
State(s): MI
Contact: Ms. Nancy Phillips
   U.S. Environmental Protection Agency
   Region V (WQW-16J)
   77 W. Jackson Boulevard
   Chicago, IL 60604
   Phone: (312)886-9376
   Fax: (312)886-7804

L504

Salmon Habitat and River Enhancement
   (SHARE)
State(s): ME
Contact:
   Champion International Corp.
   Bucksport, ME

L620

Salt Valley Lakes Project--Wildwood Lake
*State(s):*NE
Contact: Mr. Paul Zillig
   Lower Platte NRD
   P.O. Box 83581
   Lincoln, NE 68501
   Phone: (402)476-2729
   Fax: (402)476-6454

L505

San Francisco Bay Plan
State(s): CA
Contact:
   San Francisco Bay Conservation &
      Development Commission
   San Francisco, CA

L506

San Francisco Bay/Sacramento-San Joaquin
   Delta Estuary
State(s): CA
Contact: Mr. Patrick Wright
   Chief
   U.S. Environmental Protection Agency
   Region IX
   Bay/Delta Section (W-2-4)
   75 Hawthorne Street
   San Francisco, CA 94105-3901
   Phone: (415)744-1989
   Fax: (415)744-1078

L507

San Joaquin River Management Program
State(s): CA
Contact: Mr. Dale Hoffman-Floerke
   California Department of Water Resources
   3251 "S" Street
   Sacramento, CA 95816
   Phone: (916)227-7530
   Fax: (916)227-7554

L508

San Joaquin Valley Regional Ecosystem
   Protection Planning Group
State(s): CA
Contact:
   California Department of Water Resources
   Fresno, CA

L509

San Joaquin Valley: Strategy for Balancing
   Biodiversity and Economy
State(s): CA
Contact: Mr. James Abbott
   Area Manager
   Bureau of Land Management
   Bakersfield District
   Caliente Resource Area
   3801 Pegasus
   Bakersfield, CA 93308
   Phone: (805)391-6000
   Fax: (805)391-6040

L510

San Juan Basin Unlined Pit Closure and
   Remediation
State(s): NM
Contact: Mr. Bill Liess
   Bureau of Land Management
   Farmington District
   1235 La Plata Highway
   Farmington, NM 87401
   Phone: (505)599-6321
   Fax: (505)599-9889

L511

San Juan Bay Estuary Program
State(s): PR
Contact: Ms. Teresa Rodriguez
   Program Director
   U.S. Environmental Protection Agency
   1492 Ave Ponce De Leon, Apt. 417
   San Juan, PR 00907-4127
   Phone: (809)729-6931
   Fax: (809)729-7747

L512

San Luis Rey River Corridor Management
   Plan
State(s): CA
Contact: Ms. Stepanie L. Wilson
   U.S. Environmental Protection Agency
   Region IX (W-3-2)
   75 Hawthorne Street
   San Francisco, CA 94105-3901
   Phone: (415)744-1968
   Fax: (415)744-1078

P085

San Luis Valley Comprehensive Ecosystem
   Management Plan
State(s): CO
Contact:
   Refuge Manager
   U.S. Fish & Wildlife Service
   Alamosa/Monte Vista National Wildlife
      Refuge
   9383 El Rancho Lane
   Alamosa, CO 81101
   Phone: (719)589-4021
   Fax: (719)589-9184

L513

San Miguel River Multi-Objective Plan
State(s): CO
Contact: Ms. Karen Tucker
   Recreation Planner
   Bureau of Land Management
   UBRA
   2505 S. Townsend
   Montrose, CO 81401
   Phone: (970)249-6047
   Fax: (970)249-8484

P086
San Pedro River
State(s): AZ
Contact: Mr. Paul Hardy
    Program Manager
    The Nature Conservancy
    27 Ramsey Canyon Road
    Hereford, AZ 85615
    Phone: (602)378-2785

L514
San Rafael Valley Association Planning Efforts
State(s): AZ
Contact:
    VACA Ranch
    Cooper, AZ

L515
San Simon River Ecosystem Project
State(s): AZ, NM
Contact:
    Bureau of Land Management
    Safford District
    San Simon Resource Area

L516
Sand Pine-Scrub Oak
State(s): FL
Contact:
    USDA National Forest
    Silver Springs, FL

P087
Santa Catalina Island Ecological Restoration Program
State(s): CA
Contact: Mr. Allan Fone
    Santa Catalina Island Conservancy
    PO Box 2739
    Avalon, CA 90704
    Phone: (310)510-1299

P088
Santa Margarita River
State(s): CA
Contact: Mr. Cameron Barrows
    Southern California Area Director
    The Nature Conservancy
    PO Box 188
    Thousand Palms, CA 92276
    Phone: (619)343-1234
    Fax: (619)343-0393

L517
Santa Monica Bay National Estuary Program
State(s): CA
Contact: Ms. Cheryl McGovern
    U.S. Environmental Protection Agency
    Region IX
    75 Hawthorne Street
    San Francisco, CA 94105-3901
    Phone: (415)744-2013
    Fax: (415)744-1078

L518
Sarasota Bay National Estuary Program
State(s): FL
Contact: Mr. Hudson Slay
    U.S. Environmental Protection Agency
    Region IV
    Coastal Programs
    345 Courtland St., NE
    Atlanta, GA 30365
    Phone: (404)347-3555 ext. 2059
    Fax: (404)347-1797

L519
Savannah River Basin
State(s): NC, SC, GA
Contact: Ms. Meredith Anderson
    U.S. Environmental Protection Agency
    Region IV
    345 Courtland Street, NE
    Atlanta, GA 30365
    Phone: (404)347-2126 ext. 6581
    Fax: (404)347-3269

L520
Savannah River Basin Watershed Project
State(s): GA, SC
Contact: Ms. Meredith Anderson
    U.S. Environmental Protection Agency
    Region IV
    345 Courtland Street, N.E.
    Atlanta, GA 30365
    Phone: (404)347-3555 ext. 6581
    Fax: (404)347-3269

L521
Sevilleta National Wildlife Refuge Long-Term Ecological Research Site
State(s): NM
Contact: Dr. Bruce Milne
    Principal Investigator
    University of New Mexico
    Department of Biology
    Albuquerque, NM 87131
    Phone: (505)277-5356
    Fax: (505)277-0304
    E-mail: bmilne@sevilleta.unm.edu

L522
Shawangunk Ridge Biodiversity Partnership
State(s): NY
Contact: Mr. Michael S. Batcher
    Director of Science and Stewardship
    The Nature Conservancy
    Eastern New York Chapter
    251 River Street
    Troy, NY 12180
    Phone: (518)272-0195
    Fax: (518)272-0298

L523
Shoreline Management Initiative
State(s): TN
Contact: Ms. Tere McDonough
    Tennessee Valley Authority
    Ridgeway Road
    Norris, TN 37828
    Phone: (615)632-1542

P089
Sideling Hill Creek Bioreserve
State(s): MD, PA
Contact: Mr. Rodney Bartgis
    Manager
    The Nature Conservancy
    2995 Grade Road
    Martinsburg, WV 25401
    Phone: (304)754-6709

L524
Sierra Nevada Ecosystem Project
State(s): CA
Contact:
    California Department of Forestry
    Sacramento, CA

L525
Silver Creek
State(s): ID
Contact: Mr. Paul Todd
    The Nature Conservancy
    PO Box 165
    Sun Valley, ID 83353
    Phone: (208)788-0934

L526
Silverspot Butterfly Recovery Efforts
State(s): OR, WA, CA
Contact: Mr. Michael Clady
    Forest Coordinator for Silverspot Butterfly
    USDA Forest Service
    Siuslaw National Forest
    PO Box 1148
    Corvallis, OR 97339
    Phone: (503)750-7053
    Fax: (503)750-7234

L527
Silvio Conte Refuge Environmental Impact Statement
State(s): MA and others
Contact:
    U.S. Fish & Wildlife Service
    Turner Falls, MA

L528
Siouxon
State(s): WA
Contact:
    Washington Department of Natural Resources
    Olympia, WA

L529
Smoke Hole/North Fork Mountain Project
State(s): WV
Contact: Mr. Rodney Bartgis
    The Nature Conservancy
    Mid-Appalachians Field Office
    2995 Grade Road
    Martinsburg, WV 25401
    Phone: (304)754-6709

P090
Snake River Corridor Project
State(s): WY
Contact: Mr. Tim Young
    Project Facilitator
    Snake River Corridor Project
    Teton County
    PO Box 1727
    Jackson, WY 83001
    Phone: (307)733-8225
    Fax: (307)733-8034
    E-mail: tyoung@wyoming.com

L530
Soleduck Watershed Analysis
State(s): WA
Contact: Mr. John Meyer
    National Park Service
    Olympic National Park
    600 East Park avenue
    Port Angeles, WA 98362-6798
    Phone: (360)452-4501
    Fax: (360)452-0335

L531

Southeast Michigan Initiative
State(s): MI
Contact: Ms. Mardi Klevs
U.S. Environmental Protection Agency
Region V (WCC-15J)
77 W. Jackson Boulevard
Chicago, IL 60604
Phone: (312)353-5490
Fax: (312)886-0168

L532

Southern Appalachian Man and the Biosphere Program (SAMAB)
State(s): TN, NC, SC, GA, AL, WV
Contact: Mr. Hubert Hinote
Executive Director
SAMAB
Great Smoky Mountain National Park
1314 Cherokee Orchard Rd
Gatlinburg, TN 37738
Phone: (615)436-1701
Fax: (615)436-5598

L534

Southern Berkshires Bioreserve
State(s): MA, CT
Contact: Mr. Frank Lowenstein
Program Manager
The Nature Conservancy
South Berkshires Office
PO Box 268
Shefield, MA 01257
Phone: (413)229-0132
Fax: (413)229-0234
E-mail: FLOW@SIMONS.ROCK.EDU

L535

Southern California Ecoregion - U.S. Fish and Wildlife Service
State(s): CA
Contact: Mr. Marc Weitzel
Project Leader
U.S. Fish & Wildlife Service
Box 5839
Ventura, CA 93005
Phone: (805)644-5185

P091

South Florida/Everglades Ecosystem Restoration Initiative
State(s): FL
Contact: Col. Terrence Salt
Director, South Florida Ecosystem Restoration
Task Force
Florida International University
OE Building, Room 148
Miami, FL 33199
Phone: (305)348-4095
Fax: (305)348-4096

L536

Southern Forested Wetlands
State(s): MS, AL, AR, LA, SC, GA
Contact: Mr. Sammy King
National Biological Service
Southern Science Center
700 Cajundome Blvd.
Lafayette, LA 70506
Phone: (318)266-8619
Fax: (318)266-8592

L537

Southern Lake Michigan Initiative
State(s): MI
Contact:
The Nature Conservancy
Michigan City, IN

L538

Southern Phosphate District
State(s): FL
Contact: Mr. Tim King
Florida Game and Fresh Water Fish
Commission
Division of Fisheries
3928 Tenoroc Road
Lakeland, FL 33805
Phone: (941)499-2421

L539

Southlands Experimental Forest
State(s): GA
Contact: Mr. Craig Hedman Ph.D.
Forest Ecology Section Leader
International Paper
719 Southlands Road
Bainbridge, GA 31717
Phone: (912)246-3642 ext. 270
Fax: (912)243-0766

L540

Southwest Colorado Interagency LANDSAT Vegetation Classification Project
State(s): CO
Contact: Mr. James Ferguson
Wildlife Biologist
Bureau of Land Management
Uncompahgre Basin Resource Area
2505 S. Townsend Ave.
Montrose, CO 81401
Phone: (970)249-6047

L541

Southwest Wyoming Resource Evaluation
State(s): WY
Contact: Mr. Roger Wickstrom
Bureau of Land Management
Wyoming State Office
5353 Yellowstone Road
PO Box 1828
Cheyenne, WY 82003
Phone: (307)775-6011

L542

Special Ecological Stewardship
State(s): MO
Contact:
Missouri Department of Conservation
Jefferson City, MO

L543

Spring Creek Corridor Study
State(s): PA
Contact: Dr. Robert Carline
Pennsylvania State University
Pennsylvania Cooperative Fish & Wildlife
Research Unit
Merkle Laboratory
University Park, PA 16802
Phone: (814)865-4511
Fax: (814)863-4710

L544

Spring Straight and Cedar Creek Ecosystem Based Planning Projects
State(s): KS
Contact: Mr. Kenneth W. Hoffman
USDA Natural Resources Conservation
Service
Federal Building Rm. 190
Topeka, KS 66683-3569
Phone: (913)295-7630

L545

Spruce Creek and/or Logging Gulch - NEPA documents
State(s): ID
Contact: Ms. Lyn Morelan
Ecosystem Implementation Coordinator
1750 Front Street
Boise, ID 83702
Phone: (208)364-4170

L546

Squirrel River Integrated Activity Plan
State(s): AK
Contact: Ms. Susan M. Will
Bureau of Land Management
Kobuk District
1150 University Avenue
Fairbanks, AK 99709-3899
Phone: (800)437-7021
Fax: (907)474-2281

L547

St. Croix International Waterway Commission
State(s): ME
Contact: Ms. Lee Sochasky
St. Croix International Waterway Commission
PO Box 610
Calais, ME 04619
Phone: (506)466-7550
Fax: (506)466-7551

L548

St. Lawrence River Area of Concern
State(s): NY
Contact: Ms. Alice Yeh
U.S. Environmental Protection Agency
26 Federal Plaza
New York, NY 10278
Phone: (212)264-1865
Fax: (212)264-2194

P092

St. Marys River Remedial Action Plan
State(s): MI, Ont
Contact: Ms. Susan Stoddart
Coordinator
Ontario Ministry of Environment and Energy
747 Queen Street East
Sault Ste. Marie, CANADA, ONT P6A 2A8
Phone: (705)949-4640

L549

State Involvement in National Forest Plan Revisions throughout Colorado
State(s): CO
Contact: Mr. Steve Norris
Assistant Director
Colorado Department of Natural Resources
1313 Sherman Street
Denver, CO 80817
Phone: (303)866-3311

P093
State Lines Serpentine Barrens
State(s): PA, MD
Contact: Mr. James Thorne
The Nature Conservancy
1211 Chestnut St.
12th Floor
Philadelphia, PA 19107-4122
Phone: (215)963-1400
Fax: (215)963-1406
E-mail: jthorne@tnc.org

P094
Stegall Mountain Natural Area
State(s): MO
Contact: Mr. Larry Houf
District Wildlife Supervisor
Missouri Department of Conservation
Ozark District Office
Box 138
West Plains, MO 65775
Phone: (417)256-7161

L550
Stillwater Creek
State(s): OH
Contact: Ms. Sandra Chenal
Coordinator
Crossroads RC & D
10874 State Route 212 NE
Suite A
Bolivar, OH 44612
Phone: (216)874-4692
Fax: (216)874-3539

L551
Stone State Park Ecosystem Management Plan
State(s): IA
Contact: Mr. Jim Scheffler
Iowa Department of Natural Resources
Wallace State Office Building
Des Moines, IA 50319-0034
Phone: (515)281-6157
Fax: (515)281-6794

L552
Storm Lake Water Quality Protection Project
State(s): IA
Contact: Ms. Renee Braun
Iowa Department of Natural Resources
1617 North Lake Avenue
Storm Lake, IA 50588-1913
Phone: (712)732-3096
Fax: (712)732-6059

L553
Strategic Plan for the Illinois Department of Conservation
State(s): IL
Contact:
Illinois Department of Conservation
Springfield, IL

L554
Stream Protection and Management (SPAM) Program
State(s): HI
Contact: Ms. Sallie F. Edmunds
Project Manager
Hawaii Department of Land & Natural Resources
Commission on Water Resource Management
P.O. Box 621
Honolulu, HI 96809
Phone: (808)587-0252
Fax: (808)587-0219

L555
Structure and Function of Urban Forests
State(s): NY
Contact:
USDA Forest Service
Syracuse, NY

L556
Sumter and Francis Marion National Forests
State(s): SC
Contact: Mr. Forrest Starkey
USDA Forest Service
4931 Broad River Road
Columbia, SC 29210-4021
Phone: (803)561-4000

L557
Supersanctuary (Harris Center for Conservation Education)
State(s): NH
Contact: Mr. Meade Cadot
341 Kings Highway
Hancock, NH 03449
Phone: (603)525-3394
Fax: 603)357-0718

L558
Suwannee River Ecosystem Management Area
State(s): FL
Contact: Mr. Marvin Rallston
Suwannee River Water Management District
Route 3, Box 64
Live Oak, FL 32060
Phone: (904)362-1001
Fax: (904)362-1056

L559
Systems to Integrate Harvesting with Other Resource Mgmt Objectives
State(s): WV
Contact: Dr. Chris LeDoux
USDA Forest Service
180 Canfield St.
Morgantown, WV 26505
Phone: (304)285-1572

L560
TVA Regional Natural Heritage Project
State(s): TN
Contact: Dr. William Redmond
Tennessee Valley Authority
Ridgeway Road
Norris, TN 37828
Phone: (423)632-1593
E-mail: WREDMOND@mhs-tva.attmail.com

L561
Tampa Bay National Estuary Program
State(s): FL
Contact: Mr. Dean Ullock
Environmental Protection Agency
Region IV
345 Courtland Street, NE
Atlanta, GA 30365
Phone: (404)347-3555 ext. 2063

L562
Teanaway Ecosystem Demonstration Project
State(s): ID
Contact: Mr. Jon Haufler
Boise Cascade Corporation
PO Box 50
Boise, ID 83728
Phone: (208)384-6093

P095
Tensas River Basin Initiative
State(s): LA
Contact: Mr. Mike Adcock
Tensas River Basin Coordinator
Northeast Delta Resource Conservation & Development District
PO Box 848
Winnsboro, LA 71295
Phone: (318)435-7328
Fax: (318)435-7436

L563
Texas Hill Country
State(s): TX
Contact: Mr. Jim Fries
The Nature Conservancy
PO Box 164255
Austin, TX 78716-4255
Phone: (512)327-9472
Fax: (512)327-9625

L564
The Nature Conservancy Bioreserve Protection Program
State(s): MD
Contact: Mr. Nat Williams
Director
The Nature Conservancy
2 Wisconsin Circle
Suite 300
Chevy Chase, MD 20815
Phone: (301)656-8673

L565
Thousand Springs Preserve
State(s): ID
Contact: Mr. Chris & Mike O'Brien
The Nature Conservancy
1205 Thousand Springs Grade
Wendell, ID 83355
Phone: (208)536-6797
Fax: (208)726-1258

P096
Tidelands of the Connecticut River
State(s): CT
Contact: Dr. Juliana Barrett
Tidelands Program Director
The Nature Conservancy
Connecticut Chapter
55 High Street
Middletown, CT 06457
Phone: (860)344-0716
Fax: (860)344-1334

L566

Tillamook Bay National Estuary Program

State(s): OR

Contact: Ms. Marilyn Sigman
Tillamook Bay National Estuary Program
4000 Blimp Boulevard
Tillamook, OR 97141
Phone: (503)842-9922
Fax: (503)842-3680

L567

Tonasket Citizen's Council

State(s): WA

Contact: Ms. Elaine J. Zieroth
USDA Forest Service
Okanogan National Forest
1240 South Second
Okanogan, WA 98840
Phone: (509)826-3565

L568

Total Ecosystem Management Strategies

State(s): MI

Contact: Dr. Dean Premo
White Water Associates, Inc.
429 River Lane
Amasa, MI 49903
Phone: (906)822-7373
Fax: (906)822-7977

L569

Town Creek Ecosystem Stewardship Project

State(s): MD

Contact: Mr. Rick Latshaw
Maryland Department of Natural Resources
Forest Service
3 Pershing Street, Rm. 101
Cumberland, MD 21502
Phone: (301)777-2137

P097

Trail Creek Ecosystem Analysis

State(s): ID

Contact: Mr. Alan Pinkerton
USDA Forest Service
Sawtooth National Forest
Ketchum Ranger District
PO Box 2356
Ketchum, ID 83340
Phone: (208)622-5371

L570

Trapper Creek Aquatic and Riparian Restoration Project

State(s): CO

Contact: Mr. Jay Thompson
Fishery Biologist
Bureau of Land Management
Glenwood Springs Resource Area
PO Box 1009
Glenwood Springs, CO 81602

L571

Tri-County Leech Lake Watershed Project

State(s): MN

Contact: Mr. John Steward
Project Coordinator
Minnesota Department of Natural Resources
HCR 73, Box 172
Walker, MN 56484
Phone: (218)547-1770
Fax: (218)547-1887

L572

Trinity River Restoration Project

State(s): CA

Contact:
California Department of Water Resources
Red Bluff, CA

L573

Tripartite Agreement for Fish and Wildlife Resources

State(s): TN

Contact:
U.S. Air Force
Washington, DC

L574

Trout Creek Mountain Working Group

State(s): OR

Contact: Doc and Connie Hatfield
Trout Creek Mountain Working Group
Hatfield's High Desert Ranch
Brothers, OR 97712
Phone: (503)576-2455
Fax: (503)576-2238

P098

Trout Mountain Roadless Area

State(s): CO

Contact: Mr. Ron Pugh
Forest Planner
USDA Forest Service
Rio Grande National Forest
1803 West Highway 160
Monta Vista, CO 81144
Phone: (719)852-5941

L575

Tuluksak River Fish Harvest Study

State(s): AK

Contact:
U.S. Fish & Wildlife Service
Kenai, AK

L576

U.S. Fish & Wildlife Service--Pacific Islands Ecoregion

State(s): HI

Contact: Mr. Robert P. Smith
Pacific Islands Ecoregion Manager
U.S. Fish & Wildlife Service
Pacific Islands Ecoregion
Box 50167
Honolulu, HI 96850
Phone: (808)541-2749
Fax: (808)541-2756

L577

USFS Participation in Local Agency Planning

State(s): CA

Contact: Mr. Gary Earney
Lands and Recreation Officer
USDA Forest Service
San Bernardino National Forest
Cajon Ranger District
1209 Lytle Creek Road
Lytle Creek, CA 92358
Phone: (909)887-2576

L578

USFS/BLM Ecosystem Management Team

State(s): CO

Contact: Ms. Marsha Kearney
USDA Forest Service Pike
San Isabel National Forest
1920 Valley Drive
Pueblo, CO 81230
Phone: (719)545-8737

L579

Umpqua Basin Fisheries Restoration Initiative

State(s): OR

Contact: Mr. Rick Sohn
Lone Rock Timber Company
PO Box 1127
Roseburg, OR 97470
Phone: (503)673-0141
Fax: (503)440-2516

L580

Uncompahgre Riverway

State(s): CO

Contact: Mr. Gary Weiner
RMR-PPO
National Park Service
PO Box 25287
Denver, CO 80225

L581

Union Ridge Conservation Area

State(s): MO

Contact: Mr. Richard G. Whiteaker
District Forester
Missouri Department of Conservation
2500 S. Halliburton
Kirksville, MO 63501
Phone: (816)785-2420
Fax: (816)785-2553

L582

Upland Wildlife Ecology Program

State(s): TX

Contact: Dr. Jerry Cooke
Program Director
Texas Parks and Wildlife Department
Upland Wildlife Ecology
4200 Smith School Road
Austin, TX 78744
Phone: (512)389-4774
Fax: (512)389-4398
E-mail: jcooke@access.texas.gov

L583

Upper Arkansas Watershed Initiative and Forum

State(s): CO

Contact: Mr. Jeffrey Keidel
Upper Arkansas Watershed Coord.
PO Box 938
Buena Vista, CO 81211
Phone: (719)395-6035

L584

Upper Big Mill Creek

State(s): IA

Contact: Mr. Darcy Lee Keil
Iowa Department of Natural Resources
603 1/2 East Platt
Maquoketa, IA 52060
Phone: (319)652-2337
Fax: (319)652-4889

P099

Upper Farmington River Management Plan

State(s): CT, MA

Contact: Mr. Thomas Stanton
Chair, Farmington River Coordinating
Committee
Town of Colebrook
119 Beech Hill Road
Winsted, CT 06098
Phone: (860)379-8704

L585

Upper Feather River Watershed Projects

State(s): CA

Contact:
California Department of Water Resources
Red Bluff, CA

P100

Upper Huerfano Ecosystem

State(s): CO

Contact: Ms. Nancy Ryke
Wildlife Biologist
USDA Forest Service
San Isabel National Forest
San Carlos Ranger District
326 Dozier Avenue
Canon City, CO 81212
Phone: (719)275-4119

L586

**Upper Mississippi River System
Environmental Management Program**

State(s): MN, IA, WI, IL, MO

Contact:
U.S. Army Corps of Engineers
Chicago, IL

L587

**Upper Mississippi River/Tallgrass Prairie
Ecosystem**

State(s): MN, IA, WI, IL, MO

Contact:
U.S. Fish & Wildlife Service
Winona, MN

L588

Upper Niangua River Hydrologic Unit Area

State(s): MO

Contact: Ms. Karen Ross
PO Box 1070
Buffalo, MO 65622
Phone: (417)345-7551
E-mail: dallaco@ext.missouri.edu

L589

Upper Peace River

State(s): FL

Contact: Mr. Tim King
Florida Game and Fresh Water Fish
Commission
Division of Fisheries
3928 Teneroc Mine Road
Lakeland, FL 33805
Phone: (941)499-2421

L590

**Upper Sacramento River Riparian Habitat
Management**

State(s): CA

Contact: Ms. Stacy Cepello
2440 Main Street
Red Bluff, CA 96080-2398
Phone: (916)529-7352

L591

Upper Wabash Habitat Restoration Project

State(s): IN

Contact: Mr. Jeff Kiefer
U.S. Fish & Wildlife Service
620 S. Walker
Bloomington, IN 47403
Phone: (812)534-4261 ext. 212

L592

Upper/Middle Rio Grande Ecosystem

State(s): NM, CO

Contact:
Single Point of Contact
U.S. Fish & Wildlife Service
Sevilleta National Wildlife Refuge
PO Box 1248
Socorro, NM 87801
Phone: (505)864-4021
Fax: (505)864-7761

P101

Verde River Greenway

State(s): AZ

Contact:
Arizona State Parks
1300 West Washington Avenue
Phoenix, AZ 85007
Phone: (602)542-4147
Fax: (602)542-4180

P102

Virginia Coast Reserve

State(s): VA

Contact: Ms. Terry Thompson
The Nature Conservancy
Virginia Coast Reserve
PO Box 158
Nassanadox, VA 23413
Phone: (804)442-3049
Fax: (804)442-5418
E-mail: TATHOMPSON@AOL.COM

L593

Waikamoi Preserve

State(s): HI

Contact:
The Nature Conservancy of Hawaii
Makawao, HI

L594

**Waimea Canyon, Kokee & Polihale, and Na
Pali Coast State Parks**

State(s): HI

Contact: Mr. Wayne Souza
Kaui Parks District Supervisor
Hawaii Department of Land & Natural
Resources
3060 Eiwa Street
Room 306
Lihue, HI 96766-1875

L595

**Walnut Creek National Wildlife Refuge -
Prairie Learning Center**

State(s): IA

Contact: Mr. Richard Birger
Project Leader
U.S. Fish & Wildlife Service
PO Box 399
Prairie City, IA 50228
Phone: (515)994-2415
Fax: (515)994-2104
E-mail: BIRGER_DICK@FWS.GOV

L596

Waquoit Bay National Estuarine Reserve

State(s): MA

Contact: Ms. Christine Gault
Waquoit Bay National Estuarine Reserve
PO Box 3092
Waquoit, MA 02536
Phone: (508)457-0495
Fax: (617)727-5537

L621

Watershed Restoration Jobs Grant Program

State(s): WA

Contact: Ms. Leni Oman
Washington Department of Fish & Wildlife
Habitat Management Program
600 Capitol Way North
Olympia, WA 98501-1091
Phone: (360)902-2592
Fax: (360)902-2946

L597

**Washington State Ecosystems Conservation
Project--Riparian & Wetland Program**

State(s): WA

Contact: Mr. Dave Frederick
U.S. Fish & Wildlife Service
3704 Griffin Lane SE, Suite 102
Olympia, WA 98501-2192
Phone: (360)753-9440
Fax: (360)753-9405
E-mail: Frederickd@FWS.gov

L251

**Washington State Ecosystems Conservation
Project--Upland Wildlife Restoration**

State(s): WA

Contact: Mr. Dan Blatt
Washington Department of Fish & Wildlife
600 Capitol Way North
Olympia, WA 98501-1091
Phone: (360)902-2594

L598

Waterboro Barrens Preserve

State(s): ME

Contact: Ms. Nancy Sferra
South Maine Preserves Manager
The Nature Conservancy
160 Main St.
Sanford, ME 04073
Phone: (207)490-4012
Fax: (207)490-4012

L599

**Watershed Restoration Partnership Program
(WRPP)**

State(s): WA

Contact:
Washington Department of Fish and Wildlife
Olympia, WA

L600

Wekiva River Basin

State(s): FL

Contact: Mr. Jim Stevenson
Environmental Administrator
Florida Department of Environmental
Protection
3900 Commonwealth Blvd.
Tallahassee, FL 32399
Phone: (904)488-4892

L601

Wells Creek Watershed Partnership
State(s): MN
Contact: Beth Knudsen
Minnesota Department of Natural Resources
1801 S. Oak Street
Lake City, MN 55041
Phone: (612)345-5601
Fax: (612)345-3975
E-mail: Beth.Knudsen@dnr.state.mn.us

L602

Wells Resource Management Plan, Elk Amendment
State(s): NV
Contact: Mr. Bill Baker
Area Manager
Bureau of Land Management
Elko District Office
3900 E. Idaho Street
PO Box 831
Elko, NV 89802
Phone: (702)753-0200
Fax: (702)753-0255

L603

West Clear Creek Ecosystem Management
State(s): AZ
Contact: Mr. John Gerritsma
USDA Forest Service
Coconino National Forest
Long Valley Ranger District
Long Valley Road, HC31 Box 68
Happy Jack, AZ 86024
Phone: (520)354-2216
Fax: (520)354-2216

L604

West Elk Wilderness HRM/AMP
State(s): CO
Contact:
USDA Forest Service
Paonia, CO

L605

West Eugene Wetlands Project
State(s): OR
Contact: Mr. Johnathan T. Beall
West Eugene Wetlands Project Manager
Bureau of Land Management
Eugene District
PO Box 10226
Eugene, OR 97440-2226
Phone: (503)683-6413
Fax: (503)683-6981

L606

West Fork Bear River Ecosystem Management Project
State(s): WY
Contact:
USDA Forest Service
Evanston, WY

L607

Westvaco Corp., Timberlands Division, Southern Region/Southern Woodlands
State(s): SC
Contact: Mr. Fred W. Kinard Jr.
Westvaco
PO Box 1950
Summerville, SC 29484
Phone: (803)871-5000
Fax: (803)851-4602

L608

Wetlands Productivity Study
State(s): LA
Contact:
U.S. Air Force
Washington, DC

L609

Whole Farm/Ranch Planning
State(s): GA, ID, MN, NE, NY, PA
Contact:
USDA Natural Resources Conservation Service
Washington, DC

P103

Wild Stock Initiative
State(s): WA
Contact: Mr. Rich Lincoln
Washington Department of Fish and Wildlife
600 Capitol Way North
Olympia, WA 98501-1091
Phone: (206)902-2750

P104

Wildlife Area Planning
State(s): WA
Contact: Mr. Paul Dahmer
Wildlife Area Inv. & Planning Coordinator
Washington Department of Fish and Wildlife
600 Capitol Way North
Olympia, WA 98501-1091
Phone: (360)664-0705
Fax: (360)902-2946
E-mail: dahmepad@dfw.wa.gov

L610

Wildlife Communities and Habitat Relationships in New England
State(s): MA
Contact: Dr. Richard DeGraaf
Project Leader/Chief Research Wildlife Biologist
USDA Forest Service
Holdsworth Hall
University of Massachusetts
Amherst, MA 01003
Phone: (413)545-0357

P105

Wildlife Habitat Improvement Group
State(s): VT
Contact: Mr. David Clarkson
RR1 Box 2426
Newfane, VT 05345
Phone: (802)365-4243
E-mail: DCLARKS@LEG.STATE.VT.US

L611

Willamette River Basin
State(s): OR
Contact: Mr. Mike Rylko
U.S. Environmental Protection Agency
Region X
1200 Sixth Avenue
Seattle, WA 98101
Phone: (206)553-4014
Fax: (206)553-0165

L612

Willapa Alliance Natural Resource Program
State(s): WA
Contact: Mr. Dan'l Markham
Willapa Alliance
PO Box 278
South Bend, WA 98586
Phone: (360)875-5195
Fax: (360)875-5198
E-mail: willapanet@IGC.APC.ORG

L613

Wilson Creek National Battlefield
State(s): MO
Contact:
National Park Service
Republic, MO

L614

Xeric Oak Scrub Ecological Survey
State(s): FL
Contact:
U.S. Air Force
Washington, DC

L615

Yakima River Watershed Council
State(s): WA
Contact:
USDA Forest Service
Clellum, WA

L616

Yampa Valley Alliance
State(s): CO
Contact: Mr. Duane Holmes
National Park Service
RMRO-PPO
PO Box 25287
Denver, CO 80225
Phone: (303)969-2855
Fax: (330)987-6676

L617

Yavapai Ecosystem
State(s): AZ
Contact: Mr. Mark Johnson
District Ranger
USDA Forest Service
Chino Ranger Distrcit
Chino Valley, AZ 86313

L618

Yazoo Basin
State(s): MS
Contact:
The Nature Conservancy

L619

Yreka River - Siskiyou Forest Management Roundtable
State(s): CA

L533

Yuba Watershed Institute
State(s): CA
Contact:
Yuba Watershed Institute
17790 Tyler Foote Road
Nevada City, CA 95959
Phone: (916)292-3777; (916)478-0817

RESOURCE GUIDE:
SELECTED DOCUMENTS ON ECOSYSTEM MANAGEMENT

BIBLIOGRAPHIES AND INTERNET SOURCES

Cross, Diana, D. Jennings, R. Sojda, and C. Solomon, 1995. Ecosystem management: A bibliography. Information Transfer Center, National Biological Service, U.S. Department of Interior, Ft. Collins, CO 80525-5589. Also available on Internet: http://teal.itc.nbs.gov

University of Arizona, Water Resources Research Center. "Partnerships" web site (very comprehensive): http://ag.arizona.edu/partners/

CASE STUDIES AND CATALOGS

Colorado Ecosystem Partnership (Colorado Department of Natural Resources and USDI–Bureau of Land Management), 1994. Partnerships for sustainability. Colorado Department of Natural Resources, 1313 Sherman Street, Denver CO 80203.

Hartig, John H., and N.L. Law, 1994. Progress in Great Lakes remedial action plans: Implementing the ecosystem approach in Great Lakes Areas of Concern. EPA905-R-24-020. Environment Canada, Toronto, Canada; U.S. Environmental Protection Agency, Chicago, IL.

Thomas, Lisa, J. Geiselman, and K. Oakley, 1995. Natural resource partnership agreements: Examples. National Biological Service, Alaska Science Center, Anchorage AK 99503.

Wondolleck, Julia M. and S. Yaffee, 1994. Building bridges across agency boundaries: In search of excellence in the U.S. Forest Service. USDA–Forest Service, Pacific Northwest Research Station, Seattle, WA.

U.S. Department of Commerce, National Technical Information Service, 1993. International cooperation of the Man and the Biosphere (MAB) Programs of Europe and North America. U.S. Department of Commerce, Springfield, VA. NTIS PB93-183705.

USDI–Bureau of Land Management, 1994. Summit showcase displays and ecosystem case studies. Prepared by Richard Dworsky, Alaska State Office. GPO583 057/10007.

U.S. Environmental Protection Agency, 1995. A Phase I inventory of current EPA efforts to protect ecosystems. EPA841-S-95-001. Office of Water (4503F), U.S. Environmental Protection Agency, Washington, D.C. Available via internet: http://www.epa.gov/ecoplaces

OVERALL ASSESSMENTS AND POLICY STATEMENTS

American Forests, 1995. Inside urban ecosystems: seventh national urban forest conference proceedings, New York City, September 13–15, 1995. Washington, D.C.

The Keystone Center, 1993. National ecosystem management forum meeting summary, November 16–17, 1993, Airlie, VA. Available from The Keystone Center, Keystone, CO.

Morrissey, Wayne A., J.A. Zinn, and M.L. Corn, 1994. Ecosystem management: Federal agency activities. U.S. Congressional Research Service, 94-339 ENR.

Noss, Reed F., E.T. LaRoe, III, and J.M. Scott, 1995. Endangered ecosystems of the United States: A preliminary assessment of loss and degradation. U.S. Department of Interior, National Biological Service, Washington, D.C. Biological report 28.

Society of American Foresters, 1993. The Task Force report on sustaining long-term forest health and productivity. SAF 93-02. Bethesda, MD.

The President's Council on Sustainable Development, 1996. Sustainable America: A new consensus for prosperity, opportunity, and a healthy environment for the future.

Thomas, Jack Ward, 1994. Statement before the Subcommittee on National Parks, Forests and Public Lands, and the Subcommittee on Oversight and Investigations, Committee on Natural Resources, U.S. House of Representatives, Washington, D.C., February 3.

U.S. Congress, House of Representatives, 1994. Ecosystem management: Sustaining the nation's natural resources trust. Majority Staff Report, Committee on Natural Resources, 103-2, Committee Print No. 6.

USDA–Forest Service, 1992. Ecosystem management strategies for the northeastern and midwestern national forests: A report to the Chief of the Forest Service. Prepared by U.S. Forest Service: Eastern Region, Northeastern Forest Experiment Station, and North Central Forest Experiment Station.

USDA–Forest Service and Morris Arboretum, 1994. An ecosystem-based approach to urban and community forestry: An ecosystem manager's workbook. USDA Forest Service (State and Private Forestry) and Morris Arboretum of the University of Pennsylvania (Center for Urban Forestry), Philadelphia, PA.

USDI–Bureau of Land Management, 1994. Ecosystem management in the BLM: From concept to commitment. U.S. Department of Interior, BLM/SC/GI-94/0057+1736. U.S. Government Printing Office.

USDI–Fish and Wildlife Service, 1994. An ecosystem approach to fish and wildlife conservation.

U.S. Environmental Protection Agency, 1994. Toward a place-driven approach: The Edgewater consensus on an EPA strategy for ecosystem protection.

U.S. General Accounting Office, 1994. Ecosystem management: Additional actions needed to adequately test a promising approach. U.S. General Accounting Office, Resources, Community, and Economic Development Division, Washington, D.C. GAO/RCED-94-111.

THEORY

Barth, Sara, R.L. Gooch, J. Havard, D. Mindell, R. Stevens, and M. Zankel, 1994. Exploring the theory and application of ecosystem management. School of Natural Resources and Environment, The University of Michigan, Ann Arbor, MI.

Grumbine, R. Edward, 1994. What is ecosystem management? Conservation Biology 8(1):27–38.

Grumbine, R. Edward, 1990. Protecting biological diversity through the greater ecosystem concept. Natural Areas Journal 10(3):114–120.

Journal of Forestry, 1994. Ecosystem management, will it work? Journal of Forestry 92(8).

Moote, Margaret A., S. Burke, H.J. Cortner, and M.G. Wallace, 1994. Principles of ecosystem management. Water Resources Research Center, The University of Arizona.

Noss, Reed F., 1987. Protecting natural areas in fragmented landscape. Natural Areas Journal 7(1):2–13.

Rowe, J. Stan, 1992. The ecosystem approach to forestland management. The Forestry Chronicle 68(1):222–224.

Slocombe, D. Scott, 1993. Implementing ecosystem-based management: Development of theory, practice, and research for planning and managing a region. BioScience 43(9):612–622.

Stanley, Thomas R., Jr., 1995. Ecosystem management and the arrogance of humanism. Conservation Biology 9(2):255–262.

ABBREVIATIONS AND GLOSSARY
OF COMMONLY USED TERMS

ABBREVIATIONS

ACOE U.S. Army Corps of Engineers.

BLM Bureau of Land Management, U.S. Department of Interior.

BMP Best Management Practices; management practices recognized by the U.S. Environmental Protection Agency as acceptable with regard to the associated level of pollutants.

DNR Department of Natural Resources; the land management or environmental quality department in many states. Also common: DEQ—Department of Environmental Quality; DOC—Department of Conservation.

DOI U.S. Department of Interior.

EPA U.S. Environmental Protection Agency.

ESA Endangered Species Act.

FACA Federal Advisory Committee Act.

FWS U.S. Fish and Wildlife Service, U.S. Department of Interior.

GIS Geographic Information Systems.

NBS National Biological Service, U.S. Department of Interior.

NEP National Estuary Program; a nationwide program authorized by the Clean Water Act and administered by the U.S. Environmental Protection Agency.

NPS National Park Service, U.S. Department of Interior.

NRCS Natural Resources Conservation Service (formerly SCS—Soil Conservation Service), U.S. Department of Agriculture.

NWR National Wildlife Refuge; managed by the U.S. Fish and Wildlife Service.

TNC The Nature Conservancy; a national non-profit land trust.

TVA Tennessee Valley Authority.

USDA U.S. Department of Agriculture.

USFS U.S. Forest Service, U.S. Department of Agriculture.

USGS U.S. Geological Survey, U.S. Department of Interior.

GLOSSARY

Biodiversity The variety of living things, including variation at the genetic, species, and landscape levels.

Community The organisms living and interacting in a given area.

Disruption of Fire Regime
Many ecosystems in North America depend on the occurrence of fires at particular intervals for ecosystem health and regeneration. However, since the turn of the century, public and private landowners alike have tended to suppress fires. As a result, many ecosystems are imperiled: characteristic species are not regenerating, and uncharacteristic species are taking over.

Ecosystem
A three-dimensional volume of space, including all its physical and biological components, which are recognized to be interconnected. An ecosystem contains soil, air, and all organisms living in the soil or air. An ecosystem is more or less homogenous both as to the form and structure of the land and as to the vegetation supported thereon. The ecosystem concept is hierarchical. For instance, large ecosystems defined by a relatively homogenous macroclimate and physiography may be subdivided into smaller ecosystems based on finer distinctions in physiography, soil, and vegetation.

Ecosystem Management
The definition of this term is the subject of much debate in the literature. For example, Edward Grumbine states that "Ecosystem management integrates scientific knowledge of ecological relationships within a complex sociopolitical and value framework toward the general goal of protecting native ecosystem integrity over the long term." (Grumbine, E.R. 1994. What is ecosystem management? Conservation Biology 8:2:27-38.)

U.S. Forest Chief Jack Ward Thomas gives the following definition: "Ecosystem management is a holistic approach to natural resource management, moving beyond a compartmentalized approach focusing on the individual parts of the forest. It is an approach that steps back from the forest stand and focuses on the forest landscape and its position in the larger environment in order to integrate the human, biological and physical dimensions of natural resource management. Its purpose is to achieve sustainability of all resources." (Thomas, J.W. 1994. New directions for the Forest Service. Statement of Jack Ward Thomas, Chief, Forest Service, U.S. Department of Agriculture, before the Subcommittee on National Parks, Forests and Public Lands, and the Subcommittee on Oversight and Investigations, Committee on Natural Resources, U.S. House of Representatives, February 3.)

This assessment does not adhere to any particular definition in the literature. Instead, natural resource management projects are considered ecosystem management if they fulfill at least one of two criteria. (See Introduction, page 4, for criteria.)

Exotic Species
Species that do not naturally occur in the ecosystem. In many instances, exotic species become invasive pests. Well-known examples include kudzu *(Pueraria lobata)* and European buckthorn *(Rhamnus cathartica).*

Federally Listed Species
Under the Endangered Species Act (ESA), the federal government, in particular the U.S. Fish and Wildlife Service, is required to list species that are considered threatened or endangered. Listed species are protected by the provisions of the ESA.

Geographic Information Systems (GIS)
A computerized system of maps and linked databases. These systems generally make use of a base map of a particular area and a set of overlays. Each overlay shows a characteristic of the area. Through combining particular overlays, a natural resource manager has a visual image of the manner in which certain area characteristics spatially relate to one another.

Hydrologic Alteration
Modification of the amount or movement of water through an ecosystem. This includes drainage of wetlands, dams, levees, aquifer mining, reser-

voirs, and channelization of rivers and streams.

Non-Point Source Pollution

Pollution from diffuse sources that do not discharge at a single location. Examples include seepage, stormwater runoff from agricultural lands and urban streets, and septic tanks leachate.

Point Source Pollution

Pollution discharged by discrete sources, including factories, power plants, and sewage treatment plants.

Riparian Areas   Areas adjacent to streams or lakes.

Stakeholders   Individuals and organizations with an interest in a particular area or project. Stakeholders may include public agencies at all levels (federal, state, and local), non-profit organizations, private landowners, industry, and others.

State-Listed Species

In addition to the federal listing system under the Endangered Species Act, many states have enacted parallel legislation pertaining to the protection of threatened and endangered species in those states. Such species are not necessarily threatened or endangered in other states, nor are they necessarily listed as federal threatened or endangered species.

Sustainability   Management practices that do not take more from an ecosystem than it can provide. Theoretically, sustainable management practices can continue in perpetuity, since they do not lead to exhaustion of natural resources.

Watershed   An area that is drained by a particular stream system. For instance, the Mississippi watershed consists of the entire area drained by the Mississippi and its tributaries.

INDEX

ACE Basin (South Carolina), 81–82
ACE Basin National Wildlife Refuge (South Carolina), 82
Adams, Steve, 198
Adaptive management, 37, 38, 199, 245
Adcock, Mike, 270
Adirondack Mountains (New York), 157
Advice, xvii, 35–38, 44, 105, 218, 290
    community relations, 35, 38
    open/collaborative process, 35, 36
    role of science, 35, 37
    securing resources and support, 35, 37
    stakeholder involvement, 35–36
Age of projects, 13, 59
Agencies:
    norms, 31
    policies and programs, 21
Agreements, cooperative, 19, 93, 123, 240
Agriculture:
    agricultural practices as a stress, 9, 10, 69–71, 91, 99,
        103, 113, 125, 127, 133, 137, 139, 141, 149, 155, 163,
        167, 169, 175, 177, 179, 181, 187, 195, 199, 203, 209,
        219, 227, 233, 237, 243, 251, 261, 269, 283, 285
    grazing, see Grazing/range management
    historic, 145, 193, 213
    land conversion to, 69–71, 89, 91, 99, 103, 113, 141, 143,
        159, 165, 167, 169, 175, 177, 179, 187, 195, 197, 199,
        217, 221, 223, 227, 237, 243, 255, 261, 269, 285
    land use, 63–65, 83, 87, 89, 91, 97, 99, 101, 103, 113, 115,
        123, 125, 157, 131, 133, 139, 141, 153, 155, 157, 159,
        163, 165, 167, 169, 175, 177, 179, 181, 195, 197, 199,
        203, 207, 209, 217, 219, 223, 227, 226, 237, 243, 249,
        257, 261, 265, 269, 277, 281, 283, 285
    removal of lands from, 176
    stakeholder representatives from, 16, 116, 126, 270
Alabama, 16, 143, 145, 155, 165
Alaska, xv, 17, 32, 171, 172, 211, 212, 239
Alaska Department of Fish and Game, 171, 211
Albany Pine Bush (New York), 7, 21, 83–84
Albany, New York, 7, 21, 83, 84
Albemarle Sound (North Carolina), 179
Alexander Archipelago (Alaska), 211
Allegan State Game Area (Michigan), 85–86
Allegheny National Forest (Pennsylvania), 245
Animas Foundation (New Mexico), 183
Animas Mountains (New Mexico), 183
Annapolis, Maryland, 113
Appalachian Mountains, 109, 219, 257
Applegate Partnership (Oregon), 13, 87–88
Appropriations, line item, 33, 112
Aransas National Wildlife Refuge (Texas), 125
Area of Concern (International Joint Commission designation), 263
Arizona, 36, 37, 38, 183, 184, 251, 281, 282
Arizona State Parks, 281, 282
Arkansas, 103, 104, 115, 223
Arkansas Wetlands and Water Resources Task Force, 104
Army Corps of Engineers (ACOE), 16, 102, 197, 198, 259, 262
Ashepoo River (South Carolina), 81
Atlanta, Georgia, 109
Atlantic Ocean/Coast, 81, 179, 265, 283
Avalon, California, 253

Babbitt, Bruce, 262, 281
Baker, Bill, 188
Baltimore, Maryland, 265
Barrett, Juliana, 272
Barrows, Cameron, 128, 256
Bartlett, Randall, 242
Bayou Lafourche (Louisiana), 89
Bear Island (State) Wildlife Management Area (South Carolina), 82
Benefits, intangible, 31
Bertie County, North Carolina, 180
Big Cypress Seminole Indian Reservation (Florida), 261
Big Darby Creek Partnership (Ohio), 91–92
Bikes, mountain, 22, 241
Binational Public Advisory Committee, 263
Bison, 18, 147, 173, 174, 221, 222, 253, 267
Black Ridge Wilderness Study Area (Colorado), 247, 248
Block Island Conservancy (Rhode Island), 95
Block Island Land Trust (Rhode Island), 95
Block Island Refuge (Rhode Island), 32, 95–96
Blue Mountains Natural Resources Institute (Oregon/Washington), 17, 97–98
Boston, Massachusetts, 207, 215
Boundaries:
    ecological, 22, 215
    property/political, 4, 26, 33, 235
Bowling, Jill, 136
Brewer, Jan, 182
Brownsville, Texas, 177
Bureau of Land Management (BLM), 9, 15, 18, 87, 88, 107, 121, 127, 128, 129, 130, 184, 187, 188, 189, 191, 192, 217, 225, 226, 233, 234, 247, 248, 251, 255, 279, 280
Buss, Sherri, 232
Butte Valley Basin (California), 99–100

Cache River (Arkansas), 103
Cache River (Illinois), 101
Cache River State Natural Area (Illinois), 101
Cache River Wetlands (Illinois), 101–102
California, 2, 99, 100, 127, 151, 253, 255, 256
California Department of Fish and Game, 99
California State Coastal Conservancy, 151
Camp Johnson Sandplain Restoration (Vermont), 13, 16
Camp Pendleton Marine Corps Base (California), 255
Campbell, Susan, 236
Canada, 133, 215, 221, 237, 243, 263, 264
Candidate sites, 4, 5
Canyon Country Partnership (Utah), 107–108
Cape May, New Jersey, 157
Carbon County, Utah, 107
Caribbean, 175, 176
Caribou National Forest (Idaho), 189
Carlson, Clint, 94
Cascade Mountains (Washington and Oregon), 87, 285, 287
Cecil County, Maryland, 265
Center for Alternative Dispute Resolution of the Judiciary of Hawaii, 201
Central flyway, 243
Charleston, South Carolina, 81
Chattahoochee National Forest (Georgia), 109

ABOUT THE AUTHORS

STEVEN YAFFEE

Dr. Steven L. Yaffee is a faculty member at the School of Natural Resources and Environment at The University of Michigan, where he teaches courses in natural resource policy and administration, negotiation skills, environmental history, and biodiversity and public policy. His research focuses on understanding and improving public decision-making processes and exploring the behavior of administrative agencies and interest groups involved in implementing public policies. He has worked for nearly twenty years on federal endangered species policy and is the author of *Prohibitive Policy: Implementing the Federal Endangered Species Act* (MIT Press, 1982) and *The Wisdom of the Spotted Owl: Policy Lessons for a New Century* (Island Press, 1994). Currently, he is collaborating with Dr. Julia Wondolleck on another book, *Building Bridges: Creating Partnerships and Resolving Conflict in Natural Resource Management*, to be published by Island Press in 1997.

Dr. Yaffee received his Ph.D. in 1979 from the Massachusetts Institute of Technology in environmental policy and planning, and has earlier degrees in natural resources. He has taught at MIT and the Kennedy School of Government at Harvard, and has been a researcher at the Oak Ridge National Laboratory and the Conservation Foundation/World Wildlife Fund.

ALI PHILLIPS

Ali Phillips is a regional development coordinator for The Wilderness Society in Denver, Colorado. His research interests have included public lands, endangered species, and ecosystem management issues. He has been a co-author of articles and reports on endangered species and federal subsidies.

Mr. Phillips received his M.S. from the School of Natural Resources and Environment at The University of Michigan in 1995 and holds a B.S. in soil science from Cornell University. He has worked in fundraising, program development, and research departments for several environmental and conservation organizations, including the The Wilderness Society, Environmental Defense Fund, and American Forests. He has also worked for Cradlerock Outdoor Network, Inc., a company specializing in adventure-based training, and the U.S. Department of Interior.

IRENE FRENTZ

Irene Frentz is a research associate at the Department of Agricultural Economics and Rural Sociology at the University of Arkansas. Her research interests include implementation of ecosystem management by public agencies, public involvement in decision-making processes of resource management agencies, and integration of community strategic planning processes with natural resource planning processes. Recently, Ms. Frentz produced a case study of the Ouachita National Forest Ecosystem Management Advisory Committee.

Ms. Frentz received her M.S. from the School of Natural Resources and Environment at The University of Michigan in 1995 and has a B.S. in microbiology from Indiana University. She has worked for the Natural Area Preservation Division of the Department of Parks and Recreation of the City of Ann Arbor, Michigan, and the Biology Department at Indiana University.

PAUL HARDY

Paul Hardy is program manager for The Nature Conservancy's (TNC) Upper San Pedro River project (see project description, page 251). He oversees TNC's operation of the Ramsey Canyon Preserve, as well as efforts to build stakeholder support for conserving the river's water resources and diverse riparian habitats. Mr. Hardy received his M.S. in natural resource policy and administration from the School of Natural Resources and Environment at The University of Michigan and an M.B.A. from the Michigan Business School in 1996. While working towards the degrees, Mr. Hardy concentrated on analyzing the strategic planning, financial management, membership recruitment, and leadership capabilities of non-profit environmental organizations.

Mr. Hardy has worked with The University of Michigan's Corporate Environmental Management Program and the Sierra Club to analyze pressure tactics, partnerships, and other strategies that non-profit environmental organizations develop toward private sector businesses. Mr. Hardy has worked at the World Wildlife Fund, where he was responsible for conducting training and technical assistance exercises that strengthen the management and planning capabilities of environmental organizations in Latin America and the Caribbean.

SUSSANNE MALEKI

Sussanne Maleki is currently employed by the Oregon Department of Fish and Wildlife as a research project coordinator assessing stream restoration and coastal salmon recovery efforts. Her research interests include effective management strategies for watershed protection, with a focus on improved monitoring and data sharing programs. She recently prepared a report on remote sensing methods for riparian monitoring for the U.S. Environmental Protection Agency in support of President Clinton's Pacific Northwest Forest Plan.

Ms. Maleki received her M.S. from The University of Michigan in 1995 with an emphasis on forest ecosystem management and land use conflict resolution. She holds a B.A. from the University of Wisconsin-Madison in environmental science and literature. Ms. Maleki has worked as an environmental specialist for industry, a fundraising coordinator for the Population Institute, an environment educator for the University of Georgia's Environmental Education Center, and as an assistant foreman and firefighter for the U.S. Forest Service.

BARBARA THORPE

Barbara Thorpe is currently working as an environmental analyst for Winslow Management Company in Boston, an investment management firm focused solely on environmentally-responsible investing. Ms. Thorpe received her M.S. in natural resource policy and administration from the School of Natural Resources and Environment at The University in Michigan in 1995, focusing on natural areas preservation and watershed management. She holds a B.S. in Political Science from St. Lawrence University.

Ms. Thorpe has worked for the U.S. Environmental Protection Agency in the Region II Office of Policy and Management (New York) on environmental policy issues, and as a communications specialist in the Headquarters Office of Air and Radiation (Washington, D.C.) on Clean Air Act issues. She has also worked for the New Jersey Public Interest Research Group.